LES SOCIÉTÉS HUMAINES FACE AUX CHANGEMENTS CLIMATIQUES

LES SOCIÉTÉS HUMAINES FACE AUX CHANGEMENTS CLIMATIQUES

Volume 2

LA PROTOHISTOIRE, DES DÉBUTS DE L'HOLOCÈNE AU DÉBUT DES TEMPS HISTORIQUES

Sous la direction de François Djindjian

ARCHAEOPRESS ARCHAEOLOGY

ARCHAEOPRESS PUBLISHING LTD
Summertown Pavilion
18–24 Middle Way
Summertown
Oxford OX2 7LG
www.archaeopress.com

ISBN 978-1-80327-262-7
ISBN 978-1-80327-263-4 (e-Pdf)

This book is available direct from Archaeopress or from our website www.archaeopress.com

Indicé

Avant Propos

Les deux présents volumes regroupent les contributions des membres de l'Union Internationale des Sciences Préhistoriques et Protohistoriques (UISPP), à un projet lancé en 2017, sous le titre « Les sociétés humaines face aux changements climatiques dans la préhistoire et la protohistoire. Des origines de l'Humanité au début des temps historiques ».

L'idée de ce projet est né de questions souvent posées aux préhistoriens de l'UISPP lors de conférences ouvertes au grand public concernant l'existence de changements climatiques dans l'Histoire de l'Humanité et la façon dont les sociétés humaines s'y étaient adaptées.

Les sociétés humaines ont connu depuis trois millions d'années une grande alternance de périodes glaciaires et interglaciaires. Quels climats ont été les plus favorables aux peuplements humains ? Quels climats ont été les plus défavorables aux peuplements humains et ont-ils entraînés des abandons de territoires et des effondrements de sociétés ? Quand et sous quels climats les groupes humains ont-ils colonisé l'ensemble des continents de la planète ? Sous quel climat et sous quelles latitudes, les innovations que représentent l'agriculture et l'élevage ont-elles réussi à se pérenniser ? Comment les sociétés agro-pastorales se sont-elles adaptées à la progression de l'aridité de l'Holocène après l'exceptionnelle période humide de ses débuts ? Le pastoralisme nomade est-il une spécialisation d'une société agro-pastorale dans un contexte d'aridité croissante ou une adaptation de la domestication animale à des zones steppiques et semi-désertiques ? Comment les sociétés agro-pastorales se sont-elles adaptées à des changements climatiques centenaires comme ceux connus des périodes protohistoriques et historiques (crise 2200 BC, crise 1200 BC, crise 800 BC, optimum climatique romain, crise du Bas-empire romain et des invasions barbares, optimum climatique médiéval, petit âge glaciaire) ? Et à des crises météorologiques sur plusieurs années à l'origine de disettes et de famines ? Une amélioration climatique avec un climat chaud et humide est-elle favorable au développement des sociétés humaines ? Le changement climatique est-il un facteur d'évolution pour les sociétés humaines, les forçant à s'adapter et à trouver des solutions durables ? Les régions du globe où les changements climatiques ont le moins d'amplitude (zones tropicales humides) ont-ils elles été favorables à l'évolution des sociétés humaines ou ont-elles eu au contraire comme conséquence une stagnation ?

Cette liste non exhaustive de questions révèle l'intérêt d'un thème de recherche que la situation actuelle de réchauffement climatique depuis le début du XX° siècle rend de plus en plus prioritaire dans le grand public, dans la jeunesse mais aussi dans la classe politique. Le succès médiatique et sociétal de cette question risque malheureusement d'en faire le sujet de manipulations, d'idéologies voire d'une nouvelle religion, avec ses faux prophètes. Aussi est-il important de la maintenir dans son contexte scientifique, où les erreurs, par définition même de la Science, peuvent être corrigées et rectifiées, alors que le laisser au mains d'idéologies ne pourra pas empêcher d'en faire un dogme définitif, brûlant pour cause de blasphèmes ou d'hérésies, tous ceux qui émettraient la moindre réserve voire la moindre demande d'explications.

Le projet a été proposé par l'UISPP à l'Union Académique Internationale qui l'approuvé en 2017 comme projet longue durée n°92 et financé. Il a également été soutenu par l'Académie Suisse des Sciences. L'UISPP remercie vivement ces deux institutions pour leur soutien.

Vingt et une contributions sont publiées ici en deux volumes. Le premier volume est consacré au Pléistocène et couvre la période allant des origines à la fin de la dernière glaciation il y a 12 000 ans, et les sociétés de chasseurs-cueilleurs. Le second volume couvre l'Holocène, les chasseurs-cueilleurs du Mésolithique et les sociétés agro-pastorales du Néolithique et des âges des métaux.

Les connaissances acquises ne permettent pas de traiter le sujet d'une façon aussi précise pour tous les continents. L'Europe est aujourd'hui la région la plus riche en sites archéologiques fouillés et en études détaillées, ce qui justifie que cette région ait été privilégiée dans les contributions reçues. Certaines régions ou continents ne permettent a pas aujourd'hui de telles synthèses comme le continent américain, l'Asie du Sud, la Chine, une partie de l'Afrique, Plusieurs contributions ont traité le thème globalement sur l'ensemble de la planète, en particulier les périodes du MIS 3 et du MIS 2. L'Asie du Sud-est et l'Afrique du Nord ont fait l'objet de contributions particulières.

Préface

Fondée en octobre 1919, l'Union Académique Internationale (désormais mieux connue par son sigle UAI) a fêté en novembre 2019 à l'Institut de France à Paris l'entrée dans son deuxième centenaire d'existence. Voulue comme un centre international d'excellence grâce à la collaboration des Académies nationales ou des grandes institutions équivalentes, elle se devait de prendre sous son aile autant les sciences humaines que les sciences sociales, et cela dans la perspective de patronner, organiser, gérer, développer, stimuler de grands projets collectifs de longue ou moyenne durée qui constitueraient et serviraient les sources de la recherche scientifique fondamentale. Ainsi, dictionnaires, corpora, atlas, index, encyclopédies, œuvres complètes ont progressivement vu le jour au fil des années, fournissant à une masse de chercheurs le produit des démarches les plus récentes dans leur discipline, accomplies très souvent dans un esprit de transdisciplinarité. En 2021, plus de cent projets avaient à cette date pu donner naissance à des produits savants de haute valeur ajoutée car réalisés par les spécialistes du moment.

Les projets s'étalant souvent sur des décennies, cette intense activité présenta toutefois - on le sait aujourd'hui encore plus clairement à la lumière des travaux d'archives menés pour célébrer le centième anniversaire de l'organisation -, une faiblesse qu'il ne convient pas de dissimuler sous les ors trompeurs des commémorations anniversaires, à savoir la faiblesse de se concentrer presque exclusivement sur la mise en valeur de problématiques relevant des sciences humaines. Le manque de projets appartenant au domaine des sciences sociales en devenait criant. On connaît les raisons de cette « discrimination », à savoir tout simplement le fait que les Académies membres proposèrent et mirent en valeur précisément des projets appartenant au champ des disciplines historiques, archéologiques et philologiques, oubliant ou négligeant de la sorte les disciplines s'inscrivant dans les sciences morales, politiques et sociales. Peu à peu, l'UAI se concentra ainsi au développement exclusif des domaines dominants appartenant aux sciences humaines, au point de n'apparaître finalement à l'extérieur que comme une organisation internationale de sciences humaines.

Une prise de conscience de l'écart existant entre les intentions et les statuts d'origine de l'Organisation et la réalité de leur application put heureusement se faire au cours des quinze dernières années. L'accent fut progressivement mis sur la nécessité de promouvoir les sciences morales, politiques et sociales et de valoriser des projets qui non seulement les mettraient en valeur mais les « serviraient » dans leurs attentes et demandes.

Dans ce contexte, il faut donc féliciter, ici et maintenant, l'Union Internationale des Sciences Pré- et Protohistoriques (UISPP). Elle qui rassemble en effet l'ensemble des préhistoriens, dont les fondements de recherche trouvent aujourd'hui leurs bases et matériaux, précisément autant dans le fond de commerce «classique» développé au sein de l'UAI que dans les empreintes et marqueurs des sciences sociales – sans oublier de mettre en évidence leur appartenance devenue naturelle et essentielle aux sciences fondamentales – a su inscrire dans la liste des projets de l'UAI, lors de l'Assemblée générale de cette dernière à Tokyo, en octobre 2017, une recherche fondamentale bousculant quelque peu les habitudes de l'Union, sous le titre « Les sociétés humaines face aux changements climatiques dans la préhistoire et la protohistoire. Des origines de l'Humanité au début des temps historiques ». Jusqu'alors, la préhistoire en tant que telle était inexistante à l'UAI, non par volonté ou ignorance, mais par simple manque de proposition de projets. Et non seulement le projet bousculait mais il arrivait à point nommé pour tenter d'analyser et d'approcher scientifiquement un questionnement planétaire et urgent... L'on projetait ainsi d'emblée l'UAI dans une démarche ultra-contemporaine. On ne pouvait espérer mieux !

A ces félicitations génériques, il convient d'ajouter en outre celles fournies par le plaisir de saluer la sortie de presse des deux premiers volumes qui donnent ainsi au projet sa première chair. Souvent en effet, le temps long laissé comme il se doit aux projets de l'UAI conduisent ceux-ci à voir leurs outils être produits lentement. Certes, les bornes sont faites pour être dépassées, ainsi que l'écrivit Antoine de Saint-Exupéry. Nous ne pouvons par conséquent qu'encourager le directeur du projet, le président François Djindjian, et l'ensemble des collègues de renom dont il a pu s'entourer pour fonder les bases de l'entreprise, à dépasser hardiment les bornes suivantes qu'ils ne manqueront pas de vouloir se fixer. L'UAI leur sait gré en tout cas maintenant de leur contribution à cette édition scientifique qu'ils ouvrent de grande et belle manière.

Jean-Luc De Paepe,
Secrétaire général adjoint de l'Union Académique Internationale

Liste des auteurs

Barbara Barich
(Dr de recherches ISMEO, Italie et Université de Rome)

Olivier Buchsenschutz
(Directeur recherches émérite CNRS)

Miguel Caparros
(Chercheur associé CNRS UMR 7194)

François Djindjian
(Président UISPP, Pf. Honoraire Université Paris 1 Panthéon Sorbonne)

Christian Dupuy
(Dr de recherches Institut des Mondes Africains Paris, UMR 8171)

Jean Guilaine
(Ancien Professeur du Collège de France)

Stefan Kozlowski
(Professeur émérite Université de Varsovie, Pologne)

Olivier Lemercier
(Professeur Université de Montpellier)

Cyril Marcigny
(Inrap Normandie, Laboratoire d'Archéologie et Histoire Merlat (LAHM), UMR 6566-CReAAH (université de Rennes, Nantes, Le Mans, CNRS, MC), Le Chaos, 14400 Longues-sur-Mer)

Marek Nowak
(Dr de recherches, Institut Jagellon, Université de Cracovie, Pologne)

Introduction au deuxième volume

François Djindjian

En hommage à Gilbert Kaenel (1949-2020)

Ce second volume est consacré pour l'essentiel à l'Holocène, la dernière période interglaciaire dans laquelle nous vivons actuellement. Les débuts de l'Holocène correspondent au retour de la forêt tempérée qui s'était réfugiée dans les régions méditerranéennes de l'Europe et qui entreprend une reconquête rapide des latitudes moyennes tandis que la forêt de conifères entreprend la conquête des altitudes et des régions septentrionales (taïga). La faune de ces forêts suit la progression végétale. La steppe froide disparait ainsi que la faune dont c'était le biotope. Face à ce changement climatique, les herbivores sont confrontés à trois situations : l'extinction (comme le cheval), la migration (par exemple vers le Nord pour le renne ou en altitude pour le bouquetin ou la marmotte) ou l'adaptation à la forêt (par exemple pour le bison). Et les carnivores partagent le même destin que leurs proies favorites.

En Europe, les chasseurs cueilleurs deviennent les hôtes de la forêt, y chassent le cerf, le chevreuil et le sanglier en inventant l'arc et la flèche et se régalent de fruits à coques et de baies. Ils pêchent dans les rivières et le long des rivages et collectent les coquillages dont ils nous ont laissé le souvenir par des grands monticules de coquilles abandonnées (amas coquilliers). La croissance démographique et la parcellisation des territoires expliquent les approvisionnements locaux, l'opportunisme saisonnier, la variété dans les ressources alimentaires et les différentiations régionales dans la géométrie des microlithes, armatures des flèches. L'art animalier disparait. C'est la période du Mésolithique (Kozlowski, ce volume).

Dans les régions méditerranéennes, la latitude plus favorable a anticipé ce processus de plusieurs milliers d'années, avec l'Epipaléolithique, dont les fines armatures devenues géométriques, favorise la chasse en milieu semi-ouvert. Au Proche-Orient, les chasseurs-cueilleurs se sont progressivement sédentarisés (Natoufien), grâce à la collecte de graminées et autres végétaux comestibles dont ils découvriront vite qu'ils sont cultivables, et à la chasse aux ovicapridés et aux gazelles dont ils découvriront aussi vite que certaines espèces sont domesticables. Ce sont les débuts de la néolithisation. Ce processus, décrit par N. Vavilov dans les années 1920, nous a fait connaître de nombreux centres d'invention de l'agriculture, dans différents continents et à différentes époques. Les espèces cultivées et les espèces domestiquées se sont ainsi répandues sur l'ensemble de la planète, transportées par les agriculteurs-éleveurs à la conquête de nouveaux espaces, poussés par les changements climatiques de l'Holocène, par la pression démographique ou relayées par d'autres groupes humains au contact des précédents. En Afrique du Nord, au repeuplement tardiglaciaire de l'Ibéromaurusien, succèdent des industries épipaléolithiques que J. Tixier avait bien décrites en son temps, dont le Capsien est l'exemple le plus abouti dont les peuplements se néolithisent progressivement pendant l'exceptionnel Holocène humide nord-africain, sur la côte méditerranéenne comme autour des grands méga-lacs du Sahara, qui resurgissent à chaque interglaciaire et qui s'enfouissent à chaque glaciaire ou à chaque poussée d'aridité en période interglaciaire comme actuellement. La préhistoire d'Afrique du Nord nous est bien connue grâce à la présence scientifique française exceptionnelle dès la fin du XIXème siècle en Algérie, au Maroc, en Tunisie, au Sahara et au Sahel grâce aux travaux de C. Arambourg, H. Alimen, L. Balout, J. Tixier, G. Camps, H.J. Hugot, H. Lhote et de bien d'autres (Caparros, ce volume). Il faut également mentionner les recherches scientifiques en Lybie de l'école italienne (Barich, ce volume) et de l'anglais McBurney dans la grotte d'Haua Fteha qui a fourni une séquence stratigraphique exceptionnelle sur les derniers 100 000 ans. L'Afrique du Nord, et tout particulièrement le Sahara, est une région particulièrement propice pour l'étude de l'adaptation des sociétés préhistoriques et protohistoriques aux changements climatiques du Pléistocène comme de l'Holocène. C'est la raison de la présence dans ce deuxième volume de quatre contributions qui se complètent en approfondissant les premières migrations humaines intra-africaines de M. Caparros, le pléistocène supérieur et les débuts de l'Holocène de B. Barich, une synthèse régionale sur le Tilemsi de Ch. Dupuy et une étude sur les relations entre pastoralisme et changements climatiques holocènes au Sahara de B. Barich.

    A l'Holocène, l'impact climatique n'est plus le même pour une économie d'agriculteur-éleveur, devenu dépendant du rendement de ses récoltes pour se nourrir et replanter et du fourrage pour ses animaux domestiques. Il devient donc nécessaire de distinguer quatre échelles de temps pour mieux comprendre l'impact des variations climatiques :

- Les temps millénaires comme celui qui sépare l'Holocène humide (à ses débuts) de l'Holocène récent dont l'aridité croit progressivement. Deux épisodes climatiques courts et intenses dont il sera beaucoup question dans ce volume ont été identifiés (et servent aujourd'hui de séparation entre les trois périodes de l'Holocène) : l'épisode froid et sec de 8200 BP et l'épisode aride de 4200 BP.
- Les temps séculaires qui voient périodiquement des périodes de péjoration climatique succéder à des périodes d'amélioration climatique, et dont les causes font encore l'objet de discussions : les glaciers de montagne s'allongent ou raccourcissent, les vendanges sont plus précoces ou plus tardives, les défrichements de nouvelles terres succèdent à l'abandon des terres les moins rentables : péjoration 1500-1200 BC, péjoration 800-700 BC, péjoration 500-400 BC, petit optimum romain, péjoration du bas empire romain

et des grandes invasions, optimum climatique médiéval, petit âge glaciaire, réchauffement actuel. Ces périodes, qui durent plusieurs siècles, entraînent des changements dans les sociétés humaines, des adaptations, parfois des effondrements (et cela d'autant plus qu'elles se sont étatisées) que l'archéologie sait déceler et étudier.

- Les temps pluriannuels qui voient se succéder plusieurs mauvaises années avec des sécheresses, des canicules, des grands froids, des pluies estivales, des gelées tardives, et qui, des origines de l'agriculture au XIX° siècle, sont causes de disette et de famines allant jusqu'à provoquer des baisses démographiques aussi dramatiques que des épidémies.

- Les temps saisonniers qui enregistrent l'instabilité de la météo, avec ses jours critiques que nos calendriers marquent de façon mnémotechnique (« les saints de glace ») et ses mauvaises surprises (gelées tardives, tempêtes de grêles, pluies d'été, canicules, cyclones, crues des rivières, invasions de sauterelles, etc.)

La réaction et l'adaptation des sociétés humaines face à ces différents temps climatiques seront forcément différentes.

Face aux aléas saisonniers et multi-annuels, que l'archéologue ne peut connaître que par des récits historiques (comme la difficulté de fourrage des légions romaines lors de l'invasion de la Gaule et les disettes et famines historiques qu'E. Le Roy Ladurie a si bien décrites du Moyen-âge à aujourd'hui), la seule solution est de puiser dans les stocks produits pendant les années fastes et conservés. Dans toute la période de la protohistoire, ces événements qui peuvent entraîner la disparition de la moitié d'une population, leur resteront inconnus.

Par contre, les effets des temps séculaires et des temps millénaires sur les sociétés humaines de l'Holocène sont observables à la fois par les enregistrements paléoclimatiques et par les changements systémiques qu'ils provoquent sur les sociétés humaines, selon leur stade d'organisation.

Si l'épisode 8200 BP ne semble avoir eu que peu de conséquences sur les dernières sociétés mésolithiques, elles ont eu un effet majeur sur les sociétés néolithiques (voir Nowak, Guilaine, Barich ce volume), effet qui va avoir des effets négatifs (baisse démographique, abandons de territoires et plus généralement choc systémique) mais aussi des rétroactions positives d'adaptation (migrations réussies, changements de systèmes de ressources alimentaires, innovations).

L'épisode aride 4200 BP va quant à lui affecter des sociétés déjà engagées dans l'urbanisation et l'étatisation au Proche-Orient comme l'ancien empire égyptien ou l'empire akkadien ou dans les iles de la Méditerranée en Crête ou à Malte et enfin le déclin des sociétés chalcolithiques d'Europe occidentale qu'accompagne le passage obscur à l'âge du Bronze (Guilaine, Lemercier, ce volume). Compte-tenu de l'organisation avancée et déjà complexe des sociétés concernées par cet épisode climatique, il n'est plus possible d'invoquer une causalité directe mais un facteur déclenchant voire des causalités convergentes dont la somme des difficultés entraine l'effondrement (disettes, déclin économique, affaiblissement militaire, montée des mécontentements, dissensions au sein du pouvoir, etc.), effondrement qui sera suivi d'un nouveau développement intégrant des innovations et des changements.

Il n'est donc pas étonnant de trouver à l'âge du Bronze et à l'âge du Fer, les mêmes débats, illustrés ici par C. Marcigny à la recherche de causalités dans les changements observées des sociétés de l'âge du Bronze et O. Buchsenschutz qui exprime des craintes face à des causalités trop caricaturales.

La disparition prématurée de Gilbert Kaenel, l'a empêché de nous remettre son manuscrit, tirée d'une communication qu'il avait donné en juin 2018 au Musée de l'Homme, dans le cadre du XIX° congrès UISPP à Paris, sous le titre : « *Les dégradations climatiques en Europe tempérée, de la fin de l'âge du Bronze au Ier siècle avant notre ère* ». Nous souhaitons citer ici *in extenso* son résumé : « *Les dégradations climatiques reconnues durant la fin de l'âge du Bronze (en croisant les résultats de différentes disciplines scientifiques) et leurs effets sur les modalités de l'occupation humaine, sont envisagées plus particulièrement ici sous l'angle de l'adaptation des communautés agricoles au milieu humide, soit les célèbres villages «lacustres» des lacs et marais subalpins, de la France de l'Est, du Plateau suisse et du sud de l'Allemagne. On observe qu'à ces dégradations (comme celle de Löbben vers 1500/1400 av. J.-C.) correspondent des transgressions lacustres et l'abandon des villages riverains. L'absence de telles occupations palafittiques entre 1500 BC et 1200 BC sur le Plateau suisse est patente. L'augmentation de l'activité solaire, dès -1100, entraîne une hausse des températures moyennes et un abaissement des niveaux des lacs qui coïncident avec plus de 2 siècles d'intenses occupations riveraines. A nouveau, une dégradation subite (la crise climatique du Subatlantique) dans la seconde moitié du IXe s. av. J.-C., signifie la fin de plus de 3 millénaires et demi de palafittes, entrecoupés d'interruptions, parfois de plusieurs siècles. L'influence du climat marque à l'évidence l'économie des sociétés, participe des changements culturels, démographiques, notamment au début de l'âge du Fer au VIIIe s. av. J.-C. Elle peut se traduire par une déstabilisation et des déplacements de populations, comme les célèbres migrations celtiques de la fin du Ve et du début du IVe s. av. J.-C. suivies de l'installation de peuples connus par l'Histoire en Italie du Nord. Inversement, la dendrochronologie et les Commentaires sur la Guerre des Gaules par Jules César nous renseignent indirectement sur le climat du milieu du Ier siècle av. J.-C.* ». Sa communication a été enregistrée et elle est heureusement accessible sur Youtube à l'adresse youtube.com/watch ?v=WTK5pDP-b9U. Elle a été vue par plus de 8 000 internautes.

Le peuplement humain au Pléistocène, avec les chapitres du premier volume, complété par deux chapitres sur la moitié Nord de l'Afrique de ce second volume, a été presque entièrement traité (avec cependant quelques manques comme l'ESA (Early Stone Age) d'Afrique ou la Chine encore trop incomplètement connue pour en tirer une synthèse). Il n'en est pas de même pour l'Holocène, car le second volume ne traite que très partiellement l'ensemble de la planète. Si l'Europe est assez bien couverte, des continents entiers sont absents comme l'Amérique du Nord au Sud, l'Australie et le Pacifique, le Proche-Orient, la zone équatoriale de la forêt tropicale humide où le thème est d'autant plus intéressant à traiter que les variations climatiques sont atténuées par rapport aux régions de haute latitude, l'Asie centrale, la Sibérie et l'Extrême-Orient. D'autres volumes nous semblent donc nécessaires pour intégrer ces études.

Extinctions animales et changements climatiques au quaternaire

Animal extinctions and climate change in the quaternary period

François Djindjian

Abstract

The causes of animal extinctions are subject to debate in Science from Buffon and Cuvier. Five large animal extinctions have been identified between 450 and 65 million years of Earth history. Then, the eras of the Tertiary and Quaternary did not show mass extinctions. However, significant variations in the climate throughout the Tertiary and Quaternary had a hard impact on evolution or disappearance of many species of animals and more systematically, on changes in the zoocenoses.

Since the beginning of the nineteenth century, once abandoned the allegory of the Noah flood, the position of naturalists for animal extinctions in the Quaternary oscillates between a climate-driven and a prehistoric human over-hunting. For the climate origin, Wallace in the nineteenth century, Hay in 1919; and for human hunting, Owen in 1860 in North America, Wallace in 1911 after changing opinion.

But the global theory, assigning animal extinctions to human action, is due to P.S. Martin between 1967 and 1984 ("Pleistocene overkill"). He concludes two main extinctions : during 50 000 - 40 000, in subtropical Africa and Eurasia and at the end of the last ice age, in Eurasia, the Americas and Australia. This model of "Pleistocene overkill" began to be questioned for each case by in-depth studies incorporating data of the archaeozoology, archaeology, more precise climate variations and more and more numerous radiocarbon dates.

The question then comes back essentially to a climate explanation, even if paradoxically, in the general public, and probably under the action of the ecological movement, the overkill hypothesis knew a strong resilience. However, there was most of the scientific work to achieve, which was to identify the processes, species by species, region by region, for each period and for each climatic environment, which are the cause of the extinctions.

This communication is intended to provide several specific examples of animal extinctions: mammoth in Eurasia and North America, large animals in Australia, large mammals in North American, etc. But we will also analyze the great migrations of fauna in latitude and altitude with glacial and interglacial climatic variations taking as an example the Western Eurasia during the stages isotopic 3, 2 and 1.

Résumé

Les causes des extinctions animales font débat dans la Science depuis Buffon et Cuvier. Cinq grandes extinctions animales ont été identifiées entre 450 et 65 millions d'années de l'histoire de la Terre. Ensuite, les ères du Tertiaire et du Quaternaire n'ont pas connu d'extinctions de masse. Cependant, les variations importantes du climat tout au long du Tertiaire et du Quaternaire ont eu un impact très fort sur l'évolution ou la disparition de nombreuses espèces animales et, plus systématiquement, sur les modifications des zoocénoses.

Depuis le début du XIX° siècle, une fois évacuée l'allégorie du Déluge, la position des naturalistes concernant les extinctions animales au Quaternaire oscille entre une origine climatique et une chasse abusive de l'homme préhistorique. Pour l'origine climatique, Wallace au XIX° siècle, Hay en 1919 ; pour la chasse humaine, Owen en 1860 en Amérique du Nord, Wallace en 1911 après avoir changé d'avis.

Mais la théorie d'ensemble attribuant les extinctions animales à l'action humaine, est due à P.S. Martin entre 1967 et 1984 (« pleistocene overkill »). Ses conclusions situaient vers 50 000 -40 000 la période majeure des extinctions, en Afrique et Eurasie subtropicale et à la fin de la dernière glaciation, les extinctions en Eurasie, dans les Amériques et en Australie.

Ce modèle du « pleistocene overkill » commença alors à être remis en cause pour chaque cas par des études approfondies intégrant les données de l'archéozoologie, de l'archéologie, des variations climatiques de plus en plus précises et des datations radiocarbone de plus en plus nombreuses. Il fallut alors revenir pour l'essentiel à une explication climatique, même si paradoxalement, dans le grand public et sans doute sous l'action des milieux écologistes, l'hypothèse de l'overkill a connu une forte résilience.

Il restait cependant l'essentiel du travail scientifique à réaliser, c'est-à-dire identifier les processus espèce par espèce, région par région, pour chaque période et pour chaque environnement climatique, qui sont à l'origine des extinctions. Mais seront également traitées les grandes migrations de faune en latitude et en altitude avec les variations climatiques glaciaires et interglaciaires en prenant comme exemple l'Eurasie occidentale au cours des stades isotopiques 3, 2 et 1.

Introduction

Les sociétés de chasseurs-cueilleurs paléolithiques sont-elles responsables à la fin du pléistocène d'exterminations d'espèces animales ayant conduit aux extinctions observées ?

Ou faut-il rechercher dans les variations climatiques les causes naturelles de ces extinctions ?

La théorie de l' « overkill pleistocene » est née en Amérique du Nord dans les années 1960-1980, influencée par les exterminations de la fin du XIXème siècle lors de la conquête de l'Ouest (tout particulièrement le bison d'Amérique). Son modèle a été progressivement appliqué à d'autres continents avec un succès variable. La théorie a été relancée dans les années 1990 en Australie en prenant modèle cette fois de l'extermination des Moas (des oiseaux coureurs géants) par les Maoris à leur arrivée en Nouvelle-Zélande. Dans tous les cas, c'est la coïncidence entre l'arrivée d'un peuplement humain (les chasseurs de Clovis en Amérique du Nord, les premiers aborigènes en Australie, les Maoris en Nouvelle-Zélande) et des extinctions d'espèces, et notamment des mégafaunes, qui est à l'origine de cette causalité. La théorie a fait l'objet de nombreuses et répétées critiques argumentées et la communauté professionnelle a pu considérer le débat comme clos dans les revues scientifiques. Mais une forte résilience peut être observée sur Internet où la théorie de l'overkill est toujours citée comme plausible voire probable. Les progrès des recherches sur le sujet depuis les années 2000 permettent d'apporter ici des arguments nouveaux pour régler définitivement la question.

Généralités

La question des extinctions de la fin du Pléistocène, que nous abordons ici, recouvre des processus en réalité très différents :
- L'extinction généralisée, que les géologues connaissent bien pour être survenue plusieurs fois sur la planète (Ordovicien, Dévonien, Permien, Trias, Crétacé), résultant de catastrophes (comme la chute d'une météorite par exemple) et de grands changements climatiques.
- L'extinction restreinte à une zone géographique plus ou moins vaste, plus ou moins oblongue pour une explosion volcanique, ou une bande en latitude plus ou moins large (zones désertiques ou zones périglaciaires) pour un changement climatique.
- Le remplacement rapide d'une espèce par une autre espèce mieux adaptée à un environnement qui a changé, résultat le plus souvent de deux migrations. Ces espèces sont particulièrement recherchées par le climatologue, pour la construction de courbes de paléotempérature et de paléohumidité, en particulier les rongeurs, les mollusques, les foraminifères et les diatomées.
- L'adaptation d'une espèce à un environnement changeant par sélection naturelle et par évolution. C'est le cas des proboscidiens (et particulièrement le mammouth et l'éléphant) dont la taille et la morphologie des molaires sont directement liées à la richesse en ressources alimentaires disponibles et à leur nature (herbes des steppes froides et des savanes, branches feuillues des espaces semi-ouverts et des forêts).
- L'extinction suite aux mécanismes de la sélection naturelle entre espèces (l'homme en faisant partie). L'extinction d'espèces endémiques suite à l'arrivée d'un prédateur dans une île en est le meilleur exemple.
- L'introduction par l'homme d'une espèce domestiquée (ou même sauvage) dans une île ou dans un nouveau continent.

- La réintroduction par l'espèce humaine d'une espèce sauvage disparue dans son biotope d'origine.
- La féralisation, c'est-à-dire le retour à la vie sauvage d'une espèce domestiquée par l'homme. Le cas de chèvres sauvages ou des mouflons dans les iles de la Méditerranée est bien connue. Le plus spectaculaire est sans doute celui des chameaux en Australie quand ils furent remplacés par des véhicules automobiles.
- La commensalité, c'est-à-dire l'adaptation des espèces non domestiquées aux sociétés humaines. Le cas du rat est celui qui a connu la plus grande explosion démographique.
- La prolifération d'une espèce, phénomène opposé à l'extinction, quand le développement démographique d'une espèce restreint la variabilité des autres espèces. Le cas de l'espèce humaine et de ses commensaux est évidemment le plus extrême.

Les extinctions ont donc des causes diverses : catastrophe géologique (météorite, éruption volcanique, tremblements de terre, déluges), changements tectoniques, changements climatiques, épidémies, irruptions d'une espèce prédatrice, surexploitation humaine et domestication.

Les animaux disparus et la préhistoire de l'Humanité

La preuve de la cohabitation de l'homme préhistorique et d'espèces animales disparues a été un des grands combats perdus de la Science au début du XIXème siècle. Les résultats des fouilles effectuées dans les remplissages des grottes et des abris, qui avaient révélé la coexistence d'outils taillés en silex par l'homme et d'animaux d'espèces disparues (comme celles de Paul Tournal à Bize dans l'Aude en 1827 ; Jules de Christol en 1828 dans les grottes de l'Hérault ; Philippe-Charles Schmerling à Engis et Engihoul en Belgique en 1829 ; Casimir Picard à Abbeville vers 1835) furent rejetés en invoquant le remaniement des remplissages qui aurait mélangé les vestiges. Les travaux de ces précurseurs se heurtèrent en France à la théorie fixiste, à Georges Cuvier, puis après sa disparition en 1832, à ses élèves, en particulier Pierre Flourens et Léonce Elie de Beaumont jusqu'en 1864, hostiles au transformisme de Jean-Baptiste de Lamarck (sans oublier les intuitions de Georges-Louis Leclerc de Buffon) et aux premiers travaux des évolutionnistes.

La première preuve indiscutable fut la découverte d'un mammouth gravé sur fragment d'ivoire de défense de mammouth à l'abri de la Madeleine, en Dordogne, publiée en 1865 par Edouard Lartet, et qui fut présentée à l'exposition universelle de 1867, où l'objet fit sensation. Elle prouvait indiscutablement non seulement la coexistence de l'homme et du mammouth, espèce éteinte mais les d'autres objets prouvaient également la coexistence de l'homme et du renne, espèce qui ne vit plus aujourd'hui que dans les régions arctiques et subarctiques. Rappelons également la moins connue sculpture féminine en ivoire « la vénus impudique » découverte par E. de Vibraye à l'abri de Laugerie-Basse, près des Eyzies, en 1863.

Assez rapidement, les paléontologues découvrent la diversité des faunes associées à l'homme dans les remplissages pléistocènes. Ce furent les débuts de l'expression « *faune chaude et faune froide* », qui devait conduire à la prise de conscience de changements climatiques importants, que les géologues devaient confirmer dès 1900 avec l'étude des glaciers alpins.

Parmi les espèces les plus fréquentes de faune froide du pléistocène supérieur, le mammouth laineux (*mammuthus primigenius*), le rhinocéros laineux (*coelodonta antiquitatis*), le renne (*rangifer tarandus*), le bœuf musqué (*ovibos moschatus*), le bison (*bison priscus*), l'antilope saïga (*saiga tatarica*), l'élan (*alces alces*), le renard polaire (*alopex lagopus*), le glouton (*gulo gulo*), le lion des cavernes (*panthera spelaea*), le loup (*canis lupus*), le bouquetin (*capra ibex*), l'ours des cavernes (*ursus spelaeus*), la marmotte (*marmota bobak*), le lièvre variable (*lepus timidus*) (figures 1 à 11).

Figure 1.[1]
Mammouth laineux
(*mammuthus primigenius*)

1 Les remarquables peintures animalières représentées ici sont l'oeuvre du peintre G.N. Glikman et proviennent du Musée de Zoologie NAS Ukraine. Elles sont publiées avec l'aimable autorisation du musée et en hommage à G.H. Glikman

Figure 2.
Rhinocéros
laineux
(*coelondonta
antiquatis*)

Figure 3. Renne
(*rangifer
tarandus*)

Figure 4. Bœuf musqué (*ovibos moschatus*)

Figure 5. Bison des steppes (*bison priscus*)

Figure 6.
Antilope saïga
(*saïga tatarica*)

Figure 7. Renard
bleu ou renard
Isatis ou renard
polaire (*vulpes
lagopus*)

Figure 8 .
Glouton (*gulo gulo*) et lièvre arctique (*lepus arcticus*)

▼ Figure 9. Loup (*canis lupus*)

Figure 10. Ours brun (*ursus arctos*)

Figure 11. Bouquetin (frise sculptée magdalénienne du Roc aux sorciers dans la Vienne). Photo L. Iakovleva

Figure 12a. Cerf élaphe (*cervus elaphus*)

Parmi les espèces les plus fréquentes de faune chaude du pléistocène supérieur, l'éléphant antique (*elephas antiquus*), le rhinocéros de Merck (*dicerorhinus mercki*), le rhinocéros de prairie (*stephanorhinus hemitoechus*), l'hippopotame (*hippopotamus major*), l'aurochs (*bos primigenius*), le daim (*dama dama*), le cerf (*cervus elaphus*), le cerf mégacéros (*mégalocéros giganteus*), le chevreuil (*capreolus capreolus*), l'âne (*equus hydruntinus*), le sanglier (*sus scrofa*), le lapin de garenne (*oryctolagus cuniculus*), le singe macaque (*macaca sylvanus*) (figure 12, 13).

Les faunes chaudes proviennent des régions méditerranéennes et d'Afrique. Les faunes froides sont des faunes de steppe froide continentale et des faunes de toundra, venant de l'Est. Une région comme le Nord de l'Europe occidentale verra donc au cours du dernier million d'années un va-et-vient de zoocénoses

Figure 12b. Cerf
élaphe à Lascaux

en fonction des variations climatiques et parallèlement une évolution des espèces dans le temps, s'adaptant aux changements d'environnement.

Remontant le temps du Quaternaire, au Pléistocène moyen et au Pléistocène inférieur, des espèces présentent ainsi des lignées évolutives comme celle des mammouths (*méridionalis, trogontherii puis primigenius*), du rhinocéros (*dicerorhinus etruscus, dicerorhinus mercki, coelodonta antiquitatis*), des ours de cavernes (*etruscus, deningeri puis speleus*), des chevaux (*stenonis, mosbachensis, ferus*) ou des bisons (*shoetensacki, priscus, bonasus*).

Les espèces apparaissent ou disparaissent. Elles se différentient suivant les continents, les géographies physiques, les environnements, les changements de ressources alimentaires. C'est le mécanisme de l'évolution et le changement climatique est un des moteurs de cette évolution.

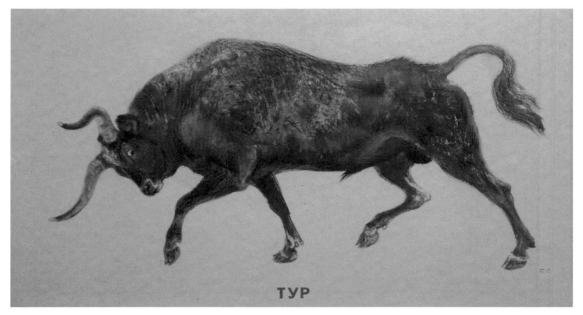

ТУР

Figure 13.
Aurochs (*bos primigenius*)

Les changements climatiques

L'évidence que l'homme préhistorique ait vécue sous un climat glaciaire fut suivie rapidement par l'évidence que le passé quaternaire avait connu des alternances de périodes glaciaires et de périodes interglaciaires. Mais il fallait estimer les amplitudes et la chronologie de ces changements, et les mettre en relations avec les occupations humaines et les zoocénoses animales. Les glaciers alpins fournirent le premier référentiel paléoclimatique au début des années 1900 (Penck et Bruckner, 1901-1909) puis la théorie de Milankovitch en expliqua la cause dans les années 1930 (Milankovitch, 1941). Au XXème siècle, dans les années 1970, les préhistoriens et les quaternaristes multiplièrent les référentiels : oiseaux, rongeurs, mollusques, diatomées, pollens, charbons de bois, séquences de lœss, terrasses fluviatiles, varves lacustres, moraines de glaciers, dendrologie (par exemple, Puyssegur, 1976 pour les mollusques ; Vernet, 1973 pour les charbons de bois ; Velitchko, 1981 pour les séquences de lœss ; Woillard 1978, pour les pollens). La paléoclimatologie moderne se constitua alors en pratiquant les carottages de plus en plus profonds dans les lacs, les océans et les calottes glaciaires, et en cherchant les meilleurs estimateurs : foraminifères, ostracodes et rapport isotopique O^{16}/O^{18} (Imbrie, Kipp, 1971 ; Dansgaard et al. 1971).

Les premières courbes paléoclimatiques sont issues de ces travaux et ces courbes sont devenues, depuis 50 ans, plus justes, plus fiables, plus précises et plus longues dans le temps. Les relations entre les sociétés humaines et les variations climatiques nécessitent cependant, outre un pixel climatique du même ordre de grandeur que le pixel archéologique, ce qui n'est pas le cas encore aujourd'hui, de construire séparément les courbes de paléotempérature et les courbes de paléohumidité. En effet, c'est la courbe de paléohumidité qui est le marqueur le plus important pour les sociétés végétales, donc animales et humaines. En période glaciaire, par exemple, l'adaptation à la sécheresse est plus difficile que l'adaptation au froid, aussi bien dans les zones septentrionales de l'Eurasie et de l'Amérique du Nord que dans les zones désertiques de l'Afrique.

La théorie de l'extermination pléistocène (« overkill pleistocene »)

La théorie de l'extermination pléistocène est née suite à la publication d'une session de communications du 7ème congrès de l'INQUA (Boulder, Colorado) en 1965 (Martin, Wright, 1967). La théorie connut alors une vogue certaine. Presque vingt ans après, un second volume avec une quarantaine de contributions eut pour but d'apporter des données nouvelles à la théorie (Martin, Klein, 1984). La théorie, initialisée et constamment soutenue par Paul S. Martin, paléontologue à l'Université d'Arizona, avec le support de nombreux chercheurs nord-américains, a été influencée par les extinctions modernes en Amérique du Nord, au XIXème siècle, notamment l'extermination du bison et du pigeon migrateur.

La théorie est née de la coïncidence, à la fin du pléistocène supérieur, en Amérique du Nord, entre l'extinction du mammouth et la présence des chasseurs de Clovis, considérés toujours par la plupart de nos collègues nord-américains comme les premiers colonisateurs du continent américain. Les chercheurs ont ensuite recherché d'autres illustrations de la théorie en Amérique (grands mammifères), en Australie (grands marsupiaux), en Eurasie (mammouth) et dans les espèces endémiques des îles de la Méditerranée.

La diffusion de la théorie a entraîné des réactions et des réfutations. Les adversaires de la théorie considèrent que c'est le changement climatique qui a entraîné au pléistocène supérieur la disparition, que les chasseurs-cueilleurs aient été ou non des prédateurs de ces animaux disparus. Un bon résumé de ces arguments, qui date des débuts des années 2000, a été donné par X. de Planhol (de Planhol, 2004, chapitre 5). A sa lecture, on ne pourrait qu'en déduire que la cause est définitivement entendue. Un des objets de ce texte est de les mettre à jour, à la lueur des données récentes.

Cependant, la consultation des sites Internet révèle toujours la grande popularité de la théorie de l'extermination pléistocène, du fait de l'amalgame fait à tort avec, au XIXème siècle, les exterminations au fusil, et avec, aux époques historiques, l'arrivée des prédateurs humains dans les îles (dont les exemples les plus connus sont les Maoris en Nouvelle-Zélande et les Européens dans les îles de l'océan indien).

La raison de cet engouement est l'interférence avec la cause écologiste, qui accueille, avec enthousiasme et de façon souvent irrationnelle, toute dénonciation de l'action destructrice de l'homme, bien que paradoxalement ce soit ici le « bon sauvage », généralement paré de toutes les vertus depuis Rousseau et Engels, qui en soit le coupable (paléo-amérindiens, Maoris, Sapiens).

Les arguments des pro- et anti-théorie de l'extermination pléistocène seront actualisés et développés ci-après, continent par continent. Mais il est déjà important de dire que les causes de ces disparitions sont systémiques intégrant tous les processus comme les changements climatiques, la déstabilisation de la dynamique des populations animales (notion de seuil critique), la durée de gestation et le nombre de petits par portée (valeurs proportionnelles à la taille des animaux et qui entraine un plus grand risque d'extinction pour les mégafaunes que pour les microfaunes en cas de changement d'environnement), et les prédations humaines et animales excessives.

Plusieurs processus, qui sont des conséquences des variations climatiques, seront approfondis :
- la sécheresse du dernier maximum glaciaire qui entraîne une forte extension des zones désertiques,
- le retour de l'humidité à la fin de la dernière glaciation pléistocène, qui entraine de fortes précipitations neigeuses en hiver dans les zones de steppe froide, condamnant à la disparition les grands mammifères brouteurs et leurs prédateurs,
- la transgression marine, en période interglaciaire, qui restreint la superficie des zoocénoses, et fait tomber la démographie de nombreuses populations animales sous le seuil critique,
- le changement de végétation au début de l'holocène, qui force la faune à s'adapter à la croissance de la forêt au détriment de la steppe froide : migrations en haute latitude (renne, bœuf musqué) et en altitude (bouquetin, marmotte), adaptations à la forêt (bisons, cervidés) ou disparitions (cheval),

Par ailleurs, la multiplication des datations radiocarbone et leur meilleure fiabilité, permettent d'obtenir l'information de la date de disparition d'une espèce en un endroit donné, voire même d'obtenir des diagrammes spatiaux de disparition dans le temps d'une espèce animale (comme par exemple l'extinction progressive du mammouth d'Europe, puis de Sibérie jusque dans l'île arctique de Wrangel).

Le dernier maximum glaciaire en Europe

Le cas du dernier maximum glaciaire, il y a vingt mille ans environ et plus précisément entre 22000 et 17 000 (en âge BP), est particulièrement instructif car la science préhistorique possède maintenant une connaissance précise (à 500 ans près) des sociétés de chasseurs-cueilleurs, de l'environnement animal et végétal, et du climat.

Si les courbes de paléotempérature mettent en évidence un maximum de froid, les courbes de paléohumidité révèlent des variations avec des épisodes très secs attendus mais également des épisodes plus humides, qui sont plus favorables à la végétation arborée, aux mammifères et aux groupes humains. Ces épisodes sont visibles dans les séquences polliniques de lacs (comme dans la séquence lacustre de Tenaghi-Philippon en Macédoine mais aussi par des sols fossiles et des gleys dans les séquences de lœss d'Europe orientale et centrale.

Ces épisodes humides sont datés entre 20 000 et 19 000 pour le premier et entre 18 500 et 17 000 pour le second (en âge BP). En Europe occidentale, où les stratigraphies en abris sous roche sont nombreuses et précises, ces peuplements correspondent à des cultures matérielles différentes dont la séquence est la suivante (Djindjian *et al.* 1999 ; Djindjian, 2018):

Fin du pléniglaciaire supérieur ancien	Gravettien final
Episode sec du Maximum glaciaire	Protosolutréen et Solutréen ancien
Episode humide 1 du Maximum glaciaire	Solutréen récent
Episode sec du Maximum glaciaire	Badegoulien ancien
Episode humide 2 du Maximum glaciaire	Badegoulien récent
Début du pléniglaciaire supérieur récent	Magdalénien inférieur

Les effets du maximum glaciaire sur la géographie de l'Europe sont spectaculaires :

- Croissance de la calotte glaciaire dont l'extension maximale arrive jusqu'à Berlin et Moscou,
- Descente du glacier alpin du Rhône jusqu'à la zone occupée aujourd'hui par la banlieue de Lyon,
- Extension de la toundra jusqu'à la vallée de la Loire,
- Baisse maximale du niveau de la mer (-120 m par rapport au niveau actuel),
- Le golfe adriatique est réduit de la moitié de sa superficie et son entrée est plus étroite,
- La mer Egée est presque fermée par le cordon de ses iles méridionales,
- La Mer Noire devient un lac.

Les changements de zoocénoses commencent à être mieux connus grâce à des datations radiocarbone d'ossements animaux dont l'espèce a été déterminée, trouvés en stratigraphie de sites préhistoriques. En Aquitaine, à partir de 27 000 BP, la disparition et/ou la migration d'espèces animales vers la péninsule ibérique subpyrénéenne et subcantabrique est observée et concerne l'aurochs, le sanglier, le chevreuil, le cerf mégacéros et l'âne. L'élan disparait de la plaine de Pannonie en Europe centrale. A l'arrivée du dernier maximum glaciaire, sont également observées l'arrivée de l'antilope saïga et la descente du bœuf musqué (en relation avec l'expansion de la toundra).

Le cas du mammouth est plus paradoxal. Présent au pléniglaciaire ancien où le climat froid et sec est favorable à la steppe froide, il préfère les zones continentales (comme le bassin de la Saône et la vallée du Rhône) aux zones océaniques plus humides (comme l'Aquitaine où sont trouvés dans les niveaux archéologiques de fragments de molaires et des fragments d'ivoire de défense mais pas l'ensemble du squelette postcrânien). Au maximum glaciaire, la proximité entre la calotte glaciaire et le glacier alpin réduit voire bloque le passage naturel des troupeaux de mammouths sur la grande plaine de l'Europe septentrionale, faisant passer la population de mammouths en Europe occidentale sous le seuil critique de reproduction. Le mammouth y reviendra au cours du pléniglaciaire supérieur récent, mais en effectif plus réduit.

Le cas de la faune de caverne est particulièrement significatif. L'ours des cavernes (*ursus spelaeus*) est un grand ursidé végétarien (descendant *d'ursus etruscus* puis *d'ursus deningeri*), qui hiberne dans des grottes d'altitude élevée (de 500 m à 2 300 m), où sont retrouvés bien conservés de nombreuses carcasses d'ours et d'oursons morts pendant l'hiver (et ayant donné sans doute naissance à la légende des dragons). Parfois, de courtes occupations humaines sont observées dans ces grottes, vestiges d'une halte de chasse pendant la bonne saison. Les ours et les humains s'y sont croisés mais ils ne se sont pas rencontrés. L'ours des cavernes (avec ses différentes sous-espèces : *ursus ingressus*, *ursus spelaeus eremus* et *ursus spelaeus ladinicus*) s'est éteint avant le début du dernier maximum glaciaire vers 24 000 – 22000 BP (Baca *et al.* 2016). La montée vers le dernier maximum glaciaire a entraîné un accroissement du permafrost et de la toundra au détriment de la steppe froide et un cloisonnement des territoires européens entraînant un déficit en nourriture végétale, auquel l'ours des cavernes, herbivore, a moins bien résisté que l'ours brun (*ursus arctos*) omnivore qui a survécu.

D'autres espèces ont disparus dans cette phase d'extinction, comme le grand ours à face courte nord-américain (*arctodus sinus*) et un éléphant, le palaeoloxodon naumani de l'archipel japonais. L'hyène des cavernes (*crocuta crocuta spelaea*) originaire d'Afrique et dont la dernière migration en Europe remonte au MIS 5, pourrait avoir perduré encore après le LGM. Le lion des cavernes (*panthera leo*) s'est éteint aussi après le LGM quand sa dernière proie habituelle, le renne, disparait de son territoire.

Pour revenir aux ursidés, l'ours à collier du Tibet (*ursus thibetanus*), une espèce végétarienne de climat tempéré, est rare mais présent en Europe dans les interglaciaires. Il a disparu au plus tard à la fin du MIS 5, alors qu'en Amérique, son cousin l'ours noir (*ursus americanus*) a survécu jusqu'à nos jours.

Il en est de même pour les groupes humains qui abandonnent les territoires de l'Europe septentrionale et de la moyenne Europe pour se réfugier dans les régions méditerranéennes : péninsule ibérique, golfe adriatique, golfe égéen, pourtour de la Mer Noire (lac). Ces groupes humains doivent alors s'adapter à des environnements différents qui les obligent à changer leur gestion des ressources alimentaires dans le cycle annuel (chasse aux animaux grégaires remplaçant la chasse aux troupeaux migrateurs) et leurs approvisionnements en matière première qui ne sont plus accessibles.

Pendant les épisodes humides, ces mêmes groupes humains remontent en Europe moyenne pendant la bonne saison pour compléter leurs ressources. C'est le cas du Solutréen récent en Europe occidentale. Le Badegoulien est ensuite une tentative de repeuplement permanent de l'Europe moyenne.

Cette précision des données à notre disposition permet de pouvoir analyser le rôle des groupes de chasseurs-cueilleurs dans les changements de zoocénoses. Ces changements de zoocénose (et c'est la même explication pour les groupes humains) sont liés au changement climatique.

L'extinction du mammouth en Eurasie

Le mammouth est un herbivore de steppe froide qui apprécie un environnement froid et sec comme le rhinocéros laineux avec qui il est souvent associé dans la même zoocénose (figure 14). Le mammouth *primigenius* (MIS 4, 3, 2) est le plus petit et le dernier de la lignée après le mammouth *trogontheri* des steppes

Figure 14. Mammouth méridional (*mammuthus meridionalis*) de Durfort (muséum National d'Histoire Naturel, Galerie de Paléontologie, Paris). Il a été découvert en 1869 à Durfort (Gard) par Paul Casalis de Fondouce et Jules Ollier de Marichard. Le « petit mammouth » à l'arrière plan est un mammouth laineux (*mammuthus primigenius*) découvert par KA Vollosovitch dans les iles Liakhov en Sibérie en 1903 et donné au muséum.

froides de la glaciation du Riss (MIS 6-8) et le mammouth *meridionalis,* un animal plus forestier des climats tempérés (MIS 11 à 17 et avant).

A partir de 14 500 BP, le climat du pléniglaciaire supérieur récent devient plus humide, les précipitations neigeuses s'accentuent et recouvrent la steppe pendant une partie de l'année, à la mauvaise saison. Faute de nourriture, l'herbe devenant inaccessible et, en l'absence d'arbres, les troupeaux sont condamnés à mourir de faim. Si tel est le cas, nous devrions découvrir les restes de ces troupeaux sous la forme de concentrations de carcasses en place, que le manteau neigeux a recouvert jusqu'à sa fonte au printemps suivant, entrainant une sédimentation rapide.

Or justement les fouilles de ces trente dernières années dans le bassin du Dniepr et les révisions des fouilles anciennes ont mis en évidence dans des ravines de versants de vallées, ces accumulations de carcasses de mammouths (Djindjian, 2015). Ces ravines sont des caractéristiques géomorphologiques bien connues des versants de vallées de la grande plaine orientale où se sont accumulées des lœss de trente à cinquante mètres de puissance. Un léger dénivelé à un endroit donné du bord du versant entraîne avec les précipitations, le creusement rapide de ces ravines dont la forme en bec de théière est caractéristique. Il est probable que les troupeaux aient essayé de se protéger des tempêtes de neige dans ces ravines bien abritées du vent. Nous savons aujourd'hui que les groupes humains recherchaient systématiquement au début du printemps, au moment de la fonte des neiges, ces accumulations et lorsqu'ils les avaient trouvées, ils les exploitaient. A ces carcasses dont une grande partie des ossements ont été prélevés, sont associés des outils en silex et en matière dure animale, ainsi que des foyers. Alors, le groupe établit son habitat à proximité et construit des cabanes en os de mammouths avec les grands ossements (crânes, mandibules, omoplates, bassins, os longs) et les défenses. Il utilise également les os frais comme combustible et comme matière première pour la fabrication d'outils. Si le groupe humain est arrivé tôt au moment de la fonte des neiges, alors le mammouth gelé peut être encore comestible et les carcasses sont dépecées, ce qui explique la présence de foyers entre les carcasses. Il ne reste alors de l'accumulation initiale des carcasses que les côtes, les vertèbres, les os de pieds et les os hyoïdes, souvent en position quasi-anatomique, ce qui permet d'identifier chaque carcasse.

Dans d'autres cas, le groupe humain est arrivé trop tard et l'exploitation en est réduite. Et parfois, l'accumulation est intacte.

Ces accumulations ont été insuffisamment étudiées pour plusieurs raisons. Les archéologues ne les ont fouillées que le temps d'une saison, car ces sites sont pauvres en industrie, et peu publiés si bien que ces accumulations ne sont connues que par les rapports de fouilles conservés en archives. Quant aux paléontologues, plus intéressés par l'étude des ossements de mammouths, ils ont négligé les artefacts humains.

Un site où ces accumulations de carcasses ont été bien étudiées, le site de Gontsy en Ukraine (Iakovleva, Djindjian, 2017), a révélé que cette accumulation, qui est le résultat d'un évènement unique, est celle d'un troupeau de mammouths, constitué de femelles, de subadultes, de jeunes et d'enfants. Aucun mâle n'est présent. Les cabanes de l'habitat, situé à proximité immédiate, ont été construites à partir des ossements récupérés de l'accumulation. La répartition en âge, en sexe et en nombre d'individus est la même. Quelques ossements d'adulte mâle ont été collectés ailleurs, recherchés pour leur grande taille, surtout les défenses, aisément reconnaissables par une altération qui montre qu'ils sont restés un certain temps à la surface du sol. Une révision des sites du bassin moyen et supérieur du Dniepr a révélé qu'à chaque habitat avec des constructions en os de mammouths, est associée une accumulation de carcasses de mammouths située à une dizaine de mètres dans une ravine voisine. Cette proximité permet de confirmer la facilité et la rapidité de construction des cabanes en quelques jours.

Il existe d'autres cas d'accumulations, qui ont été étudiées par les paléontologues. Le piège naturel en est un, comme le site Hot Springs dans le Dakota du Sud où furent trouvés des individus de l'espèce *mammuthus columbi* et de nombreuses autres espèces animales. Les 61 mammouths sont presque tous des mâles, datés d'il y a au moins 26 000 ans (Agenbroad *et al.* 1994). En milieu périglaciaire, le changement de lit des rivières érode les accumulations qui se redéposent en aval par gravité. Le site de Berelekh en Sibérie nord-orientale, fouillé par N. Vereshchagin dans les années 1970, qui est daté entre 13 700 BP et 11 600 BP, et qui a livré 140 à 200 individus sur une durée de 2 000 ans, semblerait appartenir à ce cas (Vereshagin, 1977 ; Pitulko, 2011). Par contre, à Sevsk (Briansk), en Russie européenne, fouillé et étudié par E. Maschenko (Maschenko *et al.* 2017), il s'agit bien d'un troupeau mort vers 14 000 BP. Il est constitué de 33 individus. Seize outils en silex révèlent qu'il a été exploité par un groupe humain. La distribution des parties anatomiques le confirme : 68% des os sont des côtes, des vertèbres et des os de pieds. Il reste donc à trouver l'habitat situé à proximité. Dans tous ces cas, les études taphonomiques permettent de faire le diagnostic de l'origine de ces accumulations. Le concept de « *cimetières d'éléphants* » a donc des déclinaisons multiples, mais toutes d'origines naturelles.

La question de la chasse aux mammouths a été posée dès le XIXème siècle. Les expériences de tir ont montré qu'il était possible de chasser l'éléphant avec des lances munies de pointes en silex (pointes de Clovis, pointes de Kostienki) ou en matière dure animale. Dans ce cas, l'animal saigne et parcourt plusieurs dizaines de kilomètres avant de s'effondrer obligeant les chasseurs à dépecer l'animal sur place et à revenir avec des quartiers de viande. Les archéologues nord-américains ont découvert à plusieurs reprises ce qu'ils ont appelé des piles d'ossements isolées, en fait ce qui reste d'une carcasse abandonnée après avoir été dépecée et quelques rares outils (pointes de Clovis, tranchets en os de mammouth) démontrant l'origine humaine de l'évènement. Ces chasses concernent des individus isolés, jeunes mâles et vieux mâles affaiblis ou des individus blessés. Les troupeaux, composés de femelles adultes de différents âges, de subadultes (les jeunes mâles sont expulsés du troupeau vers l'âge de douze ans), de jeunes et d'enfants, et dirigés par une matriarche, sont dangereux d'approche, les femelles chargeant pour protéger leurs petits (même au

XIXème siècle avec un fusil à longue portée). Il n'est pas possible d'imaginer la possibilité de tuer sur place un troupeau entier de quelques dizaines d'individus. Les solutions proposées ne présentent guère de faisabilité : repousser par une battue un troupeau vers une ravine (dans l'espace ouvert de la steppe froide ?), le bloquer (comment empêcher la charge des adultes ?), et lui jeter des pierres (il n'y en a pas dans la grande plaine) ou des lances du haut de la ravine (ce ne serait possible qu'à son extrémité haute car elles s'évasent rapidement le long du versant).

Il est donc important de souligner que l'exploitation des carcasses de mammouths gelés morts de famine pendant l'hiver et que la chasse aux mammouths (et aux autres mammifères comme le renne, le cheval ou le bison) pendant la bonne saison sont parfaitement complémentaires et démontrent l'intelligence de l'opportunisme des groupes humains face aux situations naturelles.

Les datations d'ossements de mammouths mettent en évidence la date de sa disparition en Eurasie à partir de 14 000 BP. Il a commencé à la fin du Dryas I mais l'épisode tempéré et humide de Bölling lui a été fatal. La disparition progressive a commencé par l'Europe vers 14 000-13 000 BP, puis la Sibérie occidentale, la Sibérie centrale jusqu'en Sibérie orientale (il y a 10 000 ans). En Amérique du Nord, la disparition est datée d'il y a 13 300 ans. Le dernier refuge dans l'île Saint-Paul en Alaska (il y a 6 400 ans) et l'île de Wrangel dans l'océan arctique (il y a 4 000 ans), isolées suite à la remontée de la mer, montre en outre un phénomène de nanisme insulaire et une dégénérescence propre à une perte de diversité génétique.

Dans le Paléolithique supérieur européen, les sites ayant livré de très nombreux ossements de mammouths sont pour l'essentiel le Gravettien ancien de Moravie, le Gravettien récent en Europe orientale, la culture de Zamiatnine sur le Don à Kostienki et le Mézinien en Ukraine. Des constructions en os de mammouths, à l'architecture différente cependant d'une culture à l'autre, sont présentes dans ces sites.

En Europe centrale et orientale, les peuplements du Paléolithique supérieur ancien et moyen, connus comme Aurignacien et Gravettien, n'ont été la cause d'aucune extermination de mammouths. En Sibérie, c'est également le cas des peuplements du Paléolithique supérieur ancien et moyen, connus comme la culture de Malta-Buret, autour de 24 000 BP.

Le Mézinien est un exemple exceptionnel. Il s'agit d'un peuplement humain occupant un territoire de 500 000 km^2 du bassin moyen et supérieur du Dniepr, pendant moins de 1000 ans. Non seulement, ce peuplement n'est pas à l'origine de la disparition des mammouths, mais il est possible d'affirmer qu'il a profité de la disparition naturelle du mammouth du fait du changement climatique en exploitant systématiquement les accumulations de carcasses congelées. Quand le mammouth a définitivement disparu, le groupement humain a également disparu après lui.

En Eurasie, les chasseurs-cueilleurs ne sont donc en aucun cas responsables de la disparition du mammouth.

Le bison en Europe et en Amérique du Nord

Le bison est un animal des grandes plaines, qui, pendant les temps glaciaires, sont occupées par la steppe froide. Au Paléolithique supérieur, au MIS 3, le bison semble avoir une large distribution géographique, mais qui, au MIS 2, se réduit à des isolats en Aquitaine, dans la plaine du Pô et sur le pourtour septentrional de la Mer Noire.

En Europe orientale, son aire de répartition est plutôt méridionale ; il ne remonte pas au-dessus du 50° parallèle. En Europe occidentale, il ne

Figure 15a. Bisons magdaléniens de la grotte du Tuc d'Audoubert (Pyrénées)

Figure 15b. Bison sculpté en ivoire du site gravettien de Zaraisk en Russie

semble pas remonter au-dessus du 47° parallèle. En Aquitaine et sur la côte cantabrique, le bison est particulièrement chassé dans la plaine au Magdalénien moyen et figuré dans les grottes avec le cheval.et le bouquetin (figures 15 a, b). Mais dans les vallées (qui descendent des Pyrénées et du Massif central), c'est la chasse spécialisée au renne en début et fin de bonne saison qui a donné l'expression « âge du renne » aux sites en abris du paléolithique supérieur. La même situation se reproduit en Europe orientale avec la chasse au bison dans la plaine et la chasse spécialisée au renne dans les vallées qui descendent des Carpates (Dniestr, Prut, Bistrita, etc.). Les restes osseux de bison, diminuent progressivement au tardiglaciaire et pourraient laisser penser que les chasseurs avaient réduit la contribution du bison à leurs ressources alimentaires. A l'Holocène, le bison (*biso priscus*) survit en s'adaptant à la forêt (*biso bonasus*). Certaines études génétiques récentes considèrent que le bison bonasus est le résultat d'un croisement entre le bison priscus et l'aurochs. Il est encore présent quoique rare à l'époque romaine et disparait au Moyen-âge. Son extinction est notifiée au Kouban en 1925 et en Pologne en 1942.

En Amérique du Nord (Speth, 2019 pour une synthèse récente), les bisons pléistocènes sont le bison antiquus et le bison latifrons, plus grands que le bison moderne. Ils se sont transformés en une espèce plus petite, le bison bison, à l'Holocène. A la différence des autres mammifères d'Amérique du Nord, le bison n'a donc pas subi d'extinction, bien qu'il ait été également chassé par les chasseurs de Clovis et de Folsom. Les études concluent que le taux de reproduction annuel est de 15 % et que les prélèvements des amérindiens à l'époque historique n'ont pas dépassé 2%. Les progrès cynégétiques et l'importance croissante de la chasse au bison dans les derniers 13 000 ans n'ont pas entrainé la disparition du bison : chasse collective à pied, chasse par battues vers des ravines, chasse à cheval (XVIII° siècle). C'est l'usage combiné de la chasse à cheval et du fusil au XIX° siècle qui conduit à l'extermination complète en trente ans (1850-1880) dont la responsabilité incombe aux Européens.

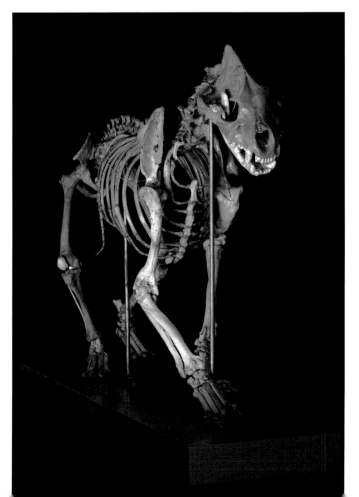

Figure 16. Hyène des cavernes (*crocuta crocuta spelaea*). Muséum d'Histoire Naturelle de Toulouse

La fin de la dernière glaciation en Europe

A partir des épisodes tardiglaciaires de Bölling mais surtout de l'Alleröd, le climat se réchauffe rapidement. Les glaciers de montagnes reculent (Alpes) ou disparaissent (Pyrénées, Massif central, Apennins, Carpates, Balkans), permettant la réoccupation ou le refuge d'espèces animales et leur développement en altitude et en conséquence des groupes humains qui les rejoignent à la bonne saison : la chasse aux rennes dans leur migration saisonnière bas vers haut de vallée au passage des gués (par exemple dans les Pyrénées, le massif central et les Alpes au Tardiglaciaire), la chasse au chamois et à la marmotte.

Le réchauffement entraîne également la disparition et/ou la migration des espèces de toundra puis de steppes froides vers le Nord-est (mammouth, rhinocéros, saïga, renne et glouton) et en corollaire l'extinction des faunes de cavernes (ours des cavernes (cf. supra), lion des cavernes (*panthera leo*), hyène (*crocuta crocuta spelaea*)) du fait de la disparition de leurs proies habituelle (figures 16, 17).

A l'inverse, le retour des faunes réfugiées dans l'Europe méridionale se concrétise dans l'Europe moyenne : c'est le retour de l'aurochs au nord des Pyrénées (Bölling), puis le sanglier, le chevreuil, le cerf mégacéros, le daim et la grande extension du cerf en Europe moyenne. Le lapin de garenne arrive, le lynx également dont le biotope est la forêt.

Le développement de la forêt holocène conduit à une adaptation progressive ou complète de certaines espèces au milieu forestier (bison, aurochs, cerf élaphe, daim, chevreuil).

Le coup de froid intense et bref du Dryas III a eu certainement des répercutions sur la faune qui avait réoccupé depuis peu, à l'Alleröd, l'Europe moyenne. Mais nous n'en savons encore que peu de choses. Les données actuelles laissent penser que cet épisode aurait été fatal au cerf mégacéros. Mais il est possible que cet animal spectaculaire ait survécu à l'Holocène dans des zones plus septentrionales ou plus continentales que la forêt n'avait pas colonisé.

Figure 17. Lion des cavernes. Sculpture aurignacienne d'homme-lion en ivoire de la grotte de Hohlenstein-Stadel (Bade-Wurtemberg)

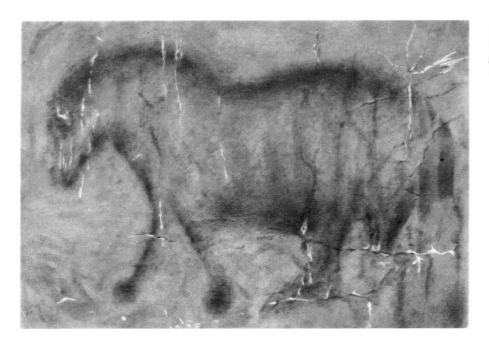

Figure 18. Cheval peint de la grotte de Kapova (Oural)

Il reste enfin à étudier une des énigmes les plus fascinantes, celle de la disparition du cheval, en Europe comme en Amérique du Nord, simultanément, à la fin des temps glaciaires. Le cheval a été aux époques paléolithiques une des principales ressources alimentaires des groupes humains (figure 18), avec le renne et dans une moindre mesure avec le bison, l'aurochs et le cerf. Nous n'avons pas encore une vue très précise sur l'extinction du cheval pléistocène en Amérique du Nord comme en Europe. Des datations radiocarbone révèlent la présence du cheval en Angleterre vers 10 000 BP, et au Canada vers 11 500 BP. En France, des restes osseux de chevaux ont été trouvés dans des niveaux du Mésolithique ou dans des niveaux du Néolithique ancien (cardial à l'abri Jean Cros) jusque vers 6 500 BP. Il disparait complètement pendant deux mille ans dans le Néolithique moyen et final pour réapparaitre au Campaniforme (2 800- 2500 BC), dans la péninsule ibérique et en France. Ces restes osseux révèlent sa consommation, à l'instar d'autres espèces animales sauvages. Serait-ce la preuve de la survivance en péninsule ibérique du cheval sauvage pléistocène ? En attendant les résultats de nouvelles études génétiques, il est possible de considérer que le cheval, animal de steppe et de prairies, n'a disparu que progressivement au début de l'Holocène, face au développement de la forêt. Pour cette raison, il aurait pu se maintenir plus facilement sur les hauts plateaux de péninsule ibérique. L'arrivée du cheval domestique en Europe, venant de l'Est, date des débuts de l'âge du Bronze à partir de 2 220 BP. Les meilleurs candidats semblent être aujourd'hui les troupeaux de chevaux des steppes du Nord Caucase. Les études génétiques ont récemment révélé que les derniers chevaux sauvages actuels des steppes asiatiques sont des chevaux féralisés.

En conclusion, en Europe, les espèces animales pléistocènes n'ont pas connu d'extinctions brutales. Certaines ont survécu à des latitudes ou des altitudes où elles ont retrouvé leur biotope naturel. D'autres se sont adaptées à l'environnement holocène forestier. D'autres sont venus recoloniser l'Europe moyenne à partir de leurs refuges méditerranéens. Les chasseurs-cueilleurs ne peuvent être tenus responsables de ces changements.

L'Amérique du Nord

La consultation des sites Wikipedia révèle, qu'encore aujourd'hui, les premiers Américains sont les chasseurs cueilleurs (Culture de Clovis) arrivés en traversant le détroit de Béring et en suivant le « corridor » entre les glaciers des Montagnes rocheuses et la calotte glaciaire des Laurentides il y a 13 500 ans. Cette théorie est toujours considérée comme la référence.

Pourtant, la multiplication des fouilles et des résultats multiplient les indices d'une plus grande ancienneté du peuplement humain en Amérique que les chasseurs de Clovis. Le site de Pedra Furada est daté de façon fiable vers 22 000 BP pour le niveau C7 associé à une industrie archaïque (Boeda *et al.* 2014). Le site de Monte Verde (Chili) est daté de 14 500 BP (Dillehay, 2015). La grotte de Bluefish fouillé par J. Cinq Mars à partir de 1979 au Yukon canadien a fourni des ossements animaux avec des marques de découpage daté de 24 000 BP qui ont été contestées en leur temps. Une récente révision du matériel aurait permis d'identifier 15 artefacts osseux avec des marques de découpage (Bourgeon *et al.* 2017).

Ces sites pré-Clovis posent alors la question du ou des moments d'arrivée des groupes humains sur le continent américain. Il n'a jamais été réellement démontré que le corridor était praticable vers 14 000 BP à la fonte du dernier glaciaire et il est possible d'en douter. On peut même considérer que cette hypothèse a été faite *a posteriori* connaissant les premières datations radiocarbone des sites de Clovis. Mais, par contre, le passage du détroit de Béring et du corridor est plus praticable pendant le MIS3 entre 38 000 et 28 000 BP et même jusqu'à 25 000 BP. Le climat est alors interpléniglaciaire, plus tempéré et humide. Les glaciers et la calotte glaciaire ont reculé. Mais le niveau de la mer n'a pas eu le temps de remonter significativement. Le passage du détroit se faisait donc à sec et le corridor est plus large.

La référence sibérienne la plus proche est le site de Yana, à l'embouchure de la rivière Yana, en Sibérie orientale à 71° de latitude Nord (Pitulko *et al.* 2004), daté de 27 800 BP, avec une industrie lithique et une industrie osseuse de type paléolithique supérieur ancien. La faune est composée du mammouth, du rhinocéros laineux, du renne, du cheval, du bison, du lièvre, du bœuf musqué, du lion, du loup, du renard polaire, de l'ours brun et du glouton. La présence du bison à une latitude aussi haute révèle que toute la Sibérie a dû être occupée par les groupes humains durant le MIS3.

Le début du MIS 2 et le retour d'un climat glaciaire va perturber l'installation des nouveaux arrivants. Car, avec le dernier maximum glaciaire, ils vont devoir refluer vers le Sud du continent. La calotte glaciaire des Laurentides va descendre jusqu'à 40° de latitude Nord. Elle va en outre détruire tous les sites de plein air des occupations humaines immédiatement antérieures au dernier LGM. Les groupes humains vont alors descendre vers le Texas, le Mexique, l'Amérique centrale et l'Amérique du Sud en quelques milliers d'années.

La culture de Clovis correspond en fait à la remontée en latitude des groupes humains à la fin de la glaciation, quand la calotte glaciaire des Laurentides amorce son retrait. Ce phénomène est l'équivalent des recolonisations du Magdalénien et du Mézinien en Europe après le dernier LGM. Elle n'est donc pas un peuplement mais un repeuplement.

Les extinctions animales en Amérique à la fin de la dernière glaciation ont été beaucoup plus nombreuses que les extinctions animales en Eurasie, faute probablement d'un espace de migration assez vaste comme en Eurasie

En Amérique du Nord, les grandes espèces ont été plus concernées que les petites espèces : 33 sur 44 grandes espèces se sont éteintes alors que seulement 6 sur 70 petites espèces se sont éteintes. La datation de cette extinction se serait faite en un millier d'années, entre 13 000 et 12 000 BP :

- Mammouth (*mammuthus columbi et mammuthus primigenius*),
- Mastodonte (*mammut americanum*),
- Gomphothère, un proboscidien à crâne allongé,
- Paresseux géant (*megalonyx jeffersonii*) (figure 19),
- Cheval (*equus curvidens*),
- Castor géant (*Castoroides nebrascensis*),
- Bison *antiquus*, l'ancêtre du bison des grandes plaines,
- Bison *latifrons*,
- *Cervalces latifrons*, un grand cervidé proche du mégacéros,
- Camelops, le chameau américain,
- Smilodon, le grand félin aux dents de sabre (figure 20),
- Lion d'Amérique (*panthera atrox*), un grand félin,
- *Canis dirus*, un grand loup,
- *Homothérium serens*, un grand carnivore de la taille d'un tigre,
- *Arctodus simus,* un grand ursidé de la taille de l'ours des cavernes européen.

Parmi les espèces survivantes, citons le caribou (le renne américain), le bœuf musqué (*ovibos moschatus*), le wapiti (cerf élaphe américain), le cerf de virginie (*odocoileus virginianus),* l'orignal (élan américain), le bison, le pronghorn (une antilope), le mouflon (*ovis canadensis, ovis dalli*), le loup (*canis lupus*), le coyote (*canis latrans*), l'ours noir (*ursus americanus*), l'ours brun (grizzli) et l'ours polaire (*ursus maritimus*).

En Amérique du Sud, la faune endémique est issue de la faune du continent Gondwana, après séparation avec l'Afrique, l'Australie et l'Antarctique. Elle a évolué de façon indépendante jusqu'à la fermeture récente de l'isthme

Figure 19.
Paresseux géant
d'Amérique du
Nord (*Megalonyx
jeffersonii*).
Muséum
d'Histoire
Naturelle de
l'Utah (USA)

du Panama il y a 3 ou 13 millions d'années, suivant les différentes théories. A partir de cette date, et sous l'effet des changements climatiques, des migrations ont été observées dans les deux sens, dont les plus importantes ont eu lieu respectivement à 2,5 millions d'années, 1.6 millions d'années, 800 000 ans et 125 000 ans.

Dans la faune endémique d'Amérique du Sud, plusieurs espèces sont connues comme les marsupiaux (superordre des Ameridelphia) qui comprend les opossums et les rats marsupiaux, les paresseux (sous-ordre des Folivora), les singes platyrrhiniens (dont l'arrivée sur le continent

Figure 20. Smilodon ou tigre à dents de sabre (*smilodon populator*) d'Amérique du Nord

américain reste sujet à débat), l'oiseau terreur, un oiseau géant non volant carnivore, disparu il y a 400 000 ans (*Titanis, Phorusrhacos et Kelenken*) dont les seuls descendants actuels sont les « petits » cariamas, un grand herbivore, le toxodon, de la famille des Notongulés et le tamanoir.

Les migrations du Sud vers le Nord ont vu l'arrivée des tatous et paresseux, puis des tamanoirs et enfin des opossums. Ces espèces ont généralement connu des difficultés d'adaptation et les plus grandes d'entre elles font partie des extinctions de la fin du Pléistocène.

Les migrations du Nord vers le Sud ont été plus nombreuses et réussies : les camélidés, les équidés, les cervidés, les pécaris, les tapirs, l'ours, les félins (puma, jaguar, smilodon), le Gomphothère, la loutre (qui a particulièrement réussi son implantation en Amazonie : la loutre géante), le loup et le procyon.

Plusieurs grandes espèces se sont éteintes en Amérique du Sud comme :
- Le Paresseux géant (*Mégathérium*),
- Le Glyptodon ou tatou géant (figure 21),
- Le Toxodon, un grand herbivore de la famille des Notongulés.

Parmi les survivants, les camélidés avec le guanaco (*lama guanicoe*), la vigogne (*vicugna vicugna*), domestiqués respectivement en lama et alpaga, le cerf des pampas (*Ozotoceros bezoarticus*), le tapir du Brésil (*Tapirus terrestris*), le pécari à collier (*Pecari Tajacu*), le tamanoir (*Myrmecophaga tridactyla*), l'ours

Figure 21.
Glyptodon ou
tatou géant
d'Amérique du
Sud

à lunettes (*Tremarcos ornatus*), les félins (jaguar, puma), le loup à crinière (*Chrysocyon brachyurus*) et un rongeur comme le capybara.

Les espèces venues d'Amérique du Nord ont donc beaucoup mieux résisté à la vague d'extinction de la fin du Pléistocène (sauf les équidés comme l'Hippidion), y ayant trouvé un espace refuge.

En conclusion, les groupes humains ne peuvent être tenus responsables de l'extinction des mégafaunes américaines pour plusieurs raisons de nature systémique :

- Les groupes humains sont arrivés bien avant l'extinction et ont cohabité ensemble pendant plus de quinze mille ans,
- Les mégafaunes d'herbivores se sont éteintes en même temps que leurs prédateurs, les grands carnivores,
- De nombreuses espèces animales du Nord du continent américain ont survécu en migrant dans le Sud du continent, comme l'espèce humaine.
- Les herbivores et carnivores de taille moyenne n'ont pas subi d'extinctions.

L'extinction de la grande faune australienne

La faune de l'Australie est une faune endémique, liée à son isolement depuis 50 millions d'années après l'éclatement du Gondwana. Elle est essentiellement caractérisée par une très grande diversité de mammifères marsupiaux, par deux espèces de monotrèmes, 350 espèces endémiques d'oiseaux et une grande diversité de reptiles.

L'arrivée des premiers groupes humains date d'il y a au moins 45 000 ans à la fin du MIS 4. Ils ont profité, pour le passage d'îles en îles selon des voies étudiées par J. Birdsell, d'un niveau de la mer bas, qui a réduit les largeurs des détroits. Ils ont bénéficié jusqu'à la fin du MIS 3 d'un climat favorable, tempéré et humide, qui a fait régresser les zones désertiques du centre du continent australien (Hiscock, 2008).

Les premiers habitants ont cohabité avec des espèces éteintes comme notamment le révèlent les ossements d'un kangourou géant (wombat) trouvé associé à la sépulture de l'homme de Mungo ou la stratigraphie du site de Cuddie Springs (cf. infra).

L'extinction de la mégafaune australienne a concerné des espèces qui avaient profité des conditions favorables de l'avant-dernier interglaciaire (MIS 5) pour se développer :

- Des kangourous géants, dont le *procoptodon* de 2,7 m de hauteur (figure 22), le *sthenurus* et le *simosthenurus*,
- Des wallabies géants, comme le *protemnodon*,
- Des wombats géants, comme le *diprotodon* de la taille d'un hippopotame (figure 23) et le *Zygomaturus*,
- Un grand koala, comme le *Phaxolarctus*,
- Des carnivores marsupiaux géants comme le *Thylacoleo*, de la taille d'une hyène, le *Sarcophilus*, ou le *Propeoplus*, un kangourou omnivore,
- Un grand monotrème (*Zaglossus hacketti*),
- Un grand varan géant, de 6 à 8 mètres de longueur, le *Magalania prisca*, ressemblant au dragon de Komodo,
- Un grand crocodile de six mètres de long, le Quinkana,
- De grands serpents constricteurs de dix mètres de long, du genre Liasis,
- Des Oiseaux non volants de la famille des Dromornithidae comme le *Dromornis Stirtoni* de trois mètres d'envergure et pesant 500 kg, ou le *Genyornis newtoni* de 2,5 mètres d'envergure et 240 kg.

Figure 22.
Procoptodon ou
Kangourou géant
d'Australie

Figure 23. Diprotodon ou wambat géant d'Australie

Cependant, il faut faire observer que la systématique des organismes vivants cache parfois les adaptations à des environnements changeants dont la réduction de taille est un des processus.

La transition du Pléistocène à l'Holocène a entraîné une tendance générale à la réduction de taille. Ce phénomène est à mettre en relation avec la disparition de la steppe et au grand développement de la forêt. Ainsi, par exemple, en Europe, le bison priscus évolue vers le bison bonasus et en Amérique du Nord, le bison antiquus et le bison latifrons vers le bison bison. La multiplication anarchique de la systématique des sous-espèces rend confuse les mécanismes de l'évolution et de l'adaptation (Buffon l'avait déjà signalé au XVIIIᵉ siècle !). Il en est de même en Australie, où ce qui est considéré comme une extinction d'une forme géante peut cacher une réduction de taille, comme par exemple pour les kangourous, les wallabies, les wombats, les koalas et les crocodiles.

Les dates de l'extinction de la mégafaune australienne, espèce par espèce, sont connues de façon trop imprécise autour de 46 000 ans. La datation radiocarbone n'étant malheureusement plus utilisable à cet horizon de date, les archéologues ont recours à l'OSL de sédiment ou l'U/Th de calcite, ce qui pose la question taphonomique de l'association entre l'ossement et la stratigraphie. Si ces dates étaient confirmées sur l'ensemble des espèces concernées, l'extinction serait survenue ou aurait débuté dans la période de sécheresse de l'épisode glaciaire MIS4. Mais certaines espèces étaient encore vivantes dans le MIS3.

Plusieurs auteurs australiens concluent à la responsabilité des chasseurs humains dans cette extinction du fait de la coïncidence des dates, avec un second argument donné que ces espèces avaient déjà survécu à d'autres variations climatiques dans le passé. L'exemple de l'extinction des Moas de Nouvelle-Zélande à l'arrivée des premiers Maoris a certainement influencé T. Flannery, qui a relancé en Australie la question la théorie de l'overkill dans les années 1990 (Flannery, Roberts, 1999). D'autres l'ont contestée (Wroe *et al.* 2004). Le lecteur trouvera dans (Hiscock, 2008, chapitre 4) une analyse critique approfondie de la question et des différents points de vue.

Plusieurs points sont importants à noter :
- Il est indispensable d'avoir des dates plus précises d'extinction, espèce par espèce, région par région et environnement par environnement, pour éviter tout présupposé théorique. Ainsi là où Flannery voit une extinction ponctuelle à 46 000 ans, en éliminant les dates qu'il juge non fiables ou les ossements non en position anatomique et donc

considérés comme remaniés, d'autres voient une extinction sur une période longue (cf. infra).

- Cette extinction ne concerne pas seulement les espèces géantes, des espèces non géantes se sont éteintes également tandis que d'autres ont survécu à l'extinction,
- La liste des espèces éteintes contient les espèces herbivores comme leurs espèces carnivores prédatrices, argumentant l'aspect systémique de cette extinction.
- Aucun kill site n'a été découvert en Australie (Hiscock, 2008, p.67), contenant des ossements de mégafaune éteinte.

Le site de Cuddie Springs fouillé par J. Field est un site clé pour résoudre la question (Field *et al.* 2001). Il s'agit d'un rivage de lac où se sont retrouvés des espèces éteintes et des groupes humains. Les espèces éteintes ont laissé leurs carcasses avant et après l'occupation humaine du site sur une durée de plus de dix millénaires de 50 000 à 35 000 ans.

Le réchauffement climatique de l'OIS 3, qui a transformé la savane en prairies, a-t-il joué un rôle dans l'extinction de la mégafaune comme le suggère Hiscock ? Ou est-ce plutôt la sécheresse de l'OIS 4 ou celle de l'OIS 2 à l'approche du dernier maximum glaciaire, mais qui implique cependant une extinction plus progressive (que contexte Flannery). Ou alors est-ce une des conséquences climatiques de la très grande éruption du volcan Toba à Sumatra il y a 73 000 ans qui concerne tous le Sundaland, et qui pourrait aussi être à l'origine de l'arrivée des premiers humains ?

La théorie de l'« *overkill pleistocene* » en Australie n'est plus crédible mais la question ne sera définitivement réglée que quand seront trouvées les explications environnementales et les dates précises de l'extinction naturelle de cette mégafaune.

Le Sahara et l'Afrique du Nord

Le Sahara, qui est actuellement un désert, a connu, avec les alternances des ères glaciaires et interglaciaires, des périodes d'humidité qui ont laissé des sols fossiles dans les séquences stratigraphiques quaternaires.

L'avant-dernier interglaciaire (MIS 5), intitulé pluvial Abbassia en Afrique du Nord, correspond à un grand développement du peuplement humain, désigné sous le nom d'Atérien, caractérisée par une industrie du paléolithique moyen à pièces pédonculées (un mode d'emmanchement caractéristique) sur l'ensemble de l'Afrique du Nord, de la Maurétanie à l'Egypte.

Au Pléistocène, les espèces animales de l'Afrique du Nord sont des espèces africaines qui effectuent un va-et-vient Nord-Sud en fonction des variations climatiques.

Quelques rares espèces sont des espèces d'origine eurasiatique venues à l'occasion d'un changement climatique, comme le rhinocéros de prairie (*stephanorhinus hemitoechus*), le cerf à joues épaisses (*mégacéroides algericus*) et le cheval (*equus algericus*).

Il est intéressant de noter que les tenants de la théorie de l'« *overkill pleistocene* » ont exclu l'Afrique des continents où cette théorie s'appliquait. Or, ces trois espèces, qui sont les seules à s'être éteintes à la fin du Pléistocène en Afrique, sont d'origine eurasiatique.

A l'inverse, des espèces africaines sont venues peupler l'espace européen dans les périodes interglaciaires. L'endroit du passage a fait longtemps débat. Si le franchissement du canal de Sicile, qui fait actuellement près de 145 km a été récusé depuis longtemps (Vaufrey, 1930), le franchissement du détroit de Gibraltar (10 km à la nage) a des défenseurs. Le trajet le plus sûr est encore la voie terrestre par le Moyen-Orient où ces faunes ont été retrouvées dans les sites pléistocènes.

Les espèces comme *elephas antiquus, hippotamus major* ou *stephanorhinus hemitoechus* (le rhinocéros de prairie) ont été retrouvées jusque sur les bords de la Tamise. Cette faune chaude est caractéristique des stades isotopiques interglaciaires 13, 11 et 9 (associée au Paléolithique inférieur : acheuléen) et du stade isotopique 5 (associée au Paléolithique moyen). Leur disparition en Europe fait toujours débat, du fait de datations encore insuffisamment précises voire inadaptées pour le radiocarbone. La date la plus probable est une disparition en Europe avec l'épisode glaciaire du MIS 4, il y a environ 50 000 ans, mais certains auteurs l'estiment plus tardive au MIS 3 vers 34 000 ans.

Revenons en Afrique. Au dernier maximum glaciaire, l'expansion du désert est maximale et les faunes nord-africaines sont descendues dans les refuges équatoriaux. C'est à ce moment, que les espèces d'origine eurasiatique ont disparu. Les espèces africaines sont revenues avec la fin de la glaciation et le passage à l'Holocène, comme par exemple la girafe (*giraffa camelopardalis*). Le début de l'Holocène fut un climat chaud et humide, à l'origine de l'expression « *le Sahara vert* » : girafes, hippopotames, crocodiles, serpents géants, varans, gnous, éléphants, rhinocéros, buffles ont été figurés par les auteurs de l'art rupestre saharien. Cette faune a disparu d'Afrique du Nord et du Sahara, en migrant vers les latitudes équatoriales, avec le retour de l'aridité il y a 6 000 ans.

Plusieurs espèces se sont éteintes au Néolithique et à l'âge du Bronze (zèbres, rhinocéros, phacochères, antilopes (élands), buffles, gnous, gazelles, damalisques, hyènes, cerfs élaphes, lycaons.

A l'époque romaine, la recherche d'animaux exotiques et d'animaux pour les jeux du cirque ont conduit à l'exploitation de nombreuses espèces en Afrique du Nord et à leurs extinctions (comme l'éléphant berbère et les fauves). A l'époque historique, les dernières extinctions datent du XIXème siècle (lions, autruches, ours, ânes) et des débuts du XXème siècle (antilopes), liées au développement de l'usage des armes à feu.

Spécificités insulaires : nanisme et gigantisme insulaire

Les îles possèdent des zoocénoses caractérisées par la présence d'espèces endémiques, l'absence de prédateurs, un phénomène de nanisme qui touche les grands mammifères (les éléphants nains devenus célèbres mais tous les grands mammifères y compris l'homme (Florès), et, à l'inverse, un gigantisme de grands rongeurs et de grands oiseaux coureurs, dont l'accroissement de taille est dû à l'absence de prédateurs.

En conséquence, les îles sont sujettes à des extinctions rapides pour des raisons, dont l'effet s'additionne :

- des raisons naturelles, le seuil critique de population et l'isolat génétique, processus qui fragilisent la population face à des évènements extérieurs puis entrainent leur extinction inéluctable. L'histoire du refuge ultime du mammouth dans l'ile arctique de Wrangel en est une des meilleures illustrations.
- par l'arrivée d'un prédateur (animal et humain, qu'il soit aborigène ou colon européen).

L'arrivée d'un prédateur dans une île a des conséquences dramatiques qui entraînent le plus souvent des extinctions rapides et massives. Car la faune endémique n'est pas préparée à la compétition entre espèces que change l'irruption brutale d'un prédateur.

L'analyse de plusieurs cas que nous allons approfondir plus loin révèle que c'est dans un contexte insulaire que surviennent les cas indiscutables et bien argumentés d'extermination. Il reste cependant à vérifier que de mauvaises datations ou des stratigraphies remaniées ne remettent pas en cause l'explication.

Par ailleurs, la colonisation humaine des îles pose l'intéressante question de la capacité de franchissement des détroits, aux différentes périodes (chasseurs-cueilleurs paléolithiques et mésolithiques, agriculteurs néolithiques) et avec différentes lignes de rivage (dépendant des régressions et des transgressions marines liées aux variations climatiques) qui augmentent ou réduisent la largeur du détroit, voire rattachent l'île au continent pour des profondeurs inférieures à 120 mètres (Djindjian, 2013). Pour les chasseurs-cueilleurs, un détroit de 30 km de large peut être franchi de jour avec des radeaux de fortune. Au Néolithique, les pirogues monoxyles permettent de transporter les personnes, les animaux domestiques, les plants, les graines et les outils, pour des navigations à vue sur des distances inférieures à 150 km. La navigation hauturière n'est pratiquée qu'à partir du 2ème millénaire BC par les Austronésiens dans l'océan pacifique et par les Minoens, les Mycéniens, les Phéniciens et les Grecs en Méditerranée.

Les chasseurs-cueilleurs espèrent que l'île va leur apporter les ressources alimentaires dans le cycle annuel suffisantes pour pouvoir s'y implanter dans la durée. Ils seront déçus quand la faune endémique, qu'ils découvrent à leur arrivée, ne leur fournit pas ces ressources.

C'est par exemple le cas de la Corse, que les Mésolithiques atteignent par l'ile d'Elbe alors rattachée à la péninsule. A leur arrivée, un petit cervidé (*mégalocéros Cazioti*), et un petit canidé, le dhole de Sardaigne (*cynothérium sardous*) venaient de s'éteindre. Ils n'y trouvent que de gros rongeurs (*prolagus corsicanus*). L'ile sera colonisée plus tard par les agriculteurs néolithiques, qui apportèrent le mouflon corse (*ovis gmelini*) puis le porc (Vigne, 1988).

Aussi, est-ce au Néolithique que les colonisations des iles furent pérennes. Car les humains apportèrent des animaux domestiques mais aussi des passagers clandestins. Ces nouvelles espèces modifièrent l'équilibre établi entrainant des extinctions et des proliférations, changements auxquels les groupes humains apportèrent leur contribution. En outre, certaines espèces domestiques se féralisèrent, c'est-à-dire retournèrent à l'état sauvage, sujet qui sera analysé plus loin.

Ces introductions suivirent les mêmes lois naturelles. Ainsi, l'introduction du cheval dans les iles Shetland créa des poneys. L'homme lui-même piégé sur l'ile de Flores par la remontée des eaux dans l'archipel indonésien devint une race naine.

Les îles de la Méditerranée

Les iles de la Méditerranée possèdent une faune endémique remarquable par sa diversité et son originalité.

Elles sont également célèbres pour ses éléphants nains, trouvés en Sicile, à Malte, en Sardaigne, en Crête, dans les iles des Cyclades (Naxos, Delos, Kythnos, Serifos, Milos) et du Dodécanèse (Rhodes, Tylos)

De petite taille ou de taille naine (figure 24), ils sont classés dans le genre *Palaeoloxodon (elephas antiquus)*, à l'exception de celui de Sardaigne classé dans le genre mammouth. La question est toujours posée pour le *mammuthus creticus* de Crête.

Figure 24. Eléphants nains des iles de la Méditerranée

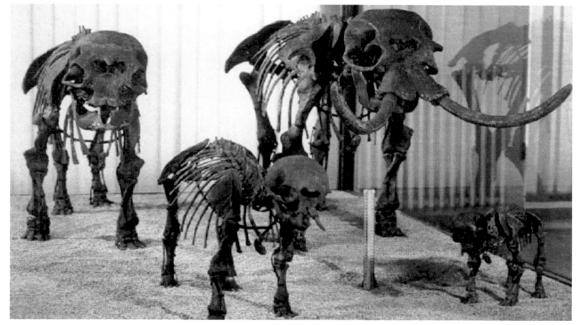

Trouvés anciennement, leur position stratigraphique précise mérite cependant d'être confirmée pour connaitre les différentes vagues de leurs arrivées et surtout la date de leur extinction. Les éléphants antiques sont arrivés en deux vagues vers 800 000 ans et vers 300 000 ans profitant des glaciations pour atteindre les îles, et grâce à leurs capacités d'excellents nageurs. Les mammouths sont arrivés plus tard vers 150 000 ans. Leur extinction est moins bien établie mais elle est pléistocène avant l'arrivée de l'homme dans les îles.

Les éléphants de Tilos, découverts par G. Theodorou dans la grotte Charkadio, ont fait couler beaucoup d'encre avec des datations radiocarbone très dispersées entre 45 000 et 3 500 BP et un contexte stratigraphique incertain (Theodorou *et al.* 2007). Derrière la belle histoire d'une disparition à l'époque historique, qui aurait été représentée sur des bas-reliefs, il n'y a sans nul doute que la pollution variable d'ossements pléistocène par du carbone récent, à l'origine de cette dispersion de dates, pour des ossements bien plus anciens.

D'autres mammifères nains sont présents dans les iles de Méditerranée comme l'hippopotame nain de Chypre (*hippopotamus minor*). Des fouilles menées sur l'île, dans les années 1980 par J. Simmons à Aetokremnos, ont mis au jour de grandes quantités d'ossements d'hippopotames, souvent brulés, mêlés à des outils et des restes de repas. Il en a déduit que les premiers humains mésolithiques de Chypre étaient responsables de son extinction. Une récente révision (Zasso *et al.* 2015) a montré ces ossements étaient antérieurs à l'occupation humaine vers 13 000 cal BP et que les mésolithiques avaient trouvé et utilisé les ossements plusieurs centaines d'années après pour alimenter leur foyer.

Le cas des îles Baléares est différent (Waldren, 1982). Situées à 92 km de la côte du Levant espagnol, elles n'ont été peuplées que tardivement vers

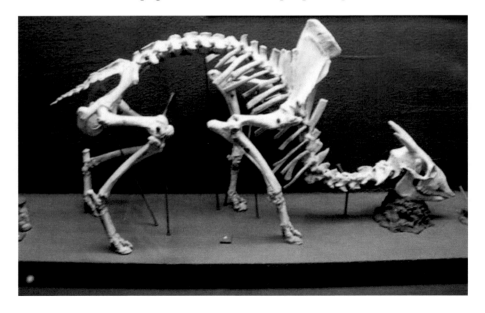

Figure 25. Antilope des Baléares (*myotragus balearicus*)

4 730 BC (Grotte de la Muleta, abri de Son Matge à Majorque). Une espèce endémique, l'antilope des Baléares (*myotragus balearicus*) y vivait encore (figure 25). Elle s'est éteinte, après une tentative avortée de domestication, à la fin du III° millénaire BC (2 143 BC).

Les autres mammifères des iles comme le daim (Chypre, Crête, Rhodes, Sardaigne), le cerf (Sardaigne, Corse, Baléares), les « sangliers, les « mouflons », les chèvres sauvages des îles grecques sont des animaux domestiques introduits par les premiers agriculteurs et qui se sont plus ou moins bien féralisés.

Les Moas de Nouvelle-Zélande ou le triste destin des oiseaux coureurs.

Les Austronésiens découvrent la Nouvelle-Zélande vers 900 AD et s'y installent. Les preuves d'agriculture, pratique qui leur était pourtant bien connue, n'existent qu'à partir de 1 200 AD.

La grande superficie des deux îles qui forment la Nouvelle-Zélande actuelle (267 000 km^2) explique la grande richesse et diversité des oiseaux coureurs (ratites), qui constituaient la totalité de la grande faune, avec toutes les échelles de taille et qui en occupaient tous les biotopes. Cette faune a presque totalement disparu, résultat d'une chasse intensive des premiers occupants, venus probablement sans leurs animaux domestiques traditionnels (porcs, poules) mais avec le chien qui fut un auxiliaire de chasse très efficace, et involontairement le rat polynésien, grand amateur des œufs de ratite.

Six sur les onze espèces de ratites furent ainsi exterminées avant 1650 AC, date de l'arrivée des Européens. Les plus grands d'entre eux étaient le *Dinornis maximus* de 3m de hauteur (figure 26), et l'*Euryapterix gravis* (1,70m de hauteur). Le modeste Kiwi (*Aperyx owenii*), haut de seulement 30 centimètres, emblème du pays, bien que raréfié, fut presque le seul survivant.

28 espèces d'oiseaux volants furent également exterminés, notamment :
- Le grand aigle (*Harpagornis moorei*), seul prédateur des ratites,
- Le busard (*Circus eylesi*),
- Le râle géant (*Aptornis cremiornis*),
- L'oie aptère (*Cnemiornis calcitrans*).

Face à ce vide biologique, presque sans équivalence sur la planète, les introductions des européens furent nombreuses, à commencer par celles du capitaine Cook : chèvres, moutons et porcs en 1774 et 1777, qui se féralisèrent rapidement. Les rats européens éliminèrent alors les rats austronésiens. 53 espèces de mammifères et 125 espèces d'oiseaux furent introduites en moins de deux siècles, principalement d'Europe mais également une faune marsupiale d'Australie qui s'adapta facilement. Certaines espèces sont devenues invasives dont le lapin et l'opossum d'Australie, échappant à tout contrôle humain. Ce sont les petites îles périphériques qui sont devenues les conservatoires de la diversité animale de la Nouvelle-Zélande (Planhol, 2004, p.877-888).

Une extinction équivalente de la faune endémique s'est produite dans trois îles de l'Océan indien (Mascareignes) au XVIIème siècle : La Réunion, Maurice et Rodriguez, dont la responsabilité est due aux navigateurs européens et aux prédateurs qui avaient été introduits. Pour les oiseaux terrestres endémiques, l'extinction a concerné 19 espèces sur 23 à La Réunion, 16 sur 21 à Maurice et 10 sur 12 à Rodriguez. Les plus connus étaient des raphidae (pigeons géants) désignés sous le nom de Dodo de Maurice, Dronte de La Réunion et Solitaire de Rodriguez.

Transgressions océaniques et extinctions

Le réchauffement climatique vers un interglaciaire entraine une transgression océanique (+120 m à l'Holocène) à l'origine d'une archipélisation de continents (Sundaland) ou de grandes îles (Antilles, Cyclades, etc.) et de l'isolement d'îles du continent (Angleterre, Tasmanie, Taiwan, Ceylan, Channel Island en Californie, Wrangel, etc.).

L'isolement active alors le processus de nanisme et gigantisme insulaire et les mécanismes d'extinction liés à la perte de taille critique, aux ressources alimentaires insuffisantes et à une compétition exacerbée sur un espace devenu trop limité. Quand s'y ajoute une transgression marine, le morcellement du territoire en chapelets d'îles accélère le processus. Les extinctions ont lieu alors au début des interglaciaires.

Le cas des Antilles est particulièrement intéressant.

La plus grande partie de la faune endémique présente au Pléistocène a disparu : 75 espèces pléistocènes se sont éteintes et parmi elles 90% des espèces hors oiseaux. Le plus spectaculaire animal disparu des Antilles était

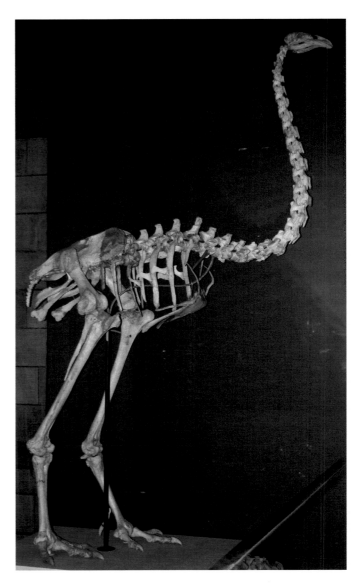

Figure 26. Dinornis maximus de Nouvelle-Zélande

un gigantesque rongeur (*Amblyrhiza inundata*) qui pesait 210 kg et avait la taille d'un cerf ! Il s'est éteint à l'avant-dernier interglaciaire (MIS 5) il y a 100 000 ans. Son territoire s'était rétréci de 2 500 km^2 à 175 km^2 avec la remontée des eaux de l'océan et ne pouvait plus assurer sa nourriture. La cause est la même pour le grand rongeur de Jamaïque (*Chlydomys jamaica*) et pour le grand paresseux (*Acractonus comes*) découvert à Haiti/Saint-Domingue. La plupart des extinctions (notamment 37 espèces de mammifères) ont eu lieu à la transition Pléistocène-Holocène, bien avant l'arrivée des premiers groupes humains vers le IV° millénaire BC à partir des côtes du Venezuela.

La féralisation ou le retour à l'état sauvage

Les colonisations néolithiques ont apporté les introductions d'espèces végétales cultivées mais aussi les introductions d'espèces animales domestiquées. Près de dix mille ans de néolithisation ont presque fait oublier que la plupart des animaux que nous côtoyons actuellement, de la nourriture végétale et animale que nous consommons, ou des fourrures ou des fibres végétales et animales dont nous nous vêtons, sont d'origines lointaines.

L'histoire ancienne nous a également laissé quelques souvenirs. Si les éléphants d'Hannibal n'ont pas fait souche en Italie, il est raisonnable de considérer que c'est à la colonisation carthaginoise que l'on doit l'introduction du petit singe *Macaca sylvanus* en Espagne (dont Gibraltar) et à Ischia. De même, la présence tardive du lion de Thessalie est sans doute à mettre en relation avec l'invasion perse de la Grèce.

Les Austronésiens qui ont peuplé les îles du Pacifique à partir du 2° millénaire BC, sont arrivés avec un cortège d'espèces qui n'est pas aussi limité que la petitesse de leurs embarcations pourrait le laisser croire : animaux (rat polynésien, chien, porc, poule) et plants (taro, igname, banane, coco, arbre à pain, etc.). Arrivés jusque sur les côtes d'Amérique du Sud, ils y ont emprunté la patate douce.

Les navigateurs européens ont procédé à une mondialisation de ces introductions, que les Portugais furent les premiers à pratiquer de façon systématique. La procédure est bien décrite dans les « *Instructions Nautiques* » d'Henri le Navigateur, qui se traduit par un rituel de « *lâcher de bêtes* », notée soigneusement sur le carnet de bord, et qui constitue à la fois une prise de possession symbolique de la terre et une réserve potentielle de nourriture pour le prochain voyage (Planhol, 2004, p.371).

En Amérique du Nord, ce sont les Espagnols qui, au XVI° siècle, ont introduit le cheval, qui avait disparu à la fin du Pléistocène et qui a révolutionné la vie de nombreuses tribus d'amérindiens des grandes plaines, en faisant d'eux des peuples cavaliers. Les amérindiens ont acquis ces chevaux soit par des razzias opérées auprès des européens soit en capturant et en dressant des animaux abandonnées revenus à l'état sauvage (mustangs). Par adaptation et par sélection, des races propres à l'Amérique du Nord sont ainsi apparues (comme la race Appaloosa des Nez-Percés).

La colonisation humaine de l'Australie fait encore débat aujourd'hui. J. Birdsell en 1967 avait proposé l'arrivée de trois vagues de colonisation sur la base de critères d'anthropologie physique des populations aborigènes actuelles. Lindsell en 2001 a rejeté ces conclusions en expliquant que ces différences étaient dues à des adaptations à des environnements différents. A.G. Thorne, entre 1971 et 2000, a étudié les squelettes et proposé de distinguer deux types de population, une forme gracile, arrivée la première et une forme robuste, arrivée ensuite. Cette explication a été rejetée par Pardoe en 1991 qui n'y a vu que la dysmorphie sexuelle, critique elle-même rejetée par Flood en 2001. En 1989, Brown a conclu que ces différences morpho-crâniennes étaient dues à des déformations crâniennes volontaires qui auraient été pratiquées à l'époque pléistocène (mais qui ne sont pas pratiquées par les Aborigènes aux époques historiques). Les études génétiques, qui ont pris le relais des études anthropologiques, ont connu des résultats contradictoires sur le mtDNA. Les différences ont été interprétées comme l'existence de plusieurs migrations, ou la divergence progressive des populations d'Australie et de Nouvelle-Guinée, une origine indienne des migrants contestée depuis, des pollutions d'échantillons par de l'ADN récent ou une seule et unique souche dont seraient issus les Aborigènes actuels. La doxa actuelle prône l'arrivée d'une population unique arrivée il y a environ 45 000 ans (cf. supra), dont les Aborigènes seraient les actuels descendants. Cette théorie de l'arrivée unique est devenue de fait une revendication politique des Aborigènes pour réclamer leur droit à la terre. Elle gène le travail scientifique en suspectant les chercheurs d'un parti pris pro-aborigènes ou pro-colonialistes suivant la conclusion proposée. Il reste la question de savoir qui donc a apporté le chien Dingo, un chien asiatique domestiqué, introduit en Australie il y environ 4 000 ans et retourné à la vie sauvage ? Ce sont probablement des Austronésiens, bien qu'il soit répété qu'ils n'aient jamais pris pied sur le continent australien. Le mystère de l'origine du Dingo demeure donc entier.

L'introduction sur le continent australien d'espèces domestiques européennes a commencé dès l'installation des Européens au XVIIIème siècle : chevaux, ânes, chèvres, porcs, dromadaires, buffles, bovins, lapins, chiens, chats, renards, mais aussi les rats et les souris. La compétition avec la faune marsupiale commença mais la différence de fécondité rendit la lutte inégale. C'est le développement d'une économie pastorale (avec l'importance de l'élevage du mouton) qui amenèrent les colons à des massacres inconsidérés de la grande faune marsupiale tandis que leurs commensaux carnivores (chiens, chats, renards) retournés à l'état sauvage faisaient de même avec la petite faune marsupiale et les oiseaux.

Une partie de cette faune domestique importée retourna à la vie sauvage : les chevaux (uniformément répartis), les ânes (concentrés au Nord-est), les chèvres (concentrées à l'Ouest et à l'Est), les porcs (concentrés au Nord et à l'Est), les bovins (au Nord), les buffles (au Nord). Le cas des dromadaires est exceptionnel. Introduits à partir de 1840 venant du Pakistan, ils ont été utilisés pour le transport de marchandises. En 1920, leur nombre a été estimé entre 10 000 et 20 000 individus. Le progrès de l'automobile les a rendus obsolètes, et, abandonnés, ils se sont féralisés. Ils se sont installés dans les régions semi-désertiques du centre de l'Australie, inutilisables pour l'élevage de moutons. En 2010, le chiffre de 1 million d'individus a été avancé.

L'état australien a organisé dans les années 1980 des campagnes annuelles d'éradication planifiée des chevaux, des ânes et des dromadaires, avec l'appui de véhicules 4x4 et d'hélicoptères. Des mouvements d'opinions et des procès l'obligèrent à réduire cette pratique. C'est aujourd'hui, dans le but « vertueux » de réduire les émissions de CO^2, que les campagnes d'éradication auraient recommencé....

Conclusions

Les sociétés de chasseurs-cueilleurs

Les sociétés de chasseurs-cueilleurs, par définition, trouvent leurs ressources alimentaires par la chasse, la pêche et le piégeage d'espèces animales et la cueillette d'espèces végétales. La démographie des territoires occupés varie de 0,01 habitants au km^2 à 0,15 habitants au km^2 suivant la superficie du territoire de circulation des groupes (Djindjian, 2014). Suivant (Redmann, 1982), la démographie des grands herbivores en espace ouvert (savanes, prairies) est de l'ordre de 0,4 individus au km^2. Les taux de prélèvement des sociétés de chasseurs-cueilleurs sont, sauf exception, très inférieurs aux taux de reproduction. Ainsi, pour le bison, aux périodes historiques, les études ont conclu que le taux de reproduction annuel est de 15 % et que les prélèvements des Amérindiens n'ont pas dépassé 2%. Est-il irrationnel de considérer que ces prélèvements étaient inférieurs aux temps préhistoriques avec des armes de chasse moins meurtrières ?

Les sociétés de chasseurs-cueilleurs n'ont pas pratiqué la domestication. Mais l'apprivoisement et la commensalité de petits carnivores charognant les restes de chasse après dépeçage (comme le loup ou le renard) a dû sans doute être à l'origine du processus.

A l'issue de l'examen approfondi qui a été fait ci-dessus, il est possible de conclure que les chasseurs-cueilleurs ne sont en rien responsables des extinctions et que le terme d'extermination pléistocène ne peut plus être sérieusement retenu aujourd'hui.

Ces extinctions sont dues aux variations climatiques pour des raisons différentes, à des moments différents et à des endroits différents. Ainsi, pour l'Europe, continent qui est le mieux connu par la production scientifique des préhistoriens depuis 150 ans, les changements de zoocénoses ont été nombreux sur une période de 20 000 ans, entre 30 000 et 10 000 BP : disparition de l'aurochs, du sanglier, du cerf mégacéros, du chevreuil des latitudes moyennes à l'approche du dernier maximum glaciaire vers 27 000 BP, extinction de l'ours des cavernes vers 24 000 BP à l'approche du LGM, disparition provisoire du mammouth en Europe occidentale mais arrivée de l'antilope saïga et descente du bœuf musqué au dernier LGM, disparition définitive du mammouth à la fin du dernier glaciaire, retour de l'aurochs, du sanglier, du cerf mégacéros, du chevreuil en Europe moyenne au Bölling, départ du renne vers les régions septentrionales à l'Alleröd, adaptation du bison à la forêt à l'Holocène, disparition du cheval. Les groupes humains n'en sont nullement responsables.

A l'arrivée des chasseurs-cueilleurs mésolithiques dans les îles de la Méditerranée, la faune endémique était déjà éteinte, en particulier les éléphants nains, les hippopotames nains et les cervidés, à l'exception de l'antilope des Baléares qui a fait l'objet d'une tentative de domestication, hélas infructueuse. La biomasse n'était tout simplement pas suffisante pour le développement d'économies de chasseurs-cueilleurs et c'est au Néolithique que ces peuplements deviendront pérennes.

La théorie de l'extermination d'une mégafaune due à l'arrivée des chasseurs-cueilleurs sur le continent américain et sur le continent australien, est également réfutée sur la base des arguments qui ont été développés ci-dessus.

Les sociétés d'agriculteurs éleveurs

Les sociétés d'agriculteurs éleveurs ont pris le relais des sociétés de chasseurs-cueilleurs il y a dix mille ans au Proche-Orient, mais à d'autres dates plus récentes dans d'autres continents (Amérique, Pacifique, Afrique) ou d'autres régions de l'Eurasie (Europe, Chine, etc.), à l'origine de cohabitations et d'acculturations avec les sociétés de chasseurs-cueilleurs, plus ou moins longues suivant les cas. Ces sociétés ont pratiqué la domestication animale, la colonisation de territoires en y introduisant les espèces domestiquées et, involontairement, des espèces commensales (rats). Certaines espèces se sont bien adaptées, voire ont proliféré et obligé les groupes humains à des campagnes d'éradication. Certaines se sont féralisées (cheval en Amérique, dingo en Australie). Entrée en contact avec la faune endémique, la compétition entres espèces a entrainé la disparition le plus souvent de cette dernière. D'autres espèces n'ont pas réussi leur adaptation et ont disparu. Pour protéger ses récoltes et ses élevages, les sociétés humaines ont considérablement réduit ou exterminé des espèces considérées comme des nuisibles et des pestes (loup, ours, rapaces, rongeurs, insectes, serpents). La transformation du paysage par anthropisation des sociétés agricoles a en outre modifié les biotopes de certaines espèces qui ont disparu localement ou totalement. C'est ainsi que la faune sauvage s'est progressivement adaptée à des environnements peu ou pas anthropisés (forêt, montagne, zones marécageuses) et à la vie nocturne. Elle s'est aussi commensalisée à l'homme non sans un grand succès pour certaines espèces.

C'est par des sociétés d'agriculteurs-éleveurs que le peuplement pérenne des îles est rendu possible par l'introduction d'espèces domestiquées et d'espèces cultivées. L'exemple le plus caractéristique est le peuplement des îles du Pacifique. Certainement, les premiers occupants ont exercé des prélèvements importants sur la faune endémique aviaire et sur les reptiles, à l'origine de nombreuses extinctions, comme à Hawaï où 50 espèces d'oiseaux ont été exterminées avant l'arrivée des Européens. La complémentarité de l'homme, du chien et du rat (pour les œufs) fut déterminante dans ces extinctions. Les grandes îles comme la Nouvelle-Calédonie (avec le *sylviornis neocaledoniae*) et la Nouvelle Zélande avaient une superficie et des ressources propres à favoriser le développement de grands oiseaux coureurs qui ont été exterminés, comme cela a été décrit supra.

Sociétés étatisées

Les sociétés étatisées ont la capacité de mettre en œuvre des infrastructures importantes, transformant le paysage, irriguant les cultures, rendant les rivières navigables, défrichant les forêts, réduisant ainsi inexorablement les biotopes des animaux sauvages. Cette exploitation a entraîné des extinctions locales qui ont conduit après le Moyen-âge à des extinctions complètes. Parallèlement à l'élevage d'animaux domestiques, la chasse a continué mais la raréfaction du gibier et la réduction des forêts a conduit à les réserver aux chasses aristocratiques, aux chasses royales, enfin à l'interdiction totale de la chasse de certaines espèces. Parallèlement à la production alimentaire, la faune était exploitée à des fins artisanales pour l'obtention d'ivoire, de corne, de dent, de bois de cervidés, d'os, de carapace, de queue, d'huile, de fourrure, de peau, de plume, de duvet, de sécrétions, de concrétions, de composants de médicaments, etc. Les déjections animales étaient utilisées comme engrais. La faune est également utilisée à des fins ludiques pour enrichir des ménageries privées, pour offrir des cadeaux diplomatiques ou pour les jeux du cirque. L'empire romain a ainsi été responsable de l'extinction de plusieurs espèces en périphérie de son empire.

Sociétés modernes

Les sociétés modernes ont intensifié les exploitations abusives en utilisant des moyens technologiques sans cesse plus efficaces (cheval, chien, fusil) qui ont entrainé au XIX° siècle des extinctions massives dont il a été parlé précédemment. La prise de conscience progressive au XXème siècle de ces désastres écologiques a entrainé la mise en œuvre de politiques de protection, d'interdictions de chasse, de conservations (réserves et zoos) permettant la reproduction d'espèces menacées et de réintroductions. Mais la faune sauvage est devenue de fait une faune commensale. Néanmoins, les rares cas connus de recul de l'occupation humaine (épidémies, catastrophe nucléaire) montre que la nature sauvage des espèces reprend vite le dessus en l'absence des humains. L'incompréhension systémique de la compétition entre espèces est malheureusement à l'origine de réintroductions ratées et, à l'inverse, de proliférations incontrôlées. Les humains, nouveaux démiurges, ont encore des progrès à faire.

Bibliographie

Agenbroad L.D., Mead J.I. (eds.) 1994. *The Hot Springs Mammoth Site: A Decade of Field and Laboratory Research in the Paleontology, Geology, and Paleoecology.* The Mammoth Site of South Dakota. Inc. Hot Springs, South Dakota.

Allen J., O'Connell J.F. 2008. Getting from Sunda to Sahel. *In:* G.R. Clark, S. O'Connor, B.F. Leach (eds.), *Islands of Inquiry: Colonization, Seafaring and the Archaeology of Maritime Landscapes".Australian National University*, p.31–46.

Auguste P. 2009. Évolution des peuplements mammaliens en Europe du Nord-ouest durant le pléistocène moyen et supérieur. Le cas de la France septentrionale. *Quaternaire*, CNRS, 2009, 20 (4), p.527-550.

Baca1 M., Popović D., Stefaniak K. Marciszak A., Urbanowski M., Nadachowski A., Mackiewicz P. 2016. Retreat and extinction of the Late Pleistocene cave bear (Ursus spelaeus sensu lato). *Sci Nat,* (2016) 103, p.92, DOI 10.1007/s00114-016-1414-8

Boëda E., Clemente-Conte I., Fontugne M., Lahaye C., Pino M., Felice Daltrini G., Guidon N., Hoeltz S., Lourdeau A., Pagli M., Pessis A-M. , Viana S., Da Costa A., Douville E. 2014. A new late Pleistocene archaeological sequence in South America: the Vale da Pedra Furada (Piauí, Brazil), *Antiquity,* 88, 2014, p. 927-955.

Bourgeon L., Burke A., Higham T. 2017. Earliest Human Presence in North America Dated to the Last Glacial Maximum: New Radiocarbon Dates from Bluefish Caves, Canada. *PLoS ONE* 12 (1): e0169486. doi:10.1371/journal.pone.0169486

Dansgaard W., Johnsen S.J., Clausen H.B., Langway E.C. 1971. Climatic Record Revealed by the Camp Century Ice Core in Late Cenozoic Glacial Ages. *In:* K.K. Turrekian (ed.) *The Late Cenozoïc glacial ages,* New Haven, Yale University Press, p.375-6.

Diedrich C.G. 2010. Disappearance of the last lions and hyenas of Europe in the Late Quaternary – a chain reaction of large mammal prey migration, extinction and human antagonism. *Geophysical Research Abstracts* Vol. 12, EGU2010-2124.

Dillehay T.D., Ocampo C., Saavedra J., Sawakuchi A.O., Vega R.M., *et al.* 2015. Correction: New Archaeological Evidence for an Early Human Presence at Monte Verde, Chile. *PLOS ONE* 10(12): e0145471.

Djindjian F. 2013. Le franchissement des détroits et des bras de mer aux périodes pré- et protohistoriques. *In:* F. Djindjian, S. Robert (eds.) *Understanding Landscapes, from land discovery to their spatial organization,* Actes du XVI° Congrès UISPP, Florianopolis, septembre 2011, sessions C19 et C22. BAR Intern. Series, n°2441, p.3-14.

Djindjian F. 2014. Contacts et déplacements des groupes humains dans le Paléolithique supérieur européen : les adaptations aux variations climatiques des stratégies de gestion des ressources dans le territoire et dans le cycle annuel. *In:* M. Otte & F. Lebrun-Ricalens (eds.) *Modes de contacts et de déplacements au Paléolithique eurasiatique.* Colloque UISPP commission 8 de Liège, mai 2012. MNHA-CNRA et Université de Liège : ERAUL 140, p.645-673.

Djindjian F. 2015. Identifying the hunter-gatherer systems behind associated mammoth bone beds and mammoth bone dwellings. *Quaternary International*, 359-360, p.47-57.

Djindjian F. (dir.) 2018. *La Préhistoire de la France.* Paris, Hermann

Djindjian F., Kozlowski J., Otte M 1999. *Le Paléolithique supérieur en Europe.* Paris, Armand Colin

Fabre V., Condemi S., Degioanni A. 2009. Genetic Evidence of Geographical Groups among Neanderthals, *PLoS ONE*, 4(4) : e5151. doi : 10.1371/journal.pone.0005151

Field J., Fullagar R., Lord G. 2001. A large area archaeological excavations at Cuddie Springs. *Antiquity*, 75, p.696-702.

Flannery T.F., Roberts R.G. 1999. Late Quaternary Extinctions in Australasia. An Overview. *In:* R.D.E. MacPhee (ed.), *Extinctions in Near Time*, New York, Springer, p.239-256.

Hays J., Imbrie J., Schackleton N.J. 1976. Variations in the Earth's Orbit : Pacemaker of the Ice Ages, *Science*, 194, p. 1121-1132.

Helgen K.M., Wells R.T., Kear B.P., Gerdtz W.R., Flannery T.F. 2006. Ecological and evolutionary significance of sizes of giant extinct kangaroos, *Australian Journal of Zoology,* 54, p.293–303 .

Hiscock P. 2008. *Archaeology of ancient Australia.* New York, Routledge.

Iakovleva L., Djindjian F. 2017. Le site paléolithique de Gontsy (Ukraine) : un habitat à cabanes en os de mammouths du paléolithique supérieur récent d'Europe orientale. Communication à l'Académie des Inscriptions et Belles-Lettres. *Comptes rendus des séances de l'Académie des Inscriptions et Belles-Lettres*, fascicule 2017-3, p.1221-1246.

Imbrie J., Kipp N.G. 1971. A New micropalaeontological method for quantitative palaeoclimatology : application to a Late Pleistocene Carribean core. *In:* K.K. Turrekian (ed.) *The Late Cenozoïc glacial ages,* New Haven, Yale University Press, p. 71-81.

Martin P.S., Wright H.E. (eds.) 1967. *Pleistocene extinctions. The search for a cause.* New-Haven.

Martin P.S., Klein R.G. (eds.) 1984. *Quaternary extinctions : a prehistoric revolution.* Tucson.

Maschenko E.N., Gablina S.S., Tesakov A.S., Simakova A.N. 2006. The Sevsk woolly mammoth (Mammuthus primigenius) site in Russia: Taphonomic, biological and behavioral interpretations *Quaternary International*, 142, p.147-165.

Milankovitch M.M. 1941. *Canon of Insolation and Ice-Age Problem,* Boegrad, Koninglich Serbische Akademie.

Penck A., Bruckner E., 1901-1909. *Die Alpen im Eiszeitalter,* Leipzig, s.n., 3 vol.

Pitulko, V.V., 2011. The Berelekh quest: a review of forty years of research in the mammoth graveyard in northeast Siberia. *Geoarchaeology,* 26, p.5–32.

Pitulko V.V., Nikolsky P.A., Girya E.Yu., Basilyan A.E., Tumskoy V.E., Koulakov S.A., Astakhov S.N., Pavlova E.Yu., Anisimov M.A. 2004. The Yana RHS Site: Humans in the Arctic Before the Last Glacial Maximum, *Science,* 2004, 303, p.52-56.

Planhol X. de 2004. *Le paysage animal.* Paris, Fayard.

Puyssegur J. 1976. *Mollusques continentaux quaternaires de Bourgogne. Significations stratigraphiques et climatologiques,* Dijon, Université de Dijon (mémoire 3).

Redmann R.E. 1982. Production and diversity in contemporary grasslands. *In:* M.D.M. Hopkins, J.V. Matthews, Jr., C.E. Schweger, S.B. Young (eds), *Paleoecology of Beringia,* Academic Press, New York, p.223–239.

Roberts R.G., Flannery T.F., Ayliffe L.K., Yoshida H., Olley J.M., Prideaux G.J., Laslett G.M., Baynes A., Smith M.A, Jones R., Smith B.L. 2001. New Ages for the Last Australian Megafauna: Continent-Wide Extinction About 46,000 Years Ago. *Science,* 292, p.1888-1892

Teschler-Nicola M. ed. 2006. *Early Humans at the Moravian Gate. The Mladec Caves and their Remains.* Springer.

Theodorou G., Symeonidis N., Stathopoulou E. 2007. Elephas tiliensis n. sp. from Tilos island (Dodecanese, Greece). *Hellenic Journal of Geosciences,* Athènes, 42, 2007, p.19–32.

Trinkaus E., Zilhão J., Rougier H., Rodrigo R., Milota S., Gherase M., Sarcinã L., Moldovan O., Bãltean I., Codrea V., Bailey S.E., Franciscus R.G., Ponce de Léon M., Zollikofer C.P.E. 2006. The Peştera cu Oase and early modern humans in Southeastern Europe. *In:* N.J. Conard (Ed.), *When Neanderthals and modern humans met,* Tübingen, Kerns Verlag, p.145-164.

Van der Kaars S., Miller G.H., Turney C.S.M., Cook L.J., Nuernberg D., Schoenfeld J., Kershaw P., Lehman S.J. 2017. Humans rather than climate the primary cause of Pleistocene megafaunal extinction in Australia. *Nature communications.* DOI 10.1038

Vaufrey, R. 1930. Les Eléphants nains des îles méditerranéennes et la question des isthmes pléistocènes. *Archiv. Inst. Paléont. Hum.,* 6, p.1-220.

Vavilov N.I. 1926. *Studies on the Origin of Cultivated Plants*, Leningrad, Inst. Appl. Plant Breeds.

Velitchko A.A. (éd.) 1981. *Archéologie et paléogéographie du paléolithique supérieur de la Plaine russe*, Moscou, Nauka (en russe).

Vereschagin N.K. 1977. Berelekh mammoth "graveyard." *Proceedings of Zoological Institute*, 72, p.5–50 (en russe).

Vernet J.-L. 1973. Étude sur l'histoire de la végétation du Sud-est de la France au Quaternaire, d'après les charbons de bois principalement, *Paléobiologie continentale*, IV.

Vigne J.-D. 1988. *Les Mammifères postglaciaires de Corse. Étude archéozoologique.* XXVI° supplément à Gallia Préhistoire, Paris, CNRS.

Waldren W.H. 1982.*Balearic prehistoric ecology and culture.* BAR Int. ser. 149, tome 1. Oxford, Archaeopress.

Woillard G.M. 1978. Grande Pile Peat Bog : A continuous pollen record for the last 140 000 years, *Quaternary research*, 9, p.121.

Wroe S., Field J., Fullagar R., Jermin L.S. 2004. Megafaunal extinction in the late Quaternary and the global overkill hypothesis, *Alcheringa*, 28, 1, p.291-331.

Zazzo A., Lebon M., Quiles A., Reiche I., Vigne J.-D. 2015. Direct Dating and Physico-Chemical Analyses Cast Doubts on the Coexistence of Humans and Dwarf Hippos in Cyprus. *PLoS ONE* 10(8): e0134429. doi:10.1371/journal.pone.0134429

The last prehistoric hunters in Europe

Stefan Karol Kozłowski

Abstract

Mesolithic is a specific, highly specialized, interglacial formation/adaptation of Early and Middle Holocene hunters, gatherers and fishers evolved from Late Pleistocene communities. It developed mostly in the forest environment. One of its technological features is the miniaturization of chipped industry, due to the inaccessibility of big size nodules of the raw material. This miniaturization concerns especially the elements (arrowheads/inserts) of hunting weapons. The miniaturization, together with geometrization of elements of arrowheads (triangles, crescents, trapezes, and rhombs) suggests certain "cultural" unification of the phenomenon but in fact Mesolithic is regionally diversified (Western Late Tanged Points, North-Eastern Maglemosian, Epigravettian and Castelnovian Complexes). Mesolithic communities were extremely conservative, highly specialized, lasting almost unchanged through millennia, living in tribal, very local system, following seasonality, and practicing individual hunting. Finally most of them invented their own pottery and participated in the neolithization of Europe.

Résumé

Le Mésolithique est une formation/adaptation interglaciaire spécifique, hautement spécialisée, de chasseurs, de cueilleurs et de pêcheurs de l'Holocène supérieur et moyen issus des communautés du Pléistocène supérieur. Il s'est développé principalement dans l'environnement forestier. L'une de ses caractéristiques technologiques est la miniaturisation de l'industrie lithique, en raison de l'inaccessibilité des rognons de matière première de grande taille. Cette miniaturisation concerne en particulier les éléments (pointes de flèches/armatures) des armes de chasse. La miniaturisation, ainsi que la géométrisation d'éléments de pointes de flèches (triangles, croissants, trapèzes et losanges) suggèrent une certaine unification « culturelle » du phénomène, mais en fait le Mésolithique est diversifié au niveau régional (complexes récents à pointes pédonculées occidentaux, Maglemosien du Nord-est, Epigravettien et Castelnovien). Les communautés mésolithiques étaient extrêmement conservatrices, hautement spécialisées ; elles ont duré presque sans changement au fil des millénaires, vivant dans un système tribal, très local, suivant les saisons, et pratiquant une chasse individuelle. A la fin, la plupart d'entre elles ont adopté la céramique et participé à la néolithisation de l'Europe.

Introduction : Mesolithic

Mesolithic - the last, big hunter-gatherer formation of the prehistoric Europe is the result of post-Pleistocene interglacial adaptation of Late Palaeolithic peoples to the new forest environment.

These last Amerindians of Europe inhabited Lowland and Highlands and seasonally the Mountains. They hunted the games (bow, arrow, spear, harpoon), gathered the fruits of forest, fished mostly in the post-glacial lake lands (nets, fish-hooks, fish-spears, fish-traps, using the canoes and paddles), living their conservative, semi-sedentary life in their forests.

They introduced and realized the extreme Mesolithic microlithisation of the lithic industries, with the geometrization of the elements of their arrowheads.

Below we propose some remarks on the imposed environmental conditions (Nature), the human response (Culture) and a synthesis of prehistory of the Mesolithic Europe.

Nature - the environmental proposal

The forest is the Mesolithic paradise (Figure 1). It created Mesolithic man, determined him and gave him the means to live. The first regional signs of a new Mesolithic formation characterized by microlithization and geometrization of armatures can be observed in the forest biotopes of the final Upper Palaeolithic of southern Europe. Although the essence of the matter escapes us, we can safely assume that to be a Mesolithic man in those times meant to live in the vast forests that covered almost the whole continent from the beginning of the Holocene.

The biotope in Europe had not been equally rich in protein mass. A.L. Kroeber estimates the capacity of the forests at much more than the tundra on one hand and of the steppe-forest on the other, so the forest is absolutely the richest biotope of the Continent.

The abundance of green mass available in this biotope stimulated growth of the animal population (cf. § Game). Of greatest economic importance were the big mammals, like elk, deer, roe, aurochs, boar, horse, etc. Added to this were numerous smaller mammals (beaver, otter, badger, fox, etc.), as well as birds. The forest was also a rich biotope for gatherers. It supplied honey, nuts and undergrowth plants (among others mushrooms !), and was an obvious source of fuel and building material.

It must have been a paradise on Earth, especially north of the Alps, following after the hardships of the Pleistocene tundra environment. Specific adaptive models were triggered, considering the specificity and habits of dominant animal species, seasonally migrating (but on short distances) within their own ranges. Hunting concerned a number of coexistent species and occurred mostly in specific season and in specific parts of the animals' own territory. It appears to have been rather individual hunting (?) with the

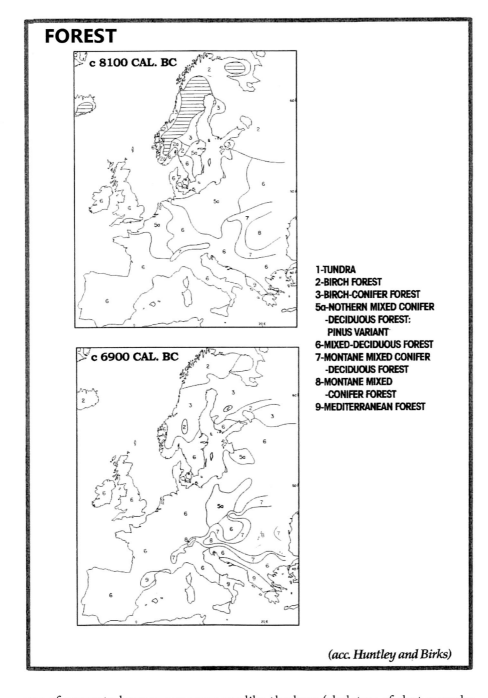

FOREST

c 8100 CAL. BC

c 6900 CAL. BC

1-TUNDRA
2-BIRCH FOREST
3-BIRCH-CONIFER FOREST
5a-NOTHERN MIXED CONIFER
    -DECIDUOUS FOREST:
    PINUS VARIANT
6-MIXED-DECIDUOUS FOREST
7-MONTANE MIXED CONIFER
    -DECIDUOUS FOREST
8-MONTANE MIXED
    -CONIFER FOREST
9-MEDITERRANEAN FOREST

(acc. Huntley and Birks)

Figure 1. . Holocene vegetation of Europe (top around 8100 cal. BC; down to 6900 cal. BC) (after Huntley & Birks, 1983)

use of accurate long-range weapons like the bow (skeleton of shot aurochs from Prejlerup, in Denmark).

The impression is of high specialization and adaptation to the environment. The apparent cost of this abundant life without stress was a resistance to novelties and a closing to the world. Mesolithic cultures and peoples were strongly conservative, living in considerable mutual isolation.

The story of forests in Europe in the Holocene started at the end of the Pleistocene and in the Early Holocene with the spreading of forest formations from southern refuges toward the north. During the climatic optimum in the Holocene (7000-4000 BC), they reached to the northern extremes of Scandinavia and Russia. In the earliest period (Preboreal), the frontier, between tundra and taiga, cut Scandinavia in two equivalent parts. Steppe formations (earlier) and forest-steppes (Atlantic period) developed along the northern Black Sea coasts and the Carpathian Basin.

Throughout the Early and Middle Holocene, the European forest featured a zonation that developed from south to north, starting with deciduous forest, followed further to the north by a mixed model and finally coniferous forest. The zones shifted gradually toward the north, in the wake of optimized climatic zones. The forest was also differentiated along an East-West line with mixed deciduous forest dominating in the west, especially in the Atlantic optimum (influence of the Gulf Stream), and simultaneously northern mixed conifer - deciduous formations in the East. Local differentiation was richer in fact, being dependent, among others, on the soils. Over time, the forest became denser and the species variability was based on the sequence: birch-pine-oak.

Variability over time and the zonation of European forest are reasons for a parallel regionalization of the forest fauna and this in turn was one of the factors determining the varied cultural behaviour of human communities observed on the European continent.

Game

The principal food resources of most Mesolithic communities were big land mammals from the forest biotopes of Europe (Figure 2). These animals were descended from the Late Pleistocene fauna minus a number of the great mammal species which disappeared shortly before, leaving room for the slightly smaller species, like reindeer, elk, aurochs, deer, roe, boar, horse, etc., sufficiently big, however, to engender the lively interest of the hunters of the time. These species have been known in Europe since a long time ago, preferring different environments, the reindeer and elk more to the North where it was colder, the forest fauna more to the South and West in a warmer climate.

The end of the Pleistocene and the beginning of the Holocene witnessed the northward expansion of forest habitats, causing the northward drift of forest fauna species which quickly occupied considerable stretches of the continent, pushing the reindeer and elk gradually further and further to the north.

The northward shift of the Early and Middle Holocene biotope resulted in at least two noteworthy phenomena (based on excavated osteological faunal remains):

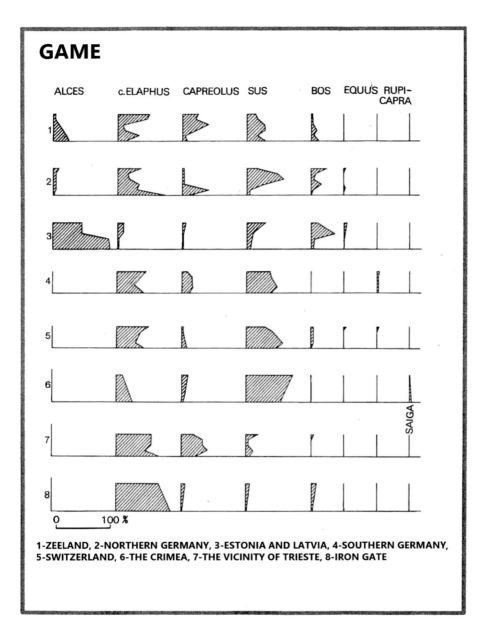

Figure 2. Fauna hunted in the Mesolithic

- Zonation of big game on the continental scale, with the tundra in the north being inhabited by reindeer, the dark coniferous forest by elk and seasonally by reindeer, and the mixed and deciduous forest by deer, roe and boar; were also subdivided along the W-E axis.
- Local fauna variability over time, corresponding to the changes of climate in the Holocene (e.g. gradual disappearance of elk from central Europe during the Boreal phase and the appearance a little later of deer and roe in the East Baltic area and even southern Scandinavia in the Atlantic period).

Mesolithic hunters also aimed for the small forest and water animals and birds. Such remains are frequently found on Mesolithic sites (cf. for example, Erik Brinch-Petersen's extensive lists for southern Scandinavia or maps edited and published by Raymond R. Newell, as well as by the author). The consumption value of these animals was low, but they were useful for their fur, musk, feathers, as well as bones for tool production.

The hunters' lifestyle depended on which animals were hunted. Greater mobility in the case of reindeer, lesser for forest fauna, herd size, specific habits of different species, skittishness of animals, etc., necessitated different adaptive models, organization of the hunt, even gear and hunting strategies.

M.R. Jarman's today forgotten theory about the Mesolithic hunters' alleged deer control is interesting to consider. Similar ideas were once expressed regarding Palaeolithic reindeer control in the ex-Soviet Union. After all, similar behaviour in the Near East led to animal domestication. In Europe, it never did.

The changing shape of a continent

The geological history of Europe in the Late and Post-Glacial periods had enormous impact on the Mesolithic of the continent. The prime mover in this case was the deglaciation of the North. The gradual melting of the Pleistocene glacier was completed in the 7th millennium BC, and it triggered at least three different processes that had direct influence on the later shape of the continent and consequently on its history in the Mesolithic.

Firstly, the permanent freeing of water from the continental ice sheet, occurring in the northern hemisphere, constantly increased the volume of the World Ocean which resulted in rising levels (maximum reached in the Atlantic Late Mesolithic) period, causing the flooding of lower-lying lands: Dogger Bank Shelf (harpoons fished out from the sea bottom), the area between the British Isles and Ireland, the Danish Straits (submerged Ertebøllian sites), the west Baltic land bridge, and finally, the Atlantic and Mediterranean coasts of Europe and the Near East (submerged PPNB site of Atlit Yam in Israel). The outcome was on one hand migrations of animals and people, and on the other, the inaccessibility of sites for research, especially sunk coastal sites.

Added to this is the planet's return to its previous roundness in parts freed from the great pressure of the continental ice sheet. This resulted in the depression in North-western Europe and the simultaneous rising of the Scandinavian plate (parallel to the rising of sea level). The first part of this mechanism resulted in the fairly rapid flooding of the northern parts of lowland Europe, the second in the appearance of modem Finland and the rising altitude of many old sites in western Norway.

The end outcome of these processes (e.g. successive stages in the development of the Baltic Sea: Yoldia, Ancylus, Littorna) is the disappearance of traditional bridges and oecumenes and the appearance of new ones. Europe was divided then, the Britain becoming separated from Ireland on one side and the Continent on the other (the English Channel) and Denmark was separated from Britain, Sweden and Poland. These processes should be kept in mind when considering the Mesolithic history of Europe (cf. the collapse of Maglemosian).

Stone raw materials

The old, Palaeolithic outcrops of flint raw materials are now almost inaccessible, they are covered by dense forests. The man is forced to gather small pieces of flint (Figure 3), among others erratic - Baltic flint from the Lowland.

This is one of the causes of the Mesolithic microlithisation (cf. infra).

The Late Mesolithic Castelnovisation (cf. infra) imposes the search for bigger nodules, mostly accessible through mining. It resulted in limited increase of the size of implements.

Territory/Culture – the human answer zonation

Mesolithic territoriality is of twofold nature: cultural and settlement, of which below. Any given cultural territory needs to be organized in one way or another, hence there needs to be a settlement pattern of some kind, most likely dependent on geographic and social factors (Figures 4, 5). The former include the water network as contrasted with the rich in game and warm river valleys as opposed to dry and poor in biomass cooler backlands. Then there are natural barriers (e.g. mountains) and passes, deposits of raw materials, differentiated flora and fauna, and a differentiated landscape (e.g. Lowland viz, Highland).

With regard to social factors, one should surely mention hierarchized segments of "tribal" organization which can be transposed into an observable segmentation of Mesolithic settlement with territories of a diameter respectively 30-50, 100-200 and 300 km constituting the principal units.

A common cemetery, uniting the people, could/should be placed in the center of the smallest structure.

The unit 30 to 50 km, is probably a closed and very isolated microcosmos, resistant to outside influence, conservative and virtually invariable (except for small and seasonal moves). These beads clusters of sites of a neckless threaded onto the river, are known from Poland, Britain, Germany, France and northern Italy. It does, however, admit occasional change of a fundamental nature in the instrumentarium as at the turn of

Figure 3.
Geometric
armatures and
microliths of the
Mesolithic

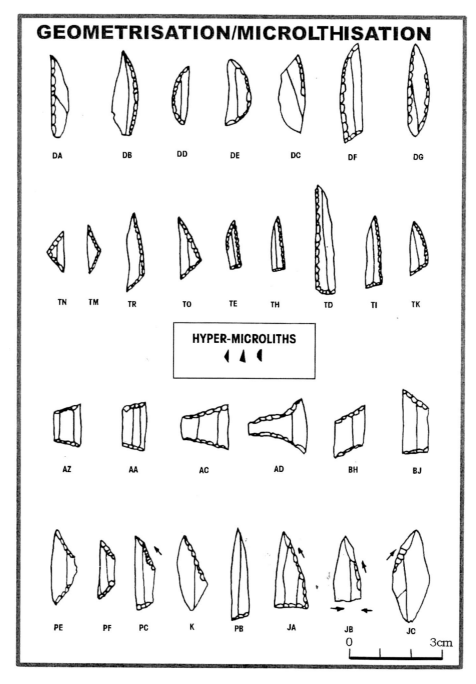

the Pleistocene and Holocene and a second time at the turn of the Middle and Late Mesolithic (cf. Castelnovisation). These changes tend to be quick (reasonably rapid mutation) rather than slow, and they are separated by usually very long periods of boring stability, not to say stagnation, in terms of tool style and the settlement pattern.

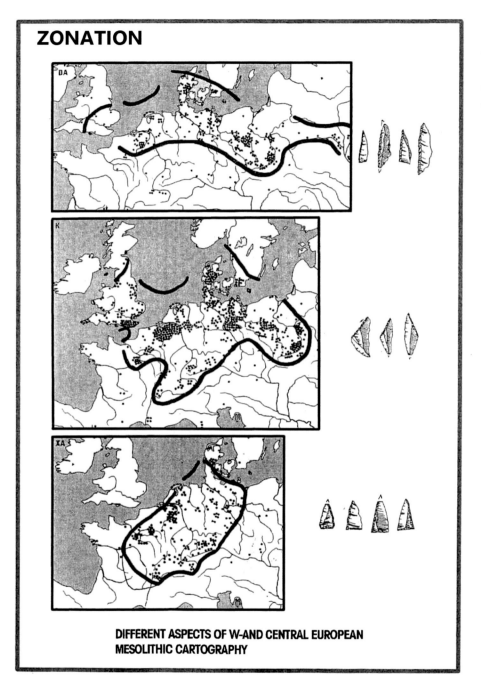

Figure 4.
Cultural zoning
from the
morphology
of geometric
armatures
(Central and
Northern
Europe)

Figure 5a. (top) Beuronian and Sauveterrien Figure 5b. (bottom). location of Mesolithic sites in western Poland

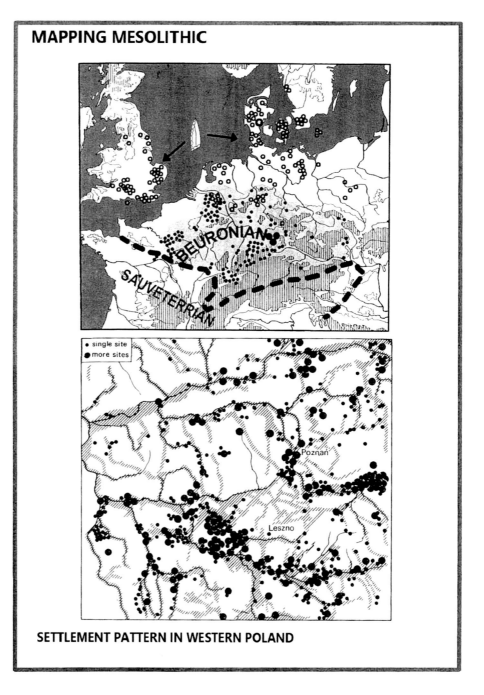

Arrow and arrowhead

The arrow is a projectile shot from the bow. Armed with the arrowhead, it enabled the exact shot, ensuring the comfortable life of its Mesolithic users.

The Mesolithic arrowhead was very often armed with geometrically shaped microlithic armatures (cf.) single or multiple.

The Mesolithic arrow had a wooden or bone shaft and an arrowhead most commonly made of stone, but also occasionally of bone (Shigirskoe points in the East - Clark's type 16). It could also be carved from a single piece of wood (e.g. blunted specimens). Some could have also been barbed ? (Wooden specimen from Veretie I in Russia).

Judging from the very few preserved stone specimens mounted on shafts, as well as stone arrowheads embedded in the bones of animals and humans, arrows can be generally divided into several groups. The first consists of arrows armed with stone arrowheads, further subdivided into composite and homogeneous examples. The latter are single points mounted at the top of the shaft. They form at least two different groups: tanged points (Stellmoor in Germany) and trapezes (Tvaermose in Denmark, vicinity of Oldenburg in northern Germany). The composite group, on the other hand, consists of a number of microliths mounted at the top of the shaft (Loshult in Sweden, Friesack in Germany) or found in sets embedded in the skeletons of hunted animals (Vig and Prejlerup, in Denmark).

The proposed classification is quite poor and surely does not exhaust all possibilities (cf. Rhythms), but finds are so rare that little can be done in this respect. There exists other evidence for the function of Mesolithic tanged points and microliths. It comes in the form of characteristic impact negatives of arrows entering an animal's body, as well as damage caused by removal from the body (broken peduncles, as for example in Stellmoor), not to mention traces of resin on the microliths (Star Carr in Britain).

Neither should the tanged points mounted on bone shafts from Olenyi Ostrov in Russian Karelia be neglected here.

A special form of arrow was the blunt wooden arrowhead forming a single piece with the shaft. Grahame Clark suggested, based on ethnographical data, that these projectiles served a very special purpose, namely, they were used to hunt birds and small fur mammals, where the objective was to protect precious feathers or fur.

Two forms of heads of these weapons can be distinguished, one like a cylinder with rounded or pointed end, the other like a reversed cone or pyramid. The former type is known from Denmark and Germany, from the Maglemosian or to be more specific, from its Svaerdborg (Holmegaard IV) and Duvensee (Hohen Viecheln) groups, while the latter type from the Ertebøllian, as well as from northeastern Europe, in association with assemblages of the Nord-Eastern Technocomplex, (e.g. Vis I in Russia).

It seems that the morphological similarity of specimens from territorially distant units, such as the Danish-German and Russian finds, is due to the nature of the raw material and the simple form of the discussed projectile.

It should be pointed out that the East European Plain has yielded undoubted counterparts in bone for the presented kinds of arrows. These are the Shigirskoe points, variant with long stem, known also as a matter of fact as a variant in wood (Veretie I).

Armatures

Despite semblances, armatures are small, mainly microlithic, but also hyper-microlithic and backed, mostly geometric (triangles, crescents and trapezes) specimens with back(s) and truncations, considered as inserts arming arrowheads (cf.), but also knives, sickles, daggers and spearheads, as proved by infrequent finds of mounted specimens from Europe, Africa and the Near East. Based on morphological characteristics, they are traditionally classified as points, backed bladelets, crescents, triangles (isosceles and scalene), microtruncations, trapezes; moreover, specific groups like the Tardenois and Sauveterre points, *feuille de gui*, Janislawician points etc. (Figure 3)

Uniformity resulting from metric (microlithization) and formal (geometrization) characteristics, armatures can vary by territory, enabling great cultural regions with evident borders to be established (cf. Territory). In the south and west of Europe and in the center of the continent common evolutionary trends have been observed (connected to arrowhead structural development) from rather broad forms to narrower ones (mainly triangles), later replaced with trapezes (cf. Rhythms). In northeastern Europe, the role of armatures was taken over by retouched and unretouched bladelets/ inserts and tanged points.

Geometrization / microlithization

Retouch technique (usually abrupt) forming tool edges (most often microliths) with the aid of two or seldom three backs or truncations (Figure 3). The resultant form is a small or very small (even only 4-5 mm!) triangle, trapezoid, rectangle or segmental piece. Such geometrically shaped microliths had been presumed initially to be specifically Mesolithic products. Sieving of deposits on pre-Holocene sites has demonstrated, however, that microliths were present also in some Late Paleolithic industries (Epi-Magdalenian, Epi-Gravettian, Ahrensburgian etc.). Formed often by microburin technique, but also by sectioning and pseudo-microburin technique, geometric microliths create the impression of a continental-wide uniformization of the European Mesolithic. This could partly be due to the limited repertory of executable formal variants. Even so, geometrics occurred only in a part of Mesolithic Europe (central, southern, southeastern and western) and moreover, there were relatively numerous specimens of a regional or even local importance, that could be differentiated regionally, by size, for example (cf. Beuronian vs. Sauveterrian cf. Territory), and testifying to a varied Mesolithic tradition. It was once thought with some naivete that microliths represented degeneration or degradation; not so today, when we are inclined to consider them as an expression of progress and effective miniaturization.

The Mesolithic groups were able to do everything from almost nothing, e.g. from easy available small raw material nodules; the outcrops of big size nodules were probably not accessible (strong forestation).

The miniaturization started in the Mediterranean area, in final palaeolithic Epigravettian cultures (Romanelli, Piancavallo, Cornille, Franchti, Cuina Turcului, Filador, Szekszard Palank), that spread to the North.

Fishing spears and harpoons for hunting

Fishing spears appear to be a specific and highly local adaptive measure, designed to take full advantage of what the environment could offer. This invention, which is known from the central and northeastern regions of the Continent, is exceptionally well represented in the lake districts circumventing the North and Baltic seas (Figure 6).

The pike skeleton with harpoon, found at the bottom of the Kunda lake in Estonia, as well as the double-mounted harpoons from Star Carr and various other mounted examples cited by Grahame Clark, not to mention fish hunted with harpoons (also cited by Clark) and hafted objects from Friesack in Germany and Veretie I in northern Russia indicate that different harpoons (mostly with small barbs) were in common use around the Baltic, neither should one forget the totally wooden leister from Skjoldnaes (Denmark).

Naturally, not all bone barbed points were fishing spears, as demonstrated by two harpoons used for hunting elk in Great Britain (High Furlong).

Spearheads and harpoons came either singly (Kunda, Friesack) or in pairs (Star Carr).

Apart from the Würmian lake district, barbed fishing spears and hunting harpoons have also been recorded in the East European Lowland. They are almost missing - and presumably not accidentally - from the rest of the continent (with the exception of flat harpoons present on the northern slopes of the Alps).

Rhythms

Change in whatever form is perhaps the characteristic trait of the European Mesolithic, regardless of how regionally it is considered. Technology was impacted and morphology of chipped products, encompassing selected groups of objects or entire industries. Evolution was either cosmetic in nature, like in the case of the microliths in the Italian Sauveterrian, or

Figure 6. Hunting
and fishing

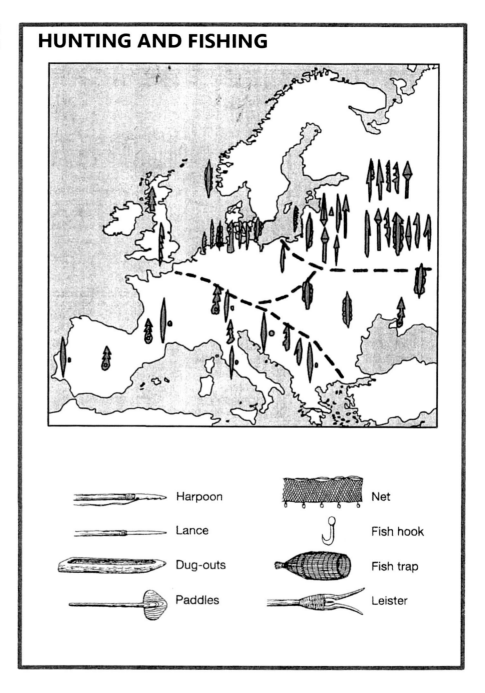

HUNTING AND FISHING

Harpoon	Net
Lance	Fish hook
Dug-outs	Fish trap
Paddles	Leister

abrupt, in which case researchers have to adapt their taxonomic descriptions (e.g. transition from Sauveterrian and Beuronian to the Castelnovian-like industries): respectively Classical Castelnovian and Montbanian (cf. Castelnovisation). In any case, this evolution concerned de facto a change in arrow construction.

It would be going too far, naturally, to suggest one evolutionary rhythm encompassing the entire Continent. Each region had its own specificity and rate of change. The most dynamic in this respect were the macroregions of the South, West and Center, which shared a certain common evolutionary trend, even in the face of distinct regionalisms (Sauveterrian, Beuronian, Maglemosian). The similarities between the three regions go beyond the presence of geometric microliths (backed points, scalenes, isosceles, crescents, rectangles, microtruncations, trapezes), covering also the changes of these stone products or at least the proportions between them.

Our reconstruction of this evolution is based on stratigraphic data from Southern Europe (Jägerhaushöhle in Germany, Romagnano III in Italy, Birsmatten-Basisgrotte in Switzerland, Rouffignac and Montclus in France, and other stratified sites), combined with numerous local or regional cultural sequences (reconstructed by Erik Brinsch-Petersen for Denmark, Hermann Schwabedissen, Bernhard Gramsch, Surendra K. Arora, Erwin Cziesla and Wolfgang Taute for Germany, Raymond R. Newell, André Gob and Pierre Vermeersch for the Benelux countries, Roger Jacobi for Britain, Max Escalon de Fonton, Jean-Georges Rozoy, André Thévenin and Michel Barbaza for France, Alberto Broglio, Paolo Biagi and Carlo Tozzi for Italy). The starting point was the broad and short microliths on irregular, short bladelets (10th-8th millennia BC): single and double truncations, isosceles and short scalenes. In the next phase (from c. 7500 BC), scalene triangles with short base appear or dominate, made on a new type of narrow bladelet; at the same time, some of the old types disappeared (together with old-fashioned "Coincy style" bladelets) completely or at least lose importance. Finally, the last phase (from c. 7000-6500 BC) is characterized by the introduction of trapezes and a new Castelnovian (cf. Castelnovisation) technology with broad, very regular bladelets ("type Montbani"), which more or less quickly pushes out the "old" microliths. Importantly, exactly the same changes of core technology and blanks were taking place in different cultural environments, but without leading to total cultural unification.

The *koine* did not include Iberia and the Italian Mezzogiorno, as well as the final Epigravettian of Balkans.

The described change of core technology, blanks (irregular, short bladelet - microbladelet - broad regular bladelet) and microliths (among others, broad triangles - narrow triangles - trapezes) is strongly linked with the evolution of arrowheads and is determined by their construction.

All this happened between Vistula river and the Atlantic Ocean, and only in this macro-region.

General and individual Culture

Should one desire to characterize it somehow in general, Mesolithic culture will have some shared traits (although hardly a proliferation of these) which are not as unique as some would like to believe.

First of all, Mesolithic culture in general is a continuation of Upper/Late Palaeolithic model of life, with or no change (local tradition without change, local tradition with change, migration of people without change of tradition, migration of people with changes). As such, it is naturally a specialized hunting-gathering formation with regional emphasis on different game hunting, intensive fishing (apparently a novelty) and a gradual increase in the significance of regionally differentiated gathering.

The instrumentarium (not taking into account the almost unknown wooden tools) is straightforward:

- different arrows (with stone geometric/microliths or tanged arrowheads, but also with wooden tips) propelled with bows,
- hurled harpoons and spears (made of bone/antler/ wood),
- "domestic" tools, mainly of small size, e.g. stone end-scrapers, retouched-use-retouched and unretouched blades and bladelets, truncations, "axes" of antler and "daggers" of antler/bone, cervid spikes, knife-scrapers made of boar tusks, wooden dugouts and paddles,
- sometimes also stone axes-adzes, burins, perforators and retouched flakes,
- stone raw material quarried or collected mainly within the framework of local autarkic economies (distance of up to 50 km); real and massive imports (radius of > 150 km) did not appear until the Late Mesolithic. Prior to that, the autarkic character of the raw material economy generally forced a considerable miniaturization of stone implements; local raw materials occurred mostly as small-size nodules or as big but internally cracked ones.

Even so, small geometries and not very big tanged points, were quite effective as arrowheads despite their micro size.

The impression of a certain stylistic uniformity of this equipment that lends itself from an analysis of the finds is also due to the simplicity of forms (mostly single or multiple backs/truncations) and techniques of execution (sectioning or microburin), as well as the small size of blanks. A real stylistic or cultural differentiation can be observed often only with regard to rare/specific types.

Supplementing this Mesolithic inventory are fishing implements - hooks, traps, nets with floats - which, however, do not need to be pan-continental in nature, even though fish were naturally caught everywhere in the Europe of the time. But not with the same intensity.

The inner structure of assemblages could reflect stylistic and/or cultural distinction, as well as their chronological position, but also the actual function of a site or complex of sites - whether it is a "base" or "satellite" camp, a workshop, raw material mine, butchering site, etc. - and whether the stone raw material was at all accessible (cf. mostly "bone" Kunda culture of Latvia and Estonia).

The said functional differentiation of assemblages is occasionally reflected in the rare, well-documented but not very surely dated Mesolithic "camps" of different size and richness; their homogeneity (meaning a short-lasting and uninterrupted occupation), however, is seldom verifiable, especially with regard to the big features. Seasonality is also confirmed.

Cultural/territorial differentiation

The techno-typological differentiation of the Mesolithic is important, at least to the present author. On the whole, it seems to correspond to the different inherited traditions separating people and communities and it appears to be verifiable cartographically. An analysis taking into consideration techno-typological and technical aspects of material/sites/assemblages demonstrates the differing territorial importance/value of various types and sets of types, which can be basically territorial, specific, individual (local or regional) or general (supra-regional, intercultural) in a territorial sense. Therefore, such cartographic characteristics can be either completely banal or particularly typical. Finally, types and sets of types can indicate chronological horizons and in such cases, they are naturally characterized by a supra-regional range (Figures 4, 5, 7).

Mapping of types and sets of types (controlled from the structural - assemblages, and chronological - datation point of view) reveals the existence of territorial structures exceeding the modern political boundaries (cf. Zonation/Territory), exhibiting own borders and featuring a common style and structure comprised of many types. It leads to generalizations on a continental scale: Europe can be divided into super-territories that are characterized by common techno-typological features (techno-complexes). These in turn are subdivided into smaller territories corresponding to particular "archaeological cultures".

Divisions of this kind are based on environmental factors like climate, fauna, flora, raw materials, geomorphology, barriers, gates, hydrography etc. Even so, they are mainly a genetic continuations, one way or another, of Terminal Palaeolithic styles reinforced by the isolationism of small microcosmoses separated one from the other by vast distances (physical and mental) and kept apart by the above-mentioned environmental factors. Of just such microcosmoses Ryszard Kapuściński wrote:

"For most people the world ends on the threshold of their own home, the outskirts of their own village, the borders of the valley they live in at the furthest"

To be brief, the Mesolithic is a very stable, one could even say petrified formation, subdivided and structured regionally, featuring definitely the longest spans of stabilization in the techno-typological, but also territorial and "cultural" sense, complete or almost complete, lasting a few millennia at a time regionally (e.g. in the Russian Lowland). Preceding and possibly

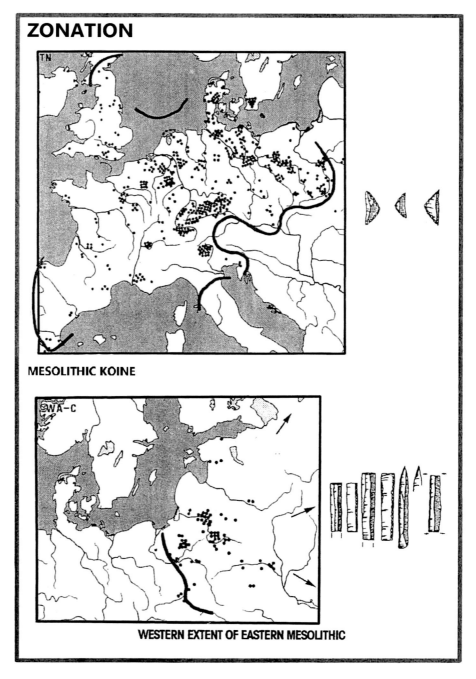

interrupting it are relatively brief moments of destabilization that result in significant techno-typological (but also behavioural) changes, instigated most frequently by drastic or at least significant ecological changes (turn of the Pleistocene and the Holocene in most areas of Europe, then the beginning of the Atlantic period), including climate, vegetation, fauna, but also geology, such as sinking of whole land formations (cf. changing shape).

The changes could be adaptive (to a new local environment) or migratory (in search of a disappearing homeland). The end effect of processes understood in this way was a gradual and multidirectional disintegration of the "old" system, occurring over a certain span of time, during which a "new" system developed appeared.

Pulsating toward the North

A gradual but regular warming of the climate in the Early and Middle Holocene pushed the vegetation zones (tundra, taiga, mixed and finally deciduous forests) to the north (cf. Forest). With these latitudinal zones went the faunal complexes (respectively, reindeer, elk, forest fauna, cf. Game).

This northward pulsating push appears to have significantly impacted several times the movements of human groups or at least the proliferation of certain ideas, as a rule from the South to the North.

1. The earliest "push" came at the end of the Pleistocene and beginning of the Holocene, which was a time of revolutionary ecological changes (10th millennium BC) marked by relatively rapid northward displacement of ecological zones and a gradual deglaciation of the north. According to the archaeological record, the Norwegian Arctic territories were settled already around this time (migration of hunters representing the Tanged Points tradition from Central and Eastern Europe). At this point a big part of northern Europe, so far uninhabited, seems to have started teeming with life.

In parallel in the same time we observe a clear proliferation to the North, of Mediterranean or more broadly speaking, Southern European elements:
- The Sauveterrian expand toward the North, following the French Rhone valley (André Thèvenin's nicely documented idea), but also comes from Slovakia to southern Poland);
- New types of geometrics known earlier from the South reach the European Lowland (important theory presented by Bernhard Gramsch);
- The southern German-French-Belgian Beuronian could have been a reminiscence of this proliferation (according to W. Taute's theory, originating from the south Germany, if not a more widely considered Epimagdalenian.

2. The second paroxysm came in the second half of the 8th millennium when the entire Northwest of Europe was "sauveterrized", that is, it underwent a certain typological uniformization that originated most likely from the southern source. This change drew from the introduction of a new types of microliths and forcing (or forced by) a change in arrowhead construction (cf. Rhythms).

It could be correlated with significant ecological change in this region at this time, mainly the spread of deciduous (oak) forests. But this change obviously did not cause the phenomenon.

Simultaneously, the real Sauveterrian also ventured north beyond the Alps (into Austrian Tyrol), the Dinaric Alps and western Slovakia. What's more, Sauveterrization does not look like actual human-group movement, contrary to the proliferation of the Sauveterrians described above.

3. The third stage of the process was the Castelnovisation (Figures 8, 9) of Europe, meaning the appearance over large parts of the continent of a

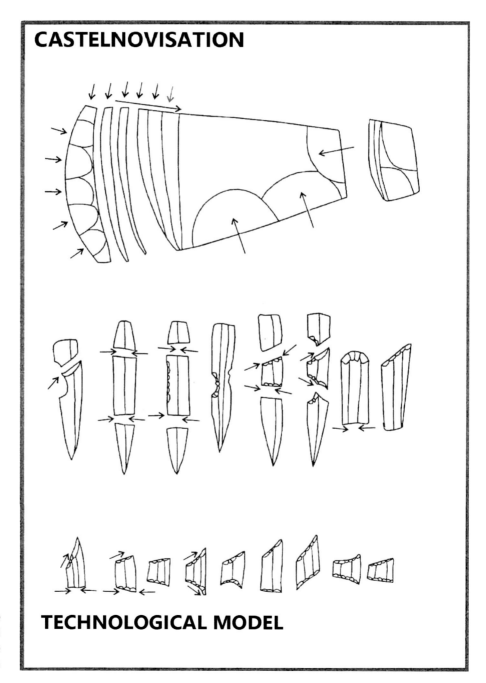

CASTELNOVISATION

TECHNOLOGICAL MODEL

Figure 8. The technological model of the Castelnovian

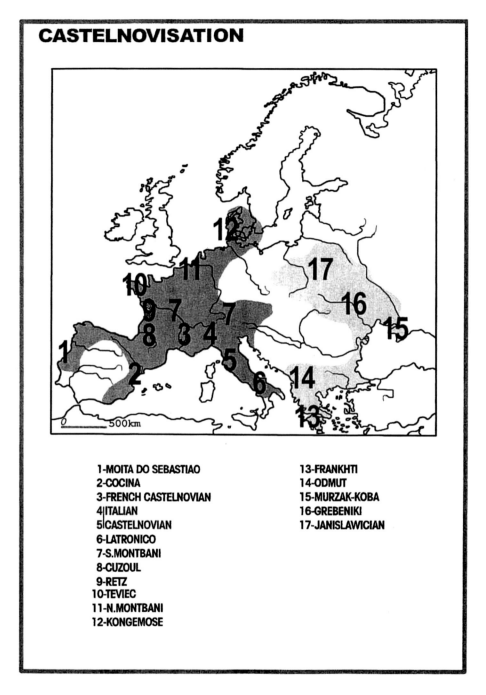

CASTELNOVISATION

1-MOITA DO SEBASTIAO
2-COCINA
3-FRENCH CASTELNOVIAN
4|ITALIAN
5|CASTELNOVIAN
6-LATRONICO
7-S.MONTBANI
8-CUZOUL
9-RETZ
10-TEVIEC
11-N.MONTBANI
12-KONGEMOSE

13-FRANKHTI
14-ODMUT
15-MURZAK-KOBA
16-GREBENIKI
17-JANISLAWICIAN

Figure 9. The Castelnovization process

new blade technology, system of long-distance distribution of raw materials and new types of arrowheads (trapezes, rhomboids). The phenomenon appears to have spread gradually from the south (Franchthi, Murzak-Koba) to the north (Kongemose, Janislawician) which nicely correlates with the northward expansion of forests in the Atlantic period (cf. Forest).

The regionally differentiated phenomenon (western and eastern zones) becomes implanted more or less deeply, but on the whole slightly later in the north. It generally spreads from the southern "center" to the north (cf. Castelnovization).

4. The phenomena that may have paralleled Castelnovization, but were at least partly caused by other reasons include:

- migration of the microbladelet tradition from southern Scandinavia to Sweden and Norway (although the causes of this migration may have been more complicated) (cf. changing shape), pushing out Fosna culture to the North;
- settling the Russian Far North, marked by such sites as Veretie I, Olei Ostrov and Vis I (assuming there had not been an earlier push paralleling the first stage in Scandinavia described above) and the emergence of the Kunda tradition in Finnish Lapland (Sujala).

The data at our disposal is still insufficient to give an understanding of which of these "pushes" can be considered as migration and which as acculturation.

Prehistory Before

Mesolithic cultures did not come from nowhere and they are naturally a continuation and adaptation of cultures from the Dryas III period, the last cooling in the Pleistocene which occurred in the 11th millennium BC. During this period Europe was divided into four separate ecological zones which extended more or less latitudinally: 1. the glacier in the north and its foreland; 2. tundra, mostly on the Plain; 3. forests, mostly birch and pine more to the north and a mixed deciduous forest in the south; and 4. steppe in the Black Sea region.

The fauna inhabiting these zones was correspondingly differentiated as well. Reindeer ruled in the tundra, the elk was found in the northern stretch of forests, while the southern forests were replete with deer, aurochs, boar, and horse; finally, in the steppe, outside the valleys where forest animals lived, the saiga antelope was hunted.

The differentiated zonal hunting economy practiced for millennia (different animal habits, different ways of hunting, different hunting *instrumentarium*, different social organization) was supplemented with other kinds of activities, especially in the south of the continent (gathering molluscs, for example, and fishing).

Topping this in the cooler Dryas III were the strongly differentiated cultural traditions corresponding roughly to ecological and geographical zones. They were to give rise in the near future to the differentiated cultures of the Mesolithic.

Following are some remarks on this subject.

In the 11th millennium, the cultures of the South of Europe (Spain, Italy, Balkans) preserved the most traditional, local Upper Palaeolithic outlook.

In Iberia, Italy and in the Balkans the final Epigravettian of Mediterranean type underwent two parallel processes: microlithization of bladelets, geometrization of arrowheads and points, and finally the miniaturization of other tools (Piancavallo in northern Italy, Montadian in Provence, Szekszard-Palank in Hungary, Molodova Ia in Moldova, Cuina Turcului in Romania, Lower-Middle Franchthi in Greece, etc.).

At the same time, the Late Upper Palaeolithic cultures of the time (among others, Western Epimagdalenian and Creswellian, as well as some variants of the Eastern Gravettian) are "Azilianized" (introduction of very short flake scrapers and arched backed points), that is, they take on western Azilian (western Europe) and northern Federmesser/Epimagdalenian (Germany), Epigravettian (Balkans) or Shan-Kobean (Ukraine) characteristics.

Just after this time the Lowland North, but also the south of Scandinavia, was settled (from where?) by the newly emerging tundra peoples/cultures using recently invented tanged points.

The archaeological map of Europe in the Dryas III and the earliest Preboreal periods is relatively well established, showing latitudinal zonation, that is, in agreement with the environmental zonation of the continent. The southern belt was occupied by Epigravettian entities: the Mediterranean type from the Balkans to the Iberia, and the Eastern type East of the southern Carpathians all the way to the Don river. In both cases, there was an "Azilianized" continuity of local development through the millennia. Both territories also reveal "azilianization" processes, in differing degrees for that matter (Italian Romanelli-type, French Azilian and German Federmesser assemblages). Both also exhibited (although in differing degrees) the phenomena of microlithization and geometrization. In both cases we are dealing with hunters interested in forest game; in the south, the fauna also included *equus hydruntinus*.

The Shan-Koban-Beloblesye culture of Ukraine and the northern piedmont of the Caucasus also belongs to this southern zone; it appears to be connected with the Caucasian-Caspian para-Gravettian tradition. This culture lives of forest game and the *saiga* antelope. Azilianization is also an observable process here, as well as partial geometrization but without microlithization pushed to the extreme.

The next zone to the north runs along the Alps from eastern France to Moravia. It is occupied by industries which refer to the Epimagdalenian (Rochèdane, Zigeunerfels, Kulna, etc.). Azilianization is also an observable phenomenon in this area (arched backed points).

The Lowland, as already mentioned above, constitutes the next zone. In the Dryas III period and the very beginning of the Preboreal, it is occupied by mostly tundra cultures/communities with tanged points as a characteristic element of their stone tool industries. These are, starting from the west: Ahrensburgian, Brommian, Swiderian and Desnian, extending from eastern England all the way to the upper Volga. The hunters preyed on tundra reindeer, occasionally also elk. Azilianization was also noted in this group (but not in northern Brommian) in the form of very short end-scrapers.

In the northwestern part of the continent the local Creswellian entered also at this time its Azilianization stage (Dead Man's Cave).

Its eastern neighbour on Dogger Bank Shelf and in northern Holland, also bearing distinct marks of Azilianization was the latest phase of the Federmesser Culture (with sites in northern Holland dated to this period! (information from Raymond R. Newell)

Summing up, the ecological and cultural zonation described above should be emphasized, because it was largely to be repeated in the Mesolithic over most of the continent (parallels: western Mediterranean Epigravettian = Sauveterrian, Balkan Epigravettian = Epigravettian Mesolithic of the Balkans, South German and Czech Epimagdalenian = Beuronian, Ahrensburgian + Cresswellian + Federmesser = Maglemosian) with the exception of the northern reaches of Europe where migratory trends were present.

Thus, a general theory about the mostly local continuations of the Upper/ Final Palaeolithic into the local/earliest Mesolithic can be assumed with some potential mobility. These continuations concern industries already azilianized in the terminal Pleistocene (almost all of Europe) and locally geometrized and microlithized (southern edge of the continent), as well as furnished with tanged points in the north. All of these characteristics will last into the Mesolithic, even if sometimes delocalized, and some of them (microlithization and geometrization) will develop and spread to larger areas, mostly to the north.

Finally, one should note that wherever significant techno-typological changes can be observed at the Palaeolithic-Mesolithic transition, they do not occur at the exact turn of the Pleistocene and Holocene, but somewhat later, which means that regionally the Final Palaeolithic traditions survived unchanged into the first half of the Preboreal, and in some regions (Balkans, Mezzogiorno, Norway, Sweden, Iberia) even much longer.

Change and continuation : the Big Change

The turn of the Pleistocene to Holocene (c. 10,500/9800 BC) in the opinion of many researchers was supposed to be the reason for the emergence of a new "cultural"/adaptive quality called the Mesolithic. This new quality was indeed revealed in many, but not all areas, and not at this time, but somewhat later (c. 9600-9500 BC).

It cannot be defined that the environmental shaking up that occurred at the beginning of a new interglacial period was sufficiently important to put the machine of history into motion, changing the fairly stable system known from Dryas III. These relatively rapid changes concerned climate foremost. In the beginnings of the Holocene it moved rapidly from a glacial to an interglacial phase, leading to a number of significant changes of biotopes in Europe. One of these was the northward proliferation of the forest zone practically up to southern Scandinavia, central England and southern Finland, thus reducing the tundra to a belt north of this zone (previously it covered practically all or at least a big part of the northern European Lowland). A parallel gradual melting of the Scandinavian glacier and perhaps the smaller one in the north of the British Isles lasted until c. 7000 BC, slowly increasing the extent of the oecumene here and decreasing there.

Forest expansion and receding of the tundra had its consequences for the migration of big mammals, constituting the main economic base of hunting-gathering societies of the period and for a long time yet. Reindeer moved up north to the areas where it lives today, while forest game - deer, elk, reindeer, aurochs, boar, horse - extended gradually its habitat toward the north, retaining an inner zonation with the elk keeping more to the coniferous forests in the north and the other species inhabiting the mixed and deciduous woods in the south. Much later, with improvement of the climate, forest fauna will expand more to the north (cf. Game).

The parallel evolution of marine reservoirs in northern Europe (Baltic and Northern Sea) changed to some extent the range of oecumenes in this period (a slow and gradual, but continuous reduction of territory, cf. Changing shape).

All these changes could not have had no any influence on the adaptive patterns of at least some Final Palaeolithic human groups, especially those located in the northern part of the Dryas III oecumene. The adaptive reactions generally fall into two basic models:
- local adaptation to new conditions (regardless of what is meant by the term) without changing the place of residence,
- migration in an effort to keep up with a receding familiar environment.

It is also possible to imagine a more complicated process in which both strategies find application (for different parts of one community, e.g. the fate of the Ahrensburgian?). Whatever the case may be, any attempt at reconstructing the process based on the archaeological record presents considerable interpretational difficulties. The same is true of the origins of the Mesolithic cultures. It is due presumably to gaps in the record; after all, most sites are not well dated, while evolutionary change/mutation in archaeology can proceed quite rapidly at times. Hence caution is recommended in reading the remarks below.

It is possible that the first model (local adaptation) was realized more frequently in the South where the Pleistocene to Holocene environmental change may not have been as bitterly felt. The second model is more likely to have characterized northern regions, where the temptation to settle newly emerging land and to follow the disappearing beloved reindeer would have been much stronger. This is generally confirmed in practice.

Wherever significant changes in culture and tool style are noted, there are usually two other, significant phenomena:

- The changes, assuming they occur, take place not in the time traditionally considered by geologists as the transition from the Pleistocene to the Holocene, but quite, a little later, that is, around 9600-9500 BC. At least some of the cultures of the terminal Pleistocene style survived into the very Early Holocene (Ahrensburgian, Swiderian, Desnian, Balkan Epi-Gravettian, Shan-Kobian, etc.). It is possible, therefore, to speak of a certain retardation of the cultural and stylistic reaction to environmental change (the oldest dates for new cultures/taxons are placed around 9500 BC)
- Classic new Mesolithic industries were preceded in at least a few cases by industries of veritable transitional style in which certain earlier characteristics disappeared (e.g. tanged points in western Ahrensburgian) with the simultaneous increase in the importance of features that are a local attribute of the Mesolithic (e.g. geometries and microliths). Examples include the Epi-Ahrensburgian of north-western Europe, the "early Mesolithic" of Belgian Wallonia, the "Mesolitico Antico" of northern Italy, etc. This transitional phase of "Early Mesolithic" cultures was usually rather short and was followed by an eruption of "Mesolithic style" industries considered classic for given regions (e.g. Early Maglemosian type Bare-Mosse, Star Carr-Klosterlund, Beuronian A and Early Sauveterrian respectively). It is another matter altogether whether this "transitional phase" means the same in all cases; in some instances it may well be nothing more than a cover for a paucity of sources in the archaeological record (small sample size translating into the limited typicality of assemblages that are de facto already of the "Mesolithic style")
- Cultures representing the classic ("new") Mesolithic of a given region started forming only after this initial "transitional" stage (?)

The above remarks clearly demonstrate the futility of searching for a single root or cause for Mesolithic Europe. Even had it existed, there is no reason to assume that it would have led every time to the same effects. Therefore, it is possible to speak of important changes taking place on a regional and even macroregional scale, but never on a continental one, this because some regions, like the South, did not experience any profound environmental/ typological/ technological transformation at this time.

This is naturally in agreement with the chronological definition of the Mesolithic, from which it draws that everything that was taking place in Europe starting from around 9600/9500 BC should be referred to as "Mesolithic" regardless of how "innovative" or "traditional" the nature of this "everything" was. Despite this there is no way of refuting that in a significant part of Europe the Mesolithic introduced an entirely new economic and cultural quality, one that is definitely easier to recognize in the north than in the south. Specific technical designs, known from earlier times but much less widespread and applied, are proliferated to the north at this time. Moreover, new territories are discovered and settled, hence the shock to stability is immense in some regions.

Models for the emergence of Mesolithic cultures can be different, as we have already said above. The following is a general presentation.

The *first model* is one of local environmental and economic, as well as stylistic continuation. It was applied at the southern edges of Europe where the transition from the Pleistocene to the Holocene had not brought any significant environmental change, the caesura provoking real changes having come earlier - in the Allerød. It initiated local adaptive patterns, including forest game hunting, development of fishing, miniaturization and local geometrization of stone tools, etc. In this way local units of the Late Mediterranean Epigravettian were formed (among others, Azilian, Montadian, Romanellian, Balkan Epigravettian of the Cuina Turcului type, Italian Piancavallo, and Iberian Maleates).

Any potential cultural change in these regions where the transition from the Pleistocene to the Holocene had not been strongly felt in the environmental sense would need to have been caused by different reasons, which is not to be fathomed at present. Neither would such change have covered the entire discussed territory. This was the case: regions that were particularly isolated, such as the southern parts of the Iberian and Apennine peninsulas, the Carpathian Basin and the northern Black Sea littoral escaped profound cultural/stylistic changes at this time. Others (south of France and central and northern Italy) passed through an evolution from the Epigravettian to the Sauveterrian.

In the *second model*, in the area to the north of these "conservative" territories (except for the Sauveterrian), in a belt from the Pyrenees to the British Isles, southern Scandinavia and Poland, the 10th millennium featured a number of important cultural and stylistic changes which took place in the early Preboreal and which led to the formation of entirely new hunting-gathering-fishing cultures that were quite different from their predecessors. They were characterized by a modified or new *instrumentarium* that had undergone serious miniaturization and geometrization, finally losing some of the "old" stylistic features as far as the flint tools were concerned. These new taxonomic units belonging to the western and northern techno-complexes (among others, the legendary western "Tardenoisian"/Beuronian and the equally famous northern "Maglemosian") have given shape to the traditional typological and stylistic definitions of the Mesolithic that are deficient, because they leave out at least a half of Europe from this time.

The described model was applied in three different climatic, environmental, landscape and cultural zones. In the South, it was France south of the Loire, central and northern Italy, part of the Spanish Levant, Italian Mezzogiorno with local Late Pleistocene Epigravettian and related industries. To a lesser (in the extreme south and west) or greater degree, these entities evolved and were transformed into the Sauveterrian, which enjoyed a very similar territorial range (from the Pyrenees to the Loire river, the Slovenian karst and Tuscany), and western Slovakia, or the para-Sauveterrian industries in the case of the Spanish Levant, and Italian Mezzogiorno. This occurred through a change of core-reduction technology (introduction of discoidal core) and an observable, rapid increase in the number of already known geometric and non-geometric microliths. The process is dated to around 9000 BC. The as yet poorly investigated transitional phase is referred to by Alberto Broglio "Mesolitico Antico".

Slightly further to the North and at the same time, the Mesolithic Beuronian A emerged in the western and central European uplands. In the South, this unit was considered by Wolfgang Taute as a continuation of a southern German and Czech Epimagdalenian of the Zigeunerfels type.

In the Western and Central European Lowland, the transition appears much more complicated as a process and it has yet to be fully understood. Even so, a few fairly certain remarks can be put forward.

In the 10th millennium BC, the territory between eastern England, Belgium, northern Germany and Denmark was the arena for the emergence of a new cultural unit called the Epi-Ahrensburgian (Zonhoven, Pinnberg I, Bare Mosse). This short-lasting entity which is devoid almost of tanged points was the possible effect of a local evolution/adaptation from the Ahresnburgian to the Maglemosian in its earliest form (Star Carr-Klosterlund-Duvensee type industries with sometimes the latest tanged points and numerous broad geometries with an Ahrensburgian flavor to them, cf. Star Carr). In the case of the Epi-Ahresburgian, the industry is a typically transitional one which will lead in time to the "true" local Mesolithic.

The fate of Late Pleistocene peoples, inhabiting the later submerged Dogger Bank Shelf of the Northern Sea is difficult to trace and understand. One should keep in mind, however, that apart from the very rare tanged points known from Epi-Ahrensburgian, there were also classic Maglemosian arched backed points (Bøllund, Broxboume, Svaerdborg) that were slender and backed microliths (e.g. Klosterlund in Denmark, Oakhanger in England), possibly originating from cultures with backed tools (e.g. Mother Groundy's Parlour, upper layer).

The last element of this still unclear puzzle (the role of backed pieces is particularly unclear to the present author) is the potential Maglemosian expansion to the East (?), to the territory of Poland (c. 9000-8000 BC), only recently left by the disappearing Swiderians. But why then should we not imagine an Early Holocene "push" of Polish Komornicians from the South to the North ?

The *third model* is fully migratory in nature and was applied in the northern extremes of Europe, that is, in the Circumbaltic. It was characterized by two things, one of which was a northward wandering of part of the Tanged Point hunters (Ahrensburgian, Brommian, Desnian) who gradually settle Scandinavia as it is deglaciated at this time, giving rise to the local Mesolithic cultures with tanged points (Fosna-Komsa-Suomusjärvi). Up to this point things appear to be clear: the moving oecumene is followed by the most closely connected specialized hunters or at least part of them. The other case is much less obvious, although just as logical. Romuald Schild has demonstrated the lack of any techno-typological ties between the Late Pleistocene Polish Swiderian and the Polish early Mesolithic (Komornician); a gap of two-three hundred radiocarbon years existed between these of two entities. It follows from this that the Swiderian reindeer hunters had abandoned their territory and the only possible route they could have taken ran along the eastern shores of the Baltic.

Some of the elements connected with the Swiderian can be traced later in the Mesolithic of Eastern Europe (a few Havel-type harpoons - some in local Kunda context, the idea for the tanged point but produced in a different technology). The matter has yet to be explained in full as the "Post-Swiderian" (= Kunda/Butovo) of Russian researchers is not connected technologically with the Swiderian, although it has some tanged points that are very much Swiderian in appearance. The Swiderians probably left their homeland and ... disappeared.

Another unique model that may have been successfully implemented in the East European Lowland in the Early Holocene was the sum of two components: pressure technology and sectioning (end-scrapers, burins on broken blade, sectioned bladelets, retouched inserts) combined with tanged points resembling Swiderian ones in appearance but not technology. The first component appears to be rooted in the Siberian Upper Paleolithic (Baikal and Angara Basins with Afontova Gora, Verkholanska Gora, Kokorievo, etc.). It is not clear when it first appeared in European Russia, meaning was it a local characteristic of the Russian Lowland from the Late Glacial or only from the Early Holocene? Russian researchers used to imagine that this "Siberian" element was influenced by Swiderian neighbours, giving birth to the Burovo and Kunda cultures.

A.N. Sorokin puts the local Resetta culture originating from the Epigravettian at the root of the Russian, Upper Volga Mesolithic.

To close these remarks, the author would like to reiterate the extensive differentiation of the beginnings of the Mesolithic in a highly regionalized Europe, the evident continuations, local adaptations/evolution, migratory models and ultimately acculturation. On the whole, however, the overall image of the continent already in the 10th millennium BC appears to be entirely new from both the environmental and cultural point of view. The new units appearing (next to surviving ones) were generated by different adaptive potential with regard to the ecological novelties of the Early Holocene, realized in different ways by an already strongly differentiated population residing on the Continent.

Stabilization

Starting with the 10th millennium BC, the more or less complicated processes of adaptation and culture creation described above resulted in the emergence of a stylistic and territorial, meaning "cultural" system which was to last for centuries or regionally (in the east) for millennia and which was not disturbed by any regional or even supraregional trend or fad. It can be assumed that the techno-complexes and their local cultures of the end of the 8th and beginning of the 7th millennium BC, took on their final shape around 9000 BC (Jagerhaushöhle for the Beuronian, Romagnano III and Predestel for the Sauveterrian, Thatcham and Star Carr for Star Carrian, Friesack and Klostelund for Duvenseean, Całowanie and Chwalim for the Komornician, Pulli for the Kunda, new far-northern Norwegian sites for the Fosna-Komsa, Padina and Franchthi for Balkanian Epigravettian, etc. etc.). Several macro-territories/techno-complexes were formed at this time only to last for hundreds/thousands of years within virtually unchanging natural borders which although penetrable, remained unpenetrated for very long periods on end. It was also at this time that the Mesolithic *koine* (geometrization and microlithization) was formed in the West, South and Centre of Europe in opposition to the Eastern and Nordic tanged point traditions.

From the very beginning this formative period in the European Mesolithic was marked by a greater regional differentiation, expressed in the division of the techno-complexes created at this time into regional entities/cultures. The situation remained relatively stable until c. 7500 BC when a transregional sauveterrization fashion occurred in the West (cf. below), in the face of a strong Mesolithic conservatism in the Centre and East. East of the Elbe river notable changes did not occur for another one thousand years, until c. 7000-6500 BC, and in Russia and the circum-Baltic area not before the 6th-5th millennium BC.

At this point one should recall the supra-regional post-Pleistocene adaptation models of the southeast of Europe, Italy and Iberia, characterized in this period by a virtually almost unchanged continuation of local Late Palaeolithic (Epigravettian) traditions/style. Moreover, the northern extremes of the European continent did not come to be settled before the Early Holocene when they were occupied by bearers of an unchanged or little changed industries of the Brommian-Ahrensburgian-Desnian-Fosna-Komsa type. Between the two extremes, the most important and most revolutionary transformations were taking place in the West and in the Centre. These were the complicated social adaptation models, both local and not local, that Late Palaeolithic communities were going through in their quest to adapt to their new environments (Ahrensburgian, Epi-Ahrensburgian, Epimagdalenian, Federmesser-Azilian, final Epigravettian, etc.). While we have still to understand the exact mechanism of these processes, we are aware of the fact that they contributed to the stabilization in the 10th-8th millennia BC.

Changes in the west

Around the mid-8th millennium BC or a little bit earlier, the flint industries of north-western Europe from central-northern France and northern Ireland to southern Scandinavia and Germany West of the Elbe river, witnessed the introduction and spread of microbladelets, which had previously been either unknown or weakly distributed in this region. Along with the microbladelets came the narrow microliths made from them, especially the scalene triangles with short base, but also regionally Sauveterrian points and rectangles (Figure 7).

Specimens of this kind appeared at this time in the Beuronian C in France, Germany and Belgium, in the Maglemosian of Britain and southern Scandinavia - mostly triangles in the latter case, causing a considerable transformation of the local and regional industries, including observable taxonomic changes like the emergence of Shippea Hill Culture in Britain and Ireland, Boberg in north-western Germany and the Netherlands and the Svaedborgian in Denmark and southernmost Sweden.

All the above-mentioned, newly introduced elements in the northern cultural environment had their earlier counterparts appearing regularly and steadily in the Sauveterrian of southern Europe. Therefore, it is difficult to avoid the impression that the entire "sauveterrization" of the northwest of the Continent, described above, originated in the south. We are incapable of reconstructing the mechanisms of this techno-typological change/phenomenon (improvement of arrowhead construction?) that swept through Europe, but at least we can describe it.

Anyhow, the previously existing latitudinal zonation in the West and Centre (starting from the south: Sauveterrian - Beuronian – Maglemosian, cf. Zonation) was disrupted at this time, mostly in the West; the new trend was limited to the West of the continent, totally avoiding the Centre where this particular triangle with short base (and only the triangle) did not appear until the very end of the 7th millennium BC.

Collapse of the Maglemosian

From the beginning the Maglemosian *sensu largo* occupied a substantial stretch of the European Lowland from Britain to Poland, including the land bridge of Dogger Bank Shelf and the Pomeranian-Danish one. Following from postglacial processes of the period, ocean levels started to rise in the Early Holocene, causing the sinking of some regions in north-western Europe (cf. Changing shape). This gradually reduced the Maglemosian oecumene, leading finally to the isolation of the British Isles and instigating natural migratory movements of the population living in the gradually submerged areas. This was first observed by Hermann Schwabedissen and later by Raymond R. Newell.

The appearance of new cultures in the Late Mesolithic of the European Lowland (De Leien-Wartena in Holland, Oldesloe-Jühnsdorf in Germany, Chojnice-Pieńki in Poland and Nostvet-Lihult in southern and central Scandinavia (this last with an entirely different raw-material base than the preceding local Fosna culture) was the probable result of the "radial", decentralizing migrations and can be dated to c. 7000 BC (?).

These new entities, mostly descendants of Svaedborgian, are distinguished foremost by the microbladelet technique (known also from southern Scandinavia, cf. Svaedborg I site, coll. Becker) which is in all likelihood rooted in the earlier sauveterrization process (cf. Pulsating). Secondly, they apparently "cancel" old structures existing in the newly acquired territories (correspondingly, Duvensee in the west and Fosna in the north, Komornician in the east). The castelnovization (cf.) of the southern Scandinavian Mesolithic progresses at this time, first weakly (model B - Ageröd I:D) around 7000 BC, then completely (model A - Konglemosian) around 6500 BC. More to the south and north, the Post-Maglemosian will continue to last for even hundreds of years to come (in Polish Pomerania even up to the 5th millennium with Ertebølle ceramics in Dąbki). The population leaving Dogger Bank must have also retreated partly to the west, into Britain, but palpable evidence of this migration has yet to be recognized in the archaeological record.

Castelnovization/technical revolution

Castelnovization was an important technical and technological change taking its name from the Castelnovian culture in southern Europe and occurring in territories around the Mediterranean, in the Pontic area and on the Atlantic coasts in the local cultures of the Middle Mesolithic. Initiated most likely at the end of the 8th millennium BC in the south of Europe (Franchthi, Murzak-Koba), it did not take long (between ca. 7000 and 6500 BC) to spread throughout territories lying roughly north of the Mediterranean and Black Seas and, in western and central Europe (but not Britain), perhaps arriving in the extreme west and north around 6500 BC. It led to considerable, but not complete, formal uniformization of a number of local flint industries (Figures 8, 9).

A new type of core appeared; it was used to produce larger than before and much more regular bladelets. The core was big in itself and it was preformed, the raw material being always of good quality and procured by regular mining (Tomaszów in Poland, Wommersom in Belgium, Melos in Greece) and transported over long distances (Melos to Franchthi - > 100 km, Tomaszów to Barycz Valley - 280 km).

Another new type was a pressure made bladelet used simply as a knife. It also served as a blank to make end-scrapers and truncations, but foremost trapezes and rhombs, formed by the microburin technique or sectioning. Relatively quickly these new arrowheads replaced the old types of microliths.

Apparently, the new techniques and new types of tools are correlated with the appearance of cemeteries or numerous graves in settlements (Moita do Sebastiao in Portugal, Hoëdic and Téviec in France, Skateholm in Sweden, Vasilyevka in Ukraine), possibly testifying to a more sedentary lifestyle of the human communities undergoing castelnovization. Whatever the case may be, it is from these "castelnovized" cultures that the early local ceramic and finally local Neolithic cultures sprang.

Consequently, the process of castelnovization can well be referred to as "Pre-neolithization", even keeping in mind the differentiation that remained in force and the fact that not all units became Neolithic at an early point, although for the most part they adopted ceramics (cf. Ceramization).

To sum up, a few remarks on the processes of proliferation of the phenomenon. There seems to be no doubt that the phenomenon spread from South to North regardless of whether it was the eastern or western part of the continent. It seems to have followed in the west sea coastlines and moved up big river valleys (Rhone, Rhine).

The western model was more extreme than the eastern, which had only limited impact on the traditional cultures of the North.

Ceramization of the Mesolithic

The last stage of the European Mesolithic is its ceramization, a technical phenomenon that preceded or paralleled Neolithization understood as an economic event. On the whole, the phenomenon can be described as Mesolithic "aborigines" absorbing usually or frequently the idea of pottery from their Neolithic neighbours (?). This first pottery is usually technically quite poor, it has a pointed or rounded base and is occasionally decorated with strongly differentiated ornaments ("impresso", pit, comb, shell impressions - "cardial" etc.). Plain ceramics are also encountered.

The earliest pottery to appear in the local Mesolithic environment in the early and mid-7th millennium BC originated from the Castelnovian complex (Italy, Balkans, Ukraine) or its local/regional successors (Spain, Portugal, France), excluding the central Balkan-Danubian "corridor", which was occupied by the earliest ceramic cultures of the 'true' Neolithic (Starčevo/Criš/Körös, Danubian). This may give a not fully justified impression of a certain marginality and possibly derivative nature of the said phenomenon (western and eastern peripheries of the "mainstream"; the fully Neolithic character of many French and Iberian early ceramic cultures is doubtful.

In the South of the continent, Mesolithic pottery appeared in Castelnovian of South-Italia (Latronico), Slovenia (Mala Triglavca) or in Dalmatia (Smilcić), Italian Dolomites (Gaban), the Po Plain (Vho, Fagnigiola). Further to the West, it will be the southern French and Iberian "cardial" type ceramics and its regional variants, accompanied by Castelnovian-type industries (e.g. pottery of the Rocadour type, following the Cuzoul Mesolithic group in southwestern France), or else local variants of very strongly locally evolved Post-Castelnovian with the even more local Muge and Cocina, II-IV or "ceramica impressa ligure" variants deriving from them.

The same is true more to the northern Central Europe where ceramization did not put in an appearance before the Post-Castelnovian (in this case, Ertebøllian, Swifterbant, and la Hoguette).

The southeast of the continent was ceramized in the Castelnovian style ("impresso" ware from Črvena Stijena, and Odmut in Montenegro).

The East of Europe partly repeats this pattern. What is observed foremost is the ceramization of Castelnovian-type cultures: Grebeniki - known as the ceramic Bug-Dniester culture, in Moldova and western Ukraine, and Janislawician further to the North, which takes on pottery of the Dubičiai-Pripiat'-Neman type.

In the same way but further to the north, ceramization occurred successively in the Dnieper and Donetz region, Upper Volga (Butovian with pottery), eastern Baltic (Post-Kundian with Narva-type ceramics) and Finland (Suomusjärvi and ceramics of Sperrings type).

In conclusion, it may be assumed that the numerous Late Mesolithic cultures (and their later but still not Neolithic mutations) adopted/invented the idea of pottery quite universally and were later (or sometimes parallel) neolithized in effect, too.

The process of ceramization can be dated, depending on the region, relatively earlier in the South (7th millennium BC) than in the North (6th/5th millennium BC).

In the central and northern part of the Continent the partly ceramized Mesolithic (cf. lasted parallel to the local (Danubian) Neolithic is even longer.

References

Bailey G., Spikins P. 2008. *Mesolithic Europe*. New York, Cambridge University Press.

Bonsall C. (ed.) 1989. *The Mesolithic in Europe*. Proceedings of the Third International Symposium, Edinburgh 1985. Edinburgh, John Donald Publishers Ltd.

Borić D., Antonović D., Mihailović B. 2021. *Foraging Assemblages*. 9th Conference: Belgrade, Serbia (14-18 September 2015). Belgrade and New York: Serbian Archaeological Society and The Italian Academy for Advanced Studies in America, Columbia University.

Cupillard C., Richard A. (ed.) 2000. *Les derniers chasseurs cueilleurs d'Europe occidentale. (13 0000- 5500 av. J.C.)*. Besançon, Presses Universitaires de Franche-Comté.

Ghesquière E., Marchand G. 2010. *Le Mésolithique en France*. Paris, La Découverte.

Huntley B., Birks H.J.B., 1983. *An atlas of past and present pollen maps for Europe: 0–13000 years ago*. Cambridge, University Press.

Kozłowski S. K. 1973. *The Mesolithic in Europe. Papers read at the 1th International Symposium on the Mesolithic in Europe. Warsaw, May 7-12, 1973*. Warsaw, Warsaw University Press.

Kozlowski S. 2009. *Thinking Mesolithic*. Oxford, Oxbow Books.

Gramsch, B. 1981. *Mesolithikum in Europa: 2th. International Symposium, Potsdam, 3. bis 8. April 1978. Berlin: Veröffentlichungen des Museums für Ur- und Frühgeschichte Potsdam 14/15*.

Larsson L., Kindgreen H., Knutsson K., Leoffler D., Akerlund A. (eds.) 2003. *Mesolithic on the move*. Papers presented at the sixth international conference on the Mesolithic in Europe (Stockholm, 2000) Oxford, Oxbow Books.

McCartan S. Schulting, R., Warren, G., Woodman, P. (eds.) 2009. *Mesolithic Horizons*. Papers Presented at the Seventh International Conference on the Mesolithic in Europe, Belfast 2005. Oxbow Books.

Milner N., Bailey G., Craig, O. (eds.) 2007. *Shell middens and coastal resources along the atlantic façade*. Oxford, Oxbow Books.

Rozoy J.G. 1978. *Les derniers chasseurs. L'Epipaléolithique en France et en Belgique. Essai de synthèse*. Bulletin de la société archéologique champenoise 3 vol.

Thévenin A. 1999. *L'Europe des derniers chasseurs: Épipaléolithique et Mésolithique. Peuplement et paléoenvironnement de l'Épipaléolithique et du Mésolithique. Actes du 5e colloque international UISPP, Commission XII. Grenoble, 18-23 septembre 1995*. Paris, Éditions du CTHS.

Vermeersch P., Van Peer Ph. 1990. *Contributions to the Mesolithic in Europe*. Papers Presented at the fourth International Symposium "The Mesolithic in Europe," Louvain, Leuven University Press.

Zvelebil M. (ed.) 1986. *Hunters in transition. Mesolithic societies of temperate Eurasia and their transition to farming*. Cambridge, Cambridge University Press.

Challenges in evaluating the role of the environment in neolithization processes. The case of South-East Europe

Marek Nowak[1]

Abstract

Neolithization processes have frequently been investigated from the environmental perspective. However, there are many ambiguities inherent in palaeoenvironmental data, especially when confronted with archaeological ones. There are also very different visions as to the role of environmental factors in development of human societies. In the contribution the most urgent problems induced by the application of an environmental approach to explain prehistoric processes, including those of neolithization, were characterised. Then, the potential role of environmental (particularly climatic) proxies in the modelling the appearance and spread of the First Neolithic in South-East Europe was demonstrated. In particular, the issue of the extent to which the Neolithic diffusions and possible adaptations were triggered or slowed down by Holocene global climate events was analysed. It was hypothesised, as one of possible scenarios, that the speed of neolithization of South-East Europe was not uniform and regular. In a measure, in some areas, this was due to the necessity to adapt to 'new' environmental conditions. However, first and foremost, these processes were evoked by the occurrence of favourable conditions ('occasions') created by global and local environmental transformations ('opportunistic colonisation').

Résumé

Les processus de néolithisation ont souvent été étudiés du point de vue de l'environnement. Cependant, il existe de nombreuses ambiguïtés inhérentes aux données paléoenvironnementales, en particulier lorsqu'elles sont confrontées à des données archéologiques. Il existe également des visions très différentes du rôle des facteurs environnementaux dans le développement des sociétés humaines. Dans cette contribution, les problèmes les plus urgents induits par l'application d'une approche environnementale pour expliquer les processus préhistoriques, y compris ceux de la néolithisation, ont été caractérisés. Ensuite, le rôle potentiel des indicateurs environnementaux (en particulier climatiques) dans la modélisation de l'apparition et de la propagation du premier Néolithique en Europe du Sud-Est a été démontré. En

1 Institute of Archaeology, Jagiellonian University, 11 Gołębia St., 31-007 Kraków, Poland. marekiauj.nowak@uj.edu.pl

particulier, la question de savoir dans quelle mesure les diffusions néolithiques et les adaptations possibles ont été déclenchées ou ralenties par les événements climatiques mondiaux de l'Holocène a été analysée. Il a été supposé, comme l'un des scénarios possibles, que la vitesse de néolithisation de l'Europe du Sud-Est n'était pas uniforme et régulière. Dans une certaine mesure, dans certaines régions, cela était dû à la nécessité de s'adapter aux « nouvelles » conditions environnementales. Cependant, ces processus ont d'abord et avant tout été évoqués par l'apparition de conditions favorables (« occasions ») créées par les transformations environnementales mondiales et locales (« colonisation opportuniste »).

Introduction

The appearance and spread of the Neolithic communities in Europe was an extremely complex process. Domestication of plants and animals as well as appearance of other elements of the so-called Neolithic Package and their involvement in the cultural system took place in the Near East. Thus, neolithization of Europe had to be a combination of different diffusions, adaptations, acculturations and most probably of generating new social and economic structures. The first, causative agents in these processes were people that migrated from the Near East (Perlès, 2001; Brandt *et al.* 2014; Hofmanová *et al.* 2016; Lipson *et al.* 2017; Mathieson *et al.* 2018) and brought with themselves novel patterns of economy, settlement as well as of social and ideological realms. The role of the local, pre-Neolithic, populations in the first neolithization seems to be minor but not nil (e.g., Nikitin *et al.* 2019).

These immensely complicated processes have been investigated from many perspectives. One of them is the environmental perspective. However, the environmental approach in archaeology is by no means universal. It has both its staunch supporters and its fervent opponents. This is due, among other things, to many ambiguities inherent in palaeoenvironmental data, especially when confronted with archaeological ones. It is also a result of very different visions of the driving causes and mechanisms of development of human societies, in which environmental factors are assigned very different roles. In the paper we would like to point out, first of all, the most vexing problems posed by the application of an environmental approach to explain prehistoric processes, including those of neolithization. Then, the role of environmental factors on development of the First Neolithic in South-East Europe[2] will be evaluated, considering the discussed problems. In particular, we will examine the controversial issue of the extent to which the diffusions and possible adaptations of the new cultural formation were triggered and/or slowed down by Holocene global climate events as well as by regional/local environmental conditions. Mostly, it will be done

---

2   This is an arbitrary term. In fact, it also includes the western coasts of Anatolia. Furthermore, just for convenience, the text distinguishes between the Aegean and Balkan Zone. Of course, most of the Aegean Zone is also part of the Balkan Peninsula.

through a demonstration of different models and scenarios, even extreme ones, which have been published in recent years. On the other hand, in the case of processes that have not been described by more or less impressive models, alternative hypotheses will be presented to explain their causes and mechanisms. To put it another way, the neolithization of South-East Europe will be used as a kind of case study for the issue of environmental determinants of neolithization processes.

The role of environment in neolithization – problems and ambiguities

Undoubtedly, the role of environmental factors in different processes within human prehistory and history is enormously hard to decipher. It is not only a matter of difficulties in analysing and summarising environmental records, i.a. because of insufficient and ambiguous sources. The point is also that much depends on views and approaches expressed by individual researchers or research schools.

We can easily find in the literature, even today, approaches very close to classical geographical determinism and processual adaptationism (e.g., Richerson *et al.* 2001; Berger *et al.* 2016: 1847; Weninger 2017; Chapman 2018: 5; Arponen *et al.* 2019: 1673; and citations therein). Some scholars are therefore of the opinion that the spread of the Neolithic, as many other processes visible in this formation, have been driven by periods of climatically induced crises and - as a consequence - by periods of instability in the economic and socio-political realms (e.g., Gronenborn 2010; Gronenborn *et al.* 2014; Berger *et al.* 2016; Sánchez Goñi *et al.* 2016: 1197; cf. also Dearing 2006, table 3). Hypotheses of this kind derive, among other things, from the demonstrated chronological correlations between stages of the Neolithic spread and environmental fluctuations, primarily climatic. The mentioned crises were also supposed to induce changes in material culture. Opinions have been expressed that the more complex ancient communities were, the more vulnerable they were to sudden climatic changes. The most vulnerable would have been those whose socio-technological systems were near equilibrium (Freeman *et al.* 2020: 8). Thus, there are reasons to expect a shift to a different mode of management (Gronneborn 2005: 9). Also acceptable is a construct in which environmental change intensified processes of economic and/or social and/or political erosion, the prime causes of which lay elsewhere and had been operating for some time.

On the other hand, it is also easy to point to approaches and studies more or less strongly denying the impact of the environment on humans and at the same time criticising deterministic approaches (e.g., Lemmen, Wirtz 2014: 66; Contreras 2017), as well as emphasising the importance of independent, active human actions that eliminate this impact (e.g. Flohr *et al.* 2016; Chapman 2018; Freeman *et al.* 2020; Nicoll, Zerboni 2020). Among others, it is argued that 'methods and concepts that assume simple causation may be invalid, yet we have few guidelines as to how to deal with the challenges presented by complex socio-ecological systems' (Dearing 2006: 188). There are also views that the productive technologies associated with agriculture (and connected social realms) that increase the carrying capacity of a system may weaken disadvantageous effects of environmental changes (Lemmen, Wirtz 2014: 66; Chapman 2018; Freeman *et al.* 2020). Transformations in material culture directly caused by environmental transformations are also doubted, emphasising that this is a matter of axiomatic belief rather than grounded knowledge (e.g., Czerniak 1992; Kukawka 1992). In short, correlations between environmental and archaeological events, even if observable in chronological terms, are entirely coincidental and irrelevant. Agricultural communities, including early agricultural ones, would not have been as susceptible to environmental turbulence as more deterministically-minded researchers suppose. As to the latter claim, it refers more to long-term environmental change. In fact, it is not possible to comment the susceptibility of these societies to short-term catastrophic events, as basically it is hardly possible to capture them in the available palaeoenvironmental proxies.

Arguably, a fair number of the constructions formulated in recent publications fall somewhere between these extremes (cf. Contreras, 2017: 8 and citations therein; Arponen *et al.* 2019). In which direction they are shifted also depends largely on the a priori views, detached to a greater or lesser extent from environmental and archaeological data. This is reflected, among other things, in the emergence of many attractively sounding terms and models in the debates (i.a. the 'resilience theory' - Gronenborn *et al.* 2014; 2017; Davies *et al.* 2018; the socio-environmental transformations/dynamics - Müller and Kirleis 2019). They are undoubtedly of great cognitive value, i.a. when they highlight the mutual, multidimensional socio-environmental relationships, dependencies, and feedbacks (Gronenborn *et al.* 2014: 75; Müller, Kirleis 2019: 1522). Nevertheless, they do not convincingly solve the fundamental difficulty of proving causation. In most cases it is still difficult to resolve whether and to what extent specific archaeological transformations were indeed caused by environmental change (Bradtmöller *et al.* 2017; Davies *et al.* 2018; Kintigh, Ingram 2018; Walsh *et al.* 2019; Freeman *et al.* 2020). Simply put, assumptions about automatic socio-environmental causal relationships are essentially the product of analyses of environmental and archaeological data and specific individual visions as well as general theories (usually borrowed from other disciplines).

There are many reasons for the uncertainty of such relationships. If we consider the significance of environmental factors for human societies, or for specific processes such as neolithization, and even if we conclude that this significance was important in a given place and time, we must bear in mind a whole series of problems which make a correct assessment of this significance difficult, both of an objective and practical nature.

Undoubtedly, one such problem is the imperceptibility of environmental change (geological time) from the perspective of a single human being, or even several human generations, except for events of a violent and catastrophic nature. Similarly, humans experience impacts of environmental change directly only at a local or regional scale (Dearing 2006: 189). Expanding on these observations, it is important to stress that for the most part socio-cultural phenomena and processes function in temporally and spatially smaller scales than environmental ones (Arponen *et al.* 2019: 1674). Even if it were otherwise, our proxy records are typically characterised by temporal resolution on the order of at least a few human generations (Chapman 2018: 11). Prehistoric man could therefore not consciously respond to environmental change (let us repeat - except for catastrophic events) because he was not aware of it and was not informed about it on television or the internet.

It is otherwise an open question whether the dating of archaeological changes and palaeoclimatic trends in the Neolithic is precise enough to demonstrate their coincidence and correlation with satisfactory probability. This is especially true for global events, which tend to be longer than archaeological events. The latter are therefore, especially if precisely dated, too short to be linked to the whole event, or even a significant part of it (cf. Krauß *et al.* 2018: 25). Is it possible, then, to suggest in all conscience a correlation and dependence? Moreover, if a 'short' archaeological change falls roughly at the beginning or end of a 'long' event, it is almost impossible to reconstruct their actual temporal relationship, given the precision of absolute dating methods. To sum up this topic, let us quote J. Chapman, according to whom 'it remains difficult to relate the effects of the 8200 BP "event" to any particular cultural development or collapse, let alone any settlement dislocation' (Chapman 2018: 5).

Of course, what has been written does not mean that such dependencies never and nowhere existed and that – contrary to the doubts expressed (Chapman 2018: 4, 11) – they could not influence human communities and channel some socio-cultural processes. If, in fact, such dependencies, to a greater or lesser extent, did take place, people were simply not aware of them, ergo they did not associate certain behaviours and transformations, perceived on a daily basis, with one or another environmental factor.

What does this imply for the archaeologist interested in this issue? Well, it follows nothing more than the imperative to search for, analyse and compare long trends and transformations, invisible to prehistoric humans but discernible to us, in the archaeological remains of the material, settlement, economic, social, ideological, and other spheres. Happily, 'archaeology with its long-term temporal perspective on human societies and landscapes, is in a unique position to trace and link comparable phenomena in the past, to study human involvement with the natural environment, human impact on nature and the consequences of the various dimensions of environmental change on human societies' (Müller, Kirleis 2019: 1518). Consequently, we may be able to identify human-environmental interrelationships and model their general, long-term, global and supra-regional patterns, those that humans have been unaware of (just as we would not have been aware of Global Warming if we had not heard and seen it in the media). These patterns can then be further detailed at regional and local levels (Dearing 2006; Müller, Kirleis 2019: 1520-1525; Nicoll, Zerboni 2020: 120) but reducing them to short- and medium-term time scales (with a temporal dimension of up to three/four human generations?) is very difficult, as correlating minor episodes, whether within global events or within regional or local processes, with specific archaeological changes is indeed fraught with a significant level of ambiguity.

The correct reading of global environmental changes, such as Rapid Climate Change (RCC) events, at regional and local scales is otherwise another problem worth mentioning. Let us remember that characteristics of global events very often come primarily from areas thousands of kilometres away from the area of interest, ergo it is not surprising that there are differences (Berger *et al.* 2016: 1850). Consequently – firstly – global environmental events do not always succeed in being captured regionally and locally, even though we would expect this to be the case due to general knowledge and belief in the importance of RCCs (Berger *et al.* 2016: 1852). This can either be a reflection of reality, i.e. the bypassing of an area by specific environmental changes, or the result of missing, imperfect or one-sided investigations. In the latter case, environmental factors are therefore unintentionally omitted from consideration (Berger *et al.* 2016: 1848). In contrast, in the former case, it is wrong to look for environmental causes of specific archaeological changes, regardless of the cost. Secondly – differences between global and regional/local levels may be observed, like e.g. chronological lag or variability in frequency and composition of change indicators. Again, there is a problem to resolve whether this is a reflection of reality or an effect of research imperfections (e.g. old-wood effect) (Berger *et al.* 2016: 1864-1865). Finally, let us remember that the response of people living in an area to specific environmental changes may be delayed.

Besides, it should also be noted that some human behaviours may themselves have led to changes and devastation of the environment, even of extra-local scales, which in turn caused negative repercussions for people who triggered them (Dearing 2006: 188-190, 197-199; Henry *et al.* 2017; Müller, Kirleis 2019: 1522). Anthropogenic environmental change can therefore be confused with natural change and vice versa, or it remains debatable what is cause and what is effect (Chapman, 2018: 5; cf. also Mercuri *et al.* 2019; Kempf, 2020).

Developing the above arguments, we feel entitled to argue that the following conclusions result from the uncertainty about the automaticity of social/cultural/environmental relationships:

1) It can hardly be expected a priori that every environmental change provoked specific human behaviour ergo resulted in cultural

transformations (Lemmen, Wirtz 2014: 70; Berger *et al.* 2016: 1870; Contreras, 2017: 8-9), whatever we understand under the latter term (cf. Müller, Kirleis 2019: 1517).

2)   To put it in a slightly opposite way, we should not look for environmental causes of every cultural change visible archaeologically, even if, in the light of chronological data, it seems that the temporal coincidence of certain archaeological changes with environmental events, of different scale, is not excluded. We may then be dealing with archaeological changes that are reflections of socio-cultural, political, economic, ideological, etc. disturbances or transformations that were not in any way triggered by and related to any environmental changes.

Regarding these two conclusions above, let us further note that climate variability during the Holocene in temperate latitudes was not staggering (cf. Benjamin *et al.* 2017: fig. 7; Finné *et al.* 2019; Walsh *et al.* 2019). Thus, one can suggest an idea that human culture, even Early Neolithic one, was able to offset these small – on a long-term scale – environmental changes. In contrast, short-term violent and catastrophic events may have been destructive.

3)   Of course, one cannot throw the baby out with the bathwater and completely, by definition, reject environmental factors when considering a given cultural change. After all, human communities have always functioned in some environmental context and must have interacted with it in some way. In the case of agricultural communities, especially fresh ones, the vulnerability to environmental turbulence may have been considerable, but on the other hand the capacity of cultural complexes to adapt and preserve their life-styles still appears as the dominant model for understanding past human ecodynamics (Lemmen, Wirtz 2014: 65, 70). The most reasonable is the statement that things may simply have been different. In any case, it is a matter of examining the available data in detail. In the archaeological realm, the greatest opportunities refer to data that reflect general, long-term trends. On the other hand, it will always be difficult to prove causation (Davies *et al.* 2018). As an illustration, let us use the much-cited article by J.-F. Berger (Berger *et al.* 2016). Despite a very meticulous and detailed discussions and analyses of four situations related to the Early Neolithic, in fact no conclusions have been offered as to the general as well as the specific model of the interaction between Early Neolithic man and the environment. What is clear is only the importance of strictly local and regional factors, not just global ones.

4)   Far-reaching, violent and destructive consequences could be caused only by exceptionally rapid, catastrophic environmental phenomena, both those one-time and short-lived (like e.g. cooling or droughts lasting for several years). These consequences, however,

usually have a short-term, ad hoc character, and may concern various areas (even political[3]). But they can have – less frequently – a long-term character (e.g. a drought of a few/a dozen or so years that caused extinction, descent or migration of populations, like migrations caused by great Sahel droughts after 1968[4]–Derrick 1977; Walther 2016).

5)   In other situations, that is to say, in the vast majority of them, if one accepts the possibility of the actual impact of environmental changes on human communities, one should not assume a priori that these changes were the only causal factors, but that they may have acted in combination with other factors, sometimes very distant from the strictly biological ones (Arponen *et al.* 2019: 1673-1674; Nicoll, Zerboni 2020: 122). For instance, adverse, disruptive changes in the cultural system, triggered by events of a political nature, may have been compounded by negative environmental changes. In a stable political situation, they would not have been so significant. Specific economic and settlement behaviours (vide e.g. the postulated destruction of irrigation systems in Central Asia by the Mongols –Andrianov, Mantellini 2016), may have initiated and/or amplified detrimental environmental changes, which then intensified even without human intervention (Kalicki 2006; Kealhofer, Marsh 2019). They in turn may have further disrupted the cultural system of the communities concerned.

6)   There is another important point to bear in mind. Namely, most publications consider the negative impact of environmental factors on human communities. That is, they examine specific reactions of the socio-cultural system to the deterioration of these environmental conditions and contexts that potentially created a situation close to equilibrium for this system. In other words, potential disturbances of this equilibrium are analysed. On the other hand, in fact, some environmental changes may be beneficial for a given socio-cultural system, i.e. they may support, facilitate and prolong its functioning. Finally, such changes may create new opportunities, e.g. in terms of the

---

3 Just as an example - 'Byzantine Emperor Michael VIII [...] had completed a series of new fortifications along the course of the river [Sakarya] by 1280. However, in the spring of 1302 the Sakarya river flooded and changed its course, with the result that the new defensive structures became useless. It is possible that it was this event that allowed Osman's men to cross the river and settle in the Byzantine province of Bithynia' (Imber 2020: 25; transl. MN).
4 'Migration has now become an inevitable method of adaptation for us ... As a means of survival for us and our animals, we are forced to continuously migrate despite all the risks involved. This is our form of adaptation. We have always mastered it, but if nothing is done to ensure the safety of our space and activities, we risk, one day, being forced to abandon our way of life and join the swelling ranks of the unemployed in the city.' Hindu Oumarou Ibrahim, Peul Mbororo of Chad - https://climatemigration.org.uk/moving-stories-the-sahel/ (viewed 5 March 2021)

efficiency of the food economy, the spatial extension of the currently practised system, or even the achievement of a new population threshold and – through this – a new socio-cultural configuration.

7) Finally, let us emphasise once again that it is impossible not to agree with the opinions that a correct approach to the annotated problematics requires an equal and comprehensive consideration of environmental and socio-cultural factors and their analysis as mutually interacted (e.g., Contreras, 2017: 8; Arponen *et al.* 2019: fig. 1; Müller, Kirleis 2019: 1524-1525). Let us also remember that human communities are more complex than many environmental-archaeological accounts would have us believe. They are not, after all, limited to subsistence issues alone. Psychological, mental, social, etc. mechanisms have also played and continue to play an important role in human behaviour and decision making. These factors are difficult to demonstrate and prove in archaeological data, but this does not mean, after all, that they did not exist and did not shape past societies.

The Aegean Zone

The first case study of this contribution is the neolithization of the broadly understood Aegean Zone. These processes have obviously also been explained in environmental terms. Their starting point is, among other things, demonstrated chronological correlations of the disappearance and appearance of certain archaeological phenomena and the limits of environmental phenomena and transformations. Several years ago, the so-called Bipartite Vulnerability Model was proposed by B. Weninger and L. Clare (Weninger, Clare 2017). The model, being the developed version of earlier constructs, generally reflects the mitigation of the RCC[5] impacts, causing biophysical and social hazards, through escape and search movements (Weninger, Clare 2017; cf. also Weninger *et al.* 2009; 2014). To some extent, this idea was revised and amended in the 2018 publication (Krauß *et al.* 2018).

Phase A (*c.* 6600-6300 BC) of this model covers the first part of the RCC identified within the 7th millennium BC (Weninger *et al.* 2014: 8-10; Weninger, Clare 2017: 70; and citations therein). Cold anomalies caused by polar air masses from Asian continent ('Siberian High') and aridification, were to be characteristic for that period. According to the quoted authors the arrival of farming communities on the Turkish West Coast and apparently – possibly with a slight delay – to some regions of present-day Greece (cf. Krauß *et al.*

---

5   In the quoted contribution the term Abrupt Climate Change has been used (Weninger and Clare 2017: 70). However, it only appears in that article, out of many written by B. Weninger and collaborators. Therefore, the more commonly used term RCC will be applied here.

2018) is coincident with the beginning of that period. This mobility was the effect of 'habitat tracking to milder regions' (Weninger, Clare 2017: 76) and can be described as 'refugium colonisation' (Weninger, Clare 2017: 84) in the coastal zone.

The existence of favourable conditions in the Aegean Zone for Neolithic settlement in the 7th millennium BC was attempted to be demonstrated in a signalled 2018 paper (Krauß *et al.* 2018). Namely, according to its authors, such conditions were present in the so-called Sub-Mediterranean-Aegean (SMA) biogeographic region. It covers the northern and north-western edge of the Aegean, including Thessaly, Macedonia, and Greek Thrace, and extends further into the Balkans, but only along the valleys of the Vardar/Axios, Struma/Strymon and Mesta/Nestos rivers. It is hard not to notice that the almost all earliest Neolithic settlements, dated to *c.* 6600/6500 BC, are concentrated within this region. Consequently, an obvious thesis has been put forward that this ecological zone attracted the earliest Neolithic groups because of its mild climatic conditions, most similar to those in the Mediterranean. It is also theorised, if understood correctly, that adaptation processes took place in these areas, to the climatically harsher (especially during winters) Balkan conditions, which must have taken place before Neolithic settlement spread to the rest of the Balkan Peninsula. This would have happened over a rather long period, up to about 6050 BC, i.e. the end of the RCC; consequently, also within the time-frames 6600-6300 BC.

Phase B of the Bipartite Vulnerability Model (Weninger, Clare 2017: 77-78, figure 2.4; cf. also Weninger *et al.* 2014: 14-17) is dated to *c.* 6300-6000 BC. The described cold conditions were amplified by the Hudson Bay (8.2 ka cal BP) event which resulted in the increased occurrence and severity of winter outbreaks over Europe and in the eastern Mediterranean. According to the model it caused societal perturbations and increased mobility in the entire eastern Mediterranean. We can infer from quoted paper that in this period further development of the Neolithic in the Marmara region (continuation of refugium colonisation?) should be postulated and perhaps an acceleration of the expansion rate between Anatolia and northern Greece. On the other hand, as mentioned above, there were no movements outside the Aegean zone. The subsequent, delayed secondary expansion into remaining parts of South-East Europe took place only around the end of the RCC conditions or rather after this end, i.e. not earlier than *c.* 6050 BC (cf. also Krauß *et al.* 2018). In other words, the quoted authors seem to imply that this expansion was possible due to amelioration of the climatic conditions and RCC under consideration stopped Neolithic expansion for *c.* 500 years.

The model proposed by B. Weninger and L. Clare in 2017, similarly to earlier publications of this research school (Weninger *et al.* 2006; 2009; 2014), to a small extent considered central and southern Greece, where some early data refer to neolithisation. Such approach underwent some change in the cited 2018 article (Krauß *et al.* 2018) where the role of maritime communications and maritime exchange networks was emphasized (Krauß *et al.* 2018: 25). These marine contacts between the Near East, Cyprus and the Aegean Sea Basin started much before the 7th millennium BC but intensified in that millennium. They brought elements which seem to be vestiges of the 'original Neolithic Package' (i.a. stone architecture, a greater role of plant foods, 'pre-domestication' of animals) to Aegean Zone which was inhabited by forager-hunter-fisher communities of the so-called Aegean Mesolithic (Sampson *et al.* 2010; Kaczanowska, Kozłowski 2014; Horejs *et al.* 2015; Horejs 2019).

Raiko Krauß and others (2018) considered the location of the Çukuriçi and Ulucak sites (perhaps the site of Uğurlu could also be added in such context), as well as the Franchti cave, as confirmation of the existence of such a southern shipping route. In doing so, they point out that the 'foundation date' of these sites is approximately the same, i.e. 6550±30 cal BC (Krauß *et al.* 2018: 25-26, 36), although in the case of Franchti the term 'foundation date' is not quite unambiguous. As they indicate 'this arrival is synchronous (within error limits of ~100 yrs) with the onset of RCC-conditions in the Eastern Mediterranean' (Krauß *et al.* 2018: 36). This means, as one might guess, that arrival was caused by negative changes in the 'home' areas. It should be noted at this point that the coastal location of Çukuriçi and Ulucak is not necessarily indicative of maritime migration. They could just as well be the end point of overland movements taking place in Anatolia. This is how it was presented in the publication from 2017 (Weninger, Clare 2017: figure 2.3).

The above-quoted authors do not specify in depth the status of southern Greece in their model (cf. Krauß *et al.* 2018: figure 4). It can be understood that it does not belong to the SMA region, as Mediterranean conditions simply prevailed and prevail there. So perhaps this was the area where it was easier for the First Neolithic migrants/sailors from the Near East to settle, as a consequence of the existence of the deeply rooted 'shipping lines'. In turn, colonisation of the SMA area would have progressed along a south-north axis.

The roughly presented ideas also give rise to other doubts, arising from some of the problems signalled in the first part of the paper. The issues of chronological matching of archaeological and climatic events should be pointed out. Bernhard Weninger and collaborators have tried to demonstrate their strong temporal correlation. However, this is only one of many possible interpretations of chronological data.

Considering the current knowledge, groups of the PPNB tradition reached the coasts of the Aegean and the Marmara Seas perhaps not later than mid-7th millennium BC (Çakırlar 2012; Horejs *et al.* 2015; Atici *et al.* 2017; Horejs 2019: 74; cf. also Stock *et al.* 2015). However, when it comes to a more precise location of the beginning of the settlement in the best-studied sites of Ulucak and Çukuriçi, we encounter different proposals in the literature. In addition to the already mentioned date of 6550 BC (Krauß *et al.* 2018: 25) there are also earlier ones, going back as far as 6850 BC (Guilbeau *et al.* 2019: 3-4, 11; Çevik, Vuruskan 2020: 97).

Given the current pattern of radiocarbon dating, it seems that Neolithic migrations did not stop very long (from our archaeological perspective) on the Turkish coasts and in southern Greece, since in Thessaly and Macedonia, the beginnings of Neolithic can be dated to c. mid-7th millennium BC (Brami, Heyd 2011; Karamitrou-Mentessidi *et al.* 2013; Lespez *et al.* 2013; Maniatis 2014; Reingruber, Thissen 2017; Reingruber *et al.* 2017; Gkouma, Karkanas 2018; Efstratiou 2019: 12-14). There have even been opinions that the Neolithic beginnings in Macedonia may have been slightly earlier compared to Thessaly (*c.* 6600 BC *vsc.* 6500 BC) (Gkouma, Karkanas 2018: 19 and citations therein), but the achievable precision of radiocarbon dating makes the plausibility of such a thesis fairly low.

As for the early dating (the first half of the 7th millennium BC and even the close of the 8th millennium BC?) of the Neolithic beginnings in southern Greece, based on few 'old' dates from Knossos (level X) and Franchti (Initial Neolithic), doubts have been put forward, in analogy with the situation in Thessaly (Brami, Heyd 2011: 173-175; Reingruber *et al.* 2017: 38-41). However, 'new' dates from Franchti, obtained from wheat grains, strongly indicate the presence of Neolithic elements at this site before the mid-7th millennium BC, perhaps even as early as 6800/6700 BC (Perlès *et al.* 2013; Reingruber, Thissen 2017). This is compatible – despite criticism (Reingruber, Thissen 2017) – with the 'new' date from Knossos, obtained from oak acorn (Efstratiou *et al.* 2004). The cultural nature of the contexts from which these dates derive remains rather an open question. It seems that they can be described as Mesolithic contexts, with a significant proportion of Neolithic elements. Nevertheless, an interpretative option emerges again that Neolithic impulses reached Crete and the north-eastern Peloponnese not only earlier than Thessaly and Macedonia, but roughly at the same time as first groups of farmers reached the West Anatolian coasts (if not earlier!). Moreover, the presence of Neolithic groups or indications of interactions with Neolithic groups before mid-7th millennium BC, perhaps again as early as 6800/6700 BC, is to be reckoned with in Boeotia, in the Sarakenos cave (layer 3 = Initial Neolithic) (Sampson *et al.* 2009; Goslar *et al.* 2016; Kaczanowska, Kozłowski 2016). In such a setting, the more acceptable is the already signalled idea that neolithisation of present-day, continental Greece progressed from south to north.

Therefore, an alternative hypothesis of Neolithic beginnings on the Turkish west coasts as well as in southern and central Greece, preceding the onset of the RCC around 6600 BC, can be constructed. Consequently, these beginnings could not be conditioned by the RCC. They had to be caused by other factors. On the other hand, the further expansion of the Neolithic northwards, into Thessaly and Macedonia (or in other words into the SMA zone) as well as into the Marmara Sea Basin, may have been conditioned by the start of the RCC in some way.

Let us also explore the issues of a possible deterioration of environment in the supposed 'home' areas of the Neolithic migrations to the Aegean Zone and of plausible milder conditions in this zone in the 7th millennium BC.

Indeed, a number of publications indicate that natural conditions in Anatolia around 6600 BC, and perhaps even slightly earlier, may not have been suitable for Neolithic settlement (e.g., Berger *et al.* 2016: 1865; Woodbridge *et al.* 2019). However, it is not clear whether they deteriorated so significantly that they caused a kind of population flight westwards, towards the Aegean coast. For instance, J. Woodbridge and collaborators (Woodbridge *et al.* 2019: 732-733), analysing the area of southern Anatolia, favours the thesis of a shift in climate seasonality beginning *c.* 6600 BC, which influenced i.a. water balance and caused water shortages. But on the other hand, in the same article it is stated that 'archaeological trends for the Early Holocene inferred from radiocarbon date densities [...] show a likely increase in population around 10,300 cal. year BP continuing until ~7500 cal. year BP' (Woodbridge *et al.* 2019: 737, figure 6). Thus, there is by no means any sign of depopulation. Moreover, it has been hypothesised that the population during this period was already actively interfering with the environment. As the cited authors wrote 'the grassland phase was followed by the development of open oak parkland (~8500 cal. year BP) that may have been managed by people' (Woodbridge *et al.* 2019: 737), although they did not rule out that the appearance of such a plant formation might also reflect climatic changes. This would imply that a systematic but mild drying of the climate was already underway at that time (cf. Woodbridge *et al.* 2019: figure 4). It took a radical turn much later, *c.* 5000/4500 BC (Finné *et al.* 2019; Walsh *et al.* 2019). If one accepts the thesis of anthropogenic transformation of vegetation, this will also indicate relatively high population levels rather than depopulation processes.

As for conditions in the Levantine Zone, the available data and their interpretations are by no means conclusive. Undoubtedly, signs and episodes of aridification have long been indicated to have taken place in the 7th millennium BC (Sanlaville *et al.* 1996; Rambeau 2010; Weninger, Lee 2017: 75-78; Palmisano *et al.* 2020).

However, firstly, in the light of some data, a more pronounced deterioration of climatic conditions began to be evident only in the second half of the 7th millennium BC, and perhaps even in the last quarter of the 7th millennium BC, as e.g. suggested by studies on the water level of the Dead Sea (Migowski *et al.* 2006). Similarly, it is emphasised that the disappearance/crisis of intensive Neolithic settlement, i.a. of the PPNB mega-sites, occurred around the mid-7th millennium BC (Rambeau, 2010: 5228). Moreover, it was around this time that Pottery Neolithic settlement on the coastal plain would have emerged to a greater extent because of the mitigation of unfavourable conditions (Weninger, Clare 2017: 75-76). The same authors point to a particularly acute crisis only in the period 6300-6000 BC (Weninger, Clare 2017: 77).

Secondly, a number of studies indicate that relatively favourable conditions were still in operation in the 7th millennium BC, at least locally. For example, according to the already quoted paper by M. Finné and others (Finné *et al.* 2019), averaged moisture conditions for most of the 7th millennium BC were still relatively positive, falling to low levels only at the end of the millennium. Moreover, it should be noted that this is mainly due to the Dead Sea data. If they were not taken into account, the decline in the overall humidity level would probably not have been recorded at all (Finné *et al.* 2019: 855; cf. also Litt, Ohlwein 2017). This means that there may have been considerable local climatic variability in the zone in question. The recent lake-level reconstructions of Lake Kinneret confirm this assumption (Vossel *et al.* 2020). Indeed, for the period *c.* 7050-6650 BC, which the cited authors somewhat hesitantly refer to as dry, quite rapid oscillations in lake-levels are apparent, but when averaged out, it appears that the levels were getting higher. This trend continues into the period *c.* 6650-5950 BC, when water levels consistently rise, but without such rapid oscillations. This period was called humid. More humid climate conditions in the second half of the 7th millennium BC were also found at other sites, such as Soreq cave (Bar-Matthews *et al.* 2000; Bar-Matthews, Ayalon 2011) and Jeita cave (Verheyden *et al.* 2008) or wetlands east of the Dead See (Rambeau 2010).

It is perhaps worth quoting at this point the results of the comparative statistical processing of radiocarbon and environmental data made by the team from University of Reading (Flohr *et al.* 2016). In their light it is impossible to demonstrate any relationship between climate and the intensity of Neolithic settlement in Levant, especially in its southern part. This is also true for various regions of Anatolia and the area of northern Mesopotamia. Interestingly, the statistical treatment of radiocarbon data by A. Palmisano and collaborators (Palmisano *et al.* 2020) came to slightly different conclusions. Namely, they indicated the existence of negative differences between the summed probability distribution (SPD) and the fitted logistic null model for the Pottery Neolithic period, from *c.* 6500/6400 BC onwards. These differences are greatest for the period *c.* 6200-5600 BC. Thus, it is theoretically possible to speak of a settlement decline and explain it by a deterioration of climatic conditions (whose role in the interpretations carried out in this paper is, however, secondary). Nevertheless, this is an averaged conclusion, resulting from the combined analysis of all data. For the different areas distinguished within that study (Lower North Levant, Cisjordan lowlands, Cisjordan highlands, Transjordan) it looks different. The most obvious negative difference appears for the Cisjordan highlands, but only for the period *c.* 6000-5500 BC. For Cisjordan lowlands, on the other hand, there is a very slight negative difference for *c.* 6700/6600 BC. It should be emphasized, however, that differences between overall result and the partial results are partially due to the criteria of the geographic divisions adopted and, as one may assume, to the degree of representativeness of the ^{14}C data for these areas. In any case, after the mid-7th millennium BC the

SPD generally – as anticipated – drops a little, except for Transjordan. The cited authors, however, do not expressly state whether this can be explained by environmental factors, i.e. mainly by aridification, signs of which they see even from the mid-8th millennium BC (Palmisano *et al.* 2020: 718).

Thus, regardless of the environmental factors themselves, the implication of which is not evident, chronological issues also make it impossible to be certain about the decisive role of environmentally driven, Levantine marine migrations for the dawn of the Neolithic in the Aegean Zone. In fact, it is possible to construct a scenario in which the beginnings of the Neolithic on the western coasts of Anatolia and in southern Greece (*c.* 6800/6700 BC) were earlier than the fundamental climatic deterioration and crisis of Neolithic communities in some regions of Levant (after 6600/6500 BC) and the resulting 'escapes', made by sea.

On the other hand, milder conditions in the Aegean Zone or at least in some parts of this zone in the earlier part of the RCC under consideration are undoubtedly suggested by some data indeed (Gkouma, Karkanas 2018: 16, and citations therein), like for example those from Lake Marmara, in central-western Turkey, ca. 100 km off the Aegean coastline (Bulkan *et al.* 2018), Lake Dojran, at the border of Greece and Northern Macedonia (Thienemann *et al.* 2017) or Lake Loudias, near the Aliakmon river delta and the site of Nea Nikomedia (Styllas, Ghillardi 2017). However, it is likely that by this time there was already a fairly strongly accentuated seasonality with warm, dry summers and more humid winters (Thienemann *et al.* 2017: 1109, and citations therein; Gkouma, Karkanas 2018: 16). Mild conditions certainly did not prevail everywhere in Greece at that time, since e.g. palaeoceanographic reconstruction based on a multi-proxy approach of the most recent sapropel event (S1) (10–7.7 kyrs BP) in a 655 m water depth record from the North Ionian Sea demonstrates a cold interruption event (S1i), which starts already *c.* 7000 BC (Checa *et al.* 2020).

Regarding the second half of the RCC, there are obvious signals of deteriorating climatic conditions, both in Anatolia (e.g., Berger *et al.* 2016: 1859) and in Greece (e.g., Berger *et al.* 2016: 1866; Gkouma, Karkanas 2018: 16-17). Nevertheless, on the other hand, there are also indications that locally (especially in the coastal zone?) the 8.2 ka event was not as severe as suggested by the global data.

As an example, the interpretations proposed by J.-F. Berger and collaborators (Berger *et al.* 2016) for Dikili Tash can be demonstrated. These authors claim that 'the decrease in trees and increase in herbs could indicate the impact of the 8.2 ka RCC, but this period also shows the first signs of human impact in the Early Neolithic. The peak in coprophilous NPPs, ruderal taxa and NPPs indicative of erosive processes are certainly due to the Early Neolithic settlement implantation in Dikili Tash [...] benefitting from pristine forested environment with multiple available resources. This is attested to in the NPP record, not only by a first coprophilous species peak but also by a decrease in deciduous forest species and increase in herbaceous taxa on the edge of the marsh. Furthermore, at the bottom of the site (Dik 12), high-percentage cereal pollen (around 9% at 8.4 ka) and the increase in ruderal taxa make it clear the anthropogenic impact on vegetation cover was associated with agropastoral activities' (Berger *et al.* 2016: 1861). They therefore assume that the changes in vegetation that could correspond to the 8.2 ka event are largely (predominantly?) due to activities of Neolithic people. The general increase in water level and probably moisture around 6200-5800 BC, visible in different data and different profiles from Dikili Tash, also did not cause a break in settlement continuity, but at most only local relocations (Berger *et al.* 2016: 1862, 1867). Besides, general simulations indicate a maximum humidification in the northern Greece at *c.* 6100-5900 BC, which is the culmination of a trend evident since *c.* 7000 BC. Such moist conditions continued to dominate for a long time, until *c.* 4300 BC (Finné *et al.* 2019: 853-855). One way or another, humid conditions in this area, during the period 6300-6000 BC were in any case more favourable than today (which of course does not exclude short-term, elusive draughts).

Let us also add that essentially no 8.2 ka event is seen in some other pollen profiles, e.g. in Attica (Elefsis Bay - Kyrikou et al. 2020) or Macedonia (Dojran Lake - Thienemann *et al.* 2017) (cf. also Marinova, Ntinou 2018: 61-62).

Summing up, the thesis can be put forward that there are not so much obvious indications of catastrophic deteriorations of natural conditions in the hypothetical Anatolian and Levantine 'home' areas, while there are indeed indications of somewhat milder conditions in some parts of the Aegean Zone, or in any case these conditions did not deteriorate as much as in the neighbouring territories. Thus, environmental conditions in the north-east Mediterranean appear to become more balanced in that time.

It seems that the recent publication by N. Gauthier (Gauthier, 2016) on the spatial pattern of climate change during the spread of farming into the Aegean is particularly worth mentioning in this context since it contains innovative ideas which largely coincide with the thesis presented above. Due to highly complex modelling of present and past climatic data he demonstrated the differences of climatic conditions and trajectories in the period roughly corresponding to RCC of the *c.* 6600-6000 BC period. In short, Greece and western Anatolia experienced more arid conditions; central Anatolia experienced no change in rainfall whereas Thrace experienced wetter winters. This heterogeneity blurred the boundaries between previously distinct climate zones as previously wetter regions received less rain, and drier regions more rain. Thus, according to the cited author, Neolithic groups moving out from South-West Asia during the 7th millennium BC encountered landscapes with a more familiar, and predictable, rainfall regime. So, we have to do here with an approach different than in case of Bipartite Model. It was not a mitigation of RCC impacts that extorted migrations ('escape' from bad conditions, 'search' for good conditions); it was a natural enlargement of 'Neolithic habitats', created by the RCC impact, which prompted migrations (cf. also Berger *et al.* 2016: 1848). We would call it 'comfortable or opportunistic colonisation', that is the colonization not caused by urgent need, but the one that made use of new opportunities.

In addition, these migrations, which took place by sea, were favoured by lower water levels in the eastern Mediterranean, since only *c.* 2000/1600 BC the coastline reached its current form there (Benjamin *et al.* 2017: 45-46; Gkouma, Karkanas 2018: 17). Another factor that some researchers believe was attractive to early farmers were the wetland environments that formed as a result of the moisture increase associated with the 8.2 ka event (Gkouma, Karkanas 2018: 17).

The Balkan Zone

The Early Neolithic groups spread northward off the Aegean Zone along at least two main axes:

1.   across the Central Balkans (sites with macroblade technology and painted ceramics), mainly along the main river valleys like Struma, Vardar, Morava, Tisa, not to speak of the section of the Danube river roughly between modern cities of Budapest and Belgrade.
2.   along the eastern Adriatic coast (sites with Impressed ware).

This spread in both cases seem to be relatively fast, at least in the archaeological time scale of the period under consideration, since opinions about the appearance of the earliest Neolithic sites and materials before 6200/6100 BC (e.g., Ciuta 2005) currently seem highly questionable, due to low number of 'old' ^{14}C dates, mostly derived from disputable contexts. It is therefore a situation somewhat like the problem of the functioning of the Greek Neolithic before the mid-7th millennium BC. Yet, on the other hand, such opinions are still expressed and maintained, like the notion of the so-called formative period in Northern Macedonia, dated to *c.* 6500-6200 BC (Fidanoski 2019: 184-186). In general, the earliest dates and other evidence seem to suggest that the beginnings of the Neolithic in Bulgaria, Northern Macedonia and Albania should be placed between 6200 and 6000 BC, perhaps with an emphasis on the 62nd century BC (Blagojević *et al.* 2017; Fidanoski 2019; cf. also Krauß *et al.* 2018: 31). Virtually within the same time-frames, in light of the available dates, the origins of the Starčevo culture in present-day Serbia can be positioned (Porčić *et al.* 2016). These beginnings must have been at least slightly later than in the aforementioned more southern territories, nevertheless it is impossible to determine their chronology precisely. We can hypothesise that they were related more to the 61st century BC, especially in northern Serbia. In contrast, the appearance of the Neolithic in Vojvodina, southern part of the Great Hungarian Plain, southern/central Transylvania, and probably also in the Wallachian Plain can be placed possibly around 6000/5900 BC (Whittle *et al.* 2002; Bronk Ramsey *et al.* 2007; Luca *et al.* 2010; Oross, Siklosi 2012; Oross *et al.* 2016; Porcić *et al.* 2016; cf. also Krauß *et al.* 2018: 37). In turn, in the north-eastern part of the Great Hungarian Plain the Neolithic appears around 5800/5700 BC (Kozłowski, Nowak 2010; Kalicz 2011), similarly to the southern Transdanubia (Oross *et al.* 2016). Finally, the First Neolithic (Starčevo) groups most probably reached the Lake Balaton area within the last two centuries of the first half of the 6th millennium BC (Regenye 2010; Kalicz *et al.* 2012).

The formation of the Impresso complex around the Ionian Sea took place around in the end of the 7th millennium BC. Probably slightly later (6000/5900 BC?) communities of this complex were already present in central and northern Dalmatia, and in the Kvarner Gulf, while most probably at the end of the first half of the 6th millennium BC they reached the Istria and coasts of the Gulf of Trieste (Forenbacher *et al.* 2013; Berger *et al.* 2016: 1864; Gaastra, Vander Linden 2018; Bunguri 2019: 63; Forenbacher 2019: 32-33; cf. also Weninger, Clare 2017: 12-13).

It is worth noting that the overall combinations of radiocarbon dates indicate a recurring pattern, at least for Serbia, Bulgaria, and Hungary. Namely, after an initially modest SPD level, it increases, reaching a clear peak around 6000 BC for Bulgaria and Serbia and around 6000/5900 BC for Hungary (Porcić *et al.* 2016; Blagojević *et al.* 2017). This phenomenon can be interpreted as reflection of quantitatively modest penetrations between 6200 and 6000 BC while around 6000 BC a crucial stage of settlement of the Early Neolithic ecumene took place. Moreover, a somewhat similar two-phase mechanism is suggested for the Adriatic Zone. The first stage of farmer-forager contacts and maritime explorations embraced almost the entire coastal zone of the eastern Adriatic region around 6000 BC, while the village stage, filling this zone more completely, begun later (Forenbacher *et al.* 2013: 603). Let us add, that in case of Bulgaria and Serbia the high level of the SPD decreases after *c.* 6000 BC (Porcić *et al.* 2016; Blagojević *et al.* 2017), but this is less relevant for the issue under discussion.

When considering the spread of the First Neolithic in the Balkans obvious attention is paid to the 8.2 ka event, primarily because of the time coincidence. There are two ways in which this can be viewed. Firstly, this event can be considered as the causal factor, i.e. this spread would have been a kind of escape and search for more favourable environmental conditions. Secondly, the 8.2 ka event could have an indirect impact slowing or blocking the Neolithic expansion until the turn from the 7th to the 6th millennia BC. Such hypothesis can be put forward because – as already mentioned – significant increase of the number of sites and dates of the First Temperate Neolithic takes place around and after 6000 BC (Brami, Heyd 2011; Porcić *et al.* 2016; Blagojević *et al.* 2017; Pilaar Birch, Van der Linden 2018). Some earlier dates (*ca.* 6200-6000 BC), therefore, can be considered as reflection of initial penetrations.

According to the already cited contribution by R. Krauß and collaborators (Krauß *et al.* 2018), the expansion northwards, but also into Thrace/Bulgaria, beyond the SMA zone, was carried out by communities that had already adapted to harsher climatic conditions of the Balkan Peninsula (especially harsh winters), during a somewhat enforced stopover within the SMA zone in the second half of the 7th millennium BC.

During these movements, some transformations of the material culture, mainly pottery, can be observed. There also are some changes visible in subsistence. For instance, recent paper by M. Ivanova and others (Ivanova *et al.* 2018) convincingly demonstrated that there was an increasing reliance on animals with dispersal of farmers beyond the zone of Mediterranean influence.

The problems we encounter in attempting to verify such hypotheses are similar to those of the Aegean Zone.

Firstly, it is a question of problematic chronological synchronisation. In fact, all the above dates should be considered as an effect of averaging of the temporal data obtained so far. The achievable precision of radiocarbon dating, in relation to the time and place under consideration, both on the archaeological and environmental side, can be estimated maximally at about 150 years. This results in two alternative scenarios. In the first one, the appearance of the First Neolithic in the area between the Aegean zone and the southern border of the Carpathian Basin would have taken place still within the RCC period, associated with the 8.2 ka event. In the second (*à la* Krauß *et al.* 2018), on the other hand, it would have happened essentially after that RCC, and thus during a period of more favourable environmental conditions. Indeed, a third scenario, which has also already been signalled, can be assumed as well. The first penetrations of the Balkan Zone occurred still in the end of the RCC in question, while the more massive migrations started after that end. It seems that the probability of these three scenarios is equivalent. Ergo, the choice of any of them depends on top-down assumptions and beliefs.

Secondly, again we find in the literature numerous studies that indicate relatively little, or none at all, impact of the 8.2 ka event in the territory of interest (e.g., Berger *et al.* 2016: 1869-1870; Thienemann *et al.* 2017; Gaastra, Vander Linden 2018: 1193; Marinova, Ntinou 2018: 61-62). As J. Chapman says, 'the vast majority of proxy records dominated by mixed deciduous forest have shown no or minimal impact from the 8200 BP "event". The most likely interpretation of these data is that the climatic impact of the 8200 BP "event" must have been small to minimal for the communities living in such a wide range of environments as were present in our study region. An alternative interpretation is that the 8200BP "event" had serious consequences for people and landscapes but that these impacts were so short-term and rapidly reversible that no impact could be detected in the vast majority of the proxy records" (Chapman 2018: 6).

Let us also quote again the study by M. Finné and collaborators (Finné *et al.* 2019), which demonstrates that in the Greek-North Macedonian-Albanian borderland, fundamental climatic changes, above all in the form of evident aridification, did not begin until the 5th millennium BC. The period of the dawn of the Neolithic still belonged definitely to the 'humid times'. The highest level of humidity can indeed be referred to *c.* 6100 BC, but thereafter there was a very slow drying, i.e. one cannot speak of dramatic differences between the declining RCC and the 6th millennium BC. Possibly, these differences were within the limits of variability acceptable for agricultural societies of the time. Incidentally, one can note differences between local multi-proxies, both in the second half of the 7th millennium BC and in the 6th millennium BC. Thus, there was no universal variation throughout the Balkan Zone.

Notwithstanding the above, the presence within the Balkan Zone of conditions supportive of Neolithic settlement has been indicated as well. Let us mention here, among others, the suggested presence of 'of natural open grasslands or steppes and a moderate to high level of natural fires which helped to maintain any existing openness of the landscape' (Chapman 2018: 11). Early Neolithic human activity may have maintained such conditions and slightly increased it. Let us also note that according to E. Marinova and M. Ntinou (2018: 62) 'the start of the climate optimum around 8000 cal. BP (6000 cal. BC) facilitated the spread of mixed oak forests at low and mid-altitudes, where the first prehistoric settlements were founded'. A similar phenomenon is demonstrated in Hungary by M. Kempf (Kempf 2020: 20-21), based on a series of palynological data. These are telling observations since the Early Neolithic groupings were strongly associated with the environment of the mixed oak forests.

The Carpathian Basin

In the northern Balkans and in the Carpathian Basin the main cultural trend with painted ceramics more distinctly separated into the western (Starčevo culture) and eastern (Körös-Criş culture) branches, in the early 6th millennium BC. These units finally reached the line running from Balaton Lake to the most northern bent of the Tisza river, around 5600 BC. 'Geoarchaeological model' of neolithisation proposed by Hungarian palaeogeographer P. Sümegi (Sümegi, Kertész 2001; Kertész, Sümegi 2001; Sümegi 2003) assumes that 'the Central European-Balkan agroecological barrier'stopped the expansion of the Neolithic in *c.* 5800-5500 BC. This invisible, environmental barrier cut across the Carpathian Basin in the Neolithic. To the south of this line sub-Mediterranean climatic influence prevailed whereas in the north the climate was characterized by oceanic influence to the west and continental in the east. One way or another, climatic conditions north of this barrier were different than sub-

Mediterranean ones, i.e. – according to P. Sümegi – they were beyond the tolerance of subsistence model based on Near-Eastern patterns. Thus, the aforementioned line became the northern limit of the Early Neolithic cultures. Therefore, in theory, we would have a repetition of the situation between the Aegean and the Balkans.

However, there are currently some doubts as to the validity of the concept of 'the Central European-Balkan agro-ecological barrier'.

Among other things, this is due to the fact, that it was generated based on modern climatic data, similarly to the SMA concept. It is also undeniable that the line of this barrier was in a great measure drawn based on archaeological rather than environmental grounds; it repeats the ranges of the First Neolithic cultures (cf. e.g., Kalicz, Makkay 1977: karte 2; Kalicz, Koós 2002), though of course this need not depreciate it.

The current chronological view of cultural development in the Carpathian Basin also does not quite fit the proposed scenario. If indeed the Starčevo culture appeared in north-western Transdanubia c. 5600 BC, it is difficult to speak of any stagnation of Neolithic expansion, since the so-called early formative phase of the Linearband Pottery culture (LBK) can be dated to 5660-5480 BC (Stadler, Kotova 2019). The model would then apply only to the eastern part of the Carpathian Basin, where one can argue about a freezing of the Neolithic expansion indeed, but not for a very long time. The barrier line would have been reached by the Körös-Criş groups around 5800/5700 BC (Whittle *et al.* 2002; Bronk Ramsey *et al.* 2007; Kozłowski, Nowak 2010; Kalicz 2011; Oross, Siklosi 2012), while the beginnings of the earliest Alföld Linear Pottery culture (ALPC), i.e. the Szatmár group, located essentially north of the range of the Körös-Criş culture, can be expected *c.* 5600 BC (Domboróczki 2009, 2010).

What is perhaps more important, discovery of some Körös-Criş sites north of the barrier in recent years does not fit to postulated model as well. At least the sites of Tiszaszőlős-Domaháza-puszta (Domboróczki 2010; Domboróczki *et al.* 2010) and Ibrány (Domboróczki, Raczky 2010) should be mentioned in this context. In theory, the case of Ibrány could in a sense corroborate the concept of agro-ecological barrier since it demonstratessome differences to more southern ones. Namely, the archaeozoological analyses carried out for this site (Kovács *et al.* 2010) demonstrate the predominance of the 'wild' sector, contrary to 'normal' Körös sites. But in terms of livestock husbandry, in ALPC, as in LBK, one does not see a greater importance of wild animals, as a potential continuation of the trend observed in Ibrány. The high proportion of wild animals therefore looks like a specific feature of a particular site.

A factor undoubtedly weakening the thesis of the existence of a barrier are some climatic reconstructions, made i.a. based on speleothems. They indicate that in the Carpathians, and therefore even more so in the Carpathian Basin, the predomination of the Mediterranean air masses circulation took place in the Early and Middle Holocene. The change to the

Atlantic-related moisture sources started as late as the early 3rd millennium BC (Perşoiu *et al.* 2017; Hercman *et al.* 2020: 844).

Conclusions

The following stages and patterns of the neolithisation of South-East Europe presented from the environmental perspective (Figure 1) can be proposed, considering reflections and conclusions demonstrated in the second chapter. This is undoubtedly one of several alternative visions of this process.

1. In the 7th millennium BC overland and maritime movements of Neolithic groups reached the Aegean Zone, where some areas of the enclave character were settled. These movements did not have to be caused by climatic factors. As demonstrated, climatic-archaeological coincidence is debatable, especially at the local level. The First Neolithic settlement on the Anatolian coasts of the Aegean and the appearance of the Neolithic elements in southern Greece could predate the beginning of the RCC period. Therefore, the former could not be conditioned by the latter, especially since in the 'home' areas it is difficult to speak of a widespread and radical deterioration of natural conditions and an equally widespread and evident decrease in the intensity of the Neolithic settlement. On the other hand, however, if we alternatively argue for correlating the beginning of the Neolithic in the whole Aegean Zone with the beginnings of the RCC around 6600/6500 BC or for an intensification of migratory movements into the Aegean Zone from that time onwards, then, it must be taken into account that the RCC phenomena in the 7th millennium BC resulted in a certain environmental homogenisation in the North-East Mediterranean. This is not to say that in places such negative phenomena as aridification, typical of the 8.2 ka event, were not severe. This is to say that at least in some places, particularly on the coasts, environments resembled those of the 'original', what undoubtedly favoured and facilitated human migrations. There was no need to adapt to altered natural conditions. Thus, aforementioned movements can be interpreted, first and foremost, as making use of newly opened settlement opportunities ('opportunistic colonisation') and not as an effect of the mitigation of detrimental conditions in the 'home' areas? In other words, not only subsistence issues played a big part as ignition causes. Neolithic colonisation is said to have been rapid in the entire Aegean Zone (e.g., Horejs 2019: 77-78). This is an acceptable scenario, although we must remember that even 100 years, from the perspective of a single individual, is a huge time span (3-4 generations). In essence, however, the above thesis is just a plausible interpretation of the radiometric data. Chronological archaeological-archaeological correlations in the Aegean Zone, similarly to climatic-archaeological

Figure 1.
Hypothesised
scenario of the
neolithization
of the Balkan
Peninsula and
Carpathian Basin.

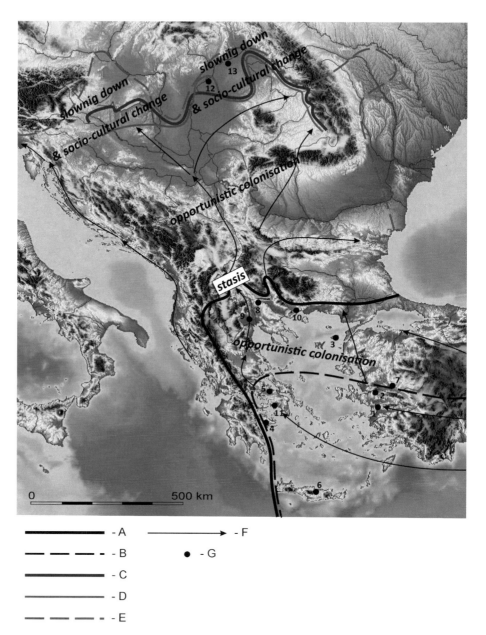

A – the line reached by the Neolithic c. mid-7th millennium BC, B – the presumable line reached by the Neolithic c. 6800/6700 BC, C – 'the Central European-Balkan agroecological barrier', D – the line reached by the Neolithic c. 5800/5700 BC, E – the line reached by the Neolithic c. 5600 BC, F – main (schematic) directions of the Neolithic spread, G – sites cited in the paper: 1 – Çukuriçi, 2 – Ulucak, 3 – Uğurlu, 4 – Franchti, 5 – Sarakenos, 6 – Knossos, 7 – Lake Marmara, 8 – Lake Dojran, 9 – Lake Loudias, 10 – Dikili Tash/ Tenaghi Philippon, 11 – Elefsis Bay, 12 - Tiszaszőlős-Domaháza, 13 - Ibrány ones, can be determined in various ways. Among other things, an alternative arrangement is acceptable in which there was a difference of about 100-300 years between Neolithic beginnings on the eastern and western coasts of the Aegean See, or between southern Greece and Thessaly/Macedonia (6800/6700 BC *vs* 6600/6500 BC). It would therefore appear that the jump across the Aegean, or from Argolid to Thessaly, was not so fast after all (if again we keep in mind the perspective of one human generation). This would mean that the neolithization of the western or north-western part of the Aegean Zone was not the result of a quick, panic-driven escape, triggered by the vision of starvation. Rather, it was a peaceful searching, finding, and exploiting those ('new') ecumenes that environmentally resembled the existing ('old') ones.

2. The RCC conditions halted further expansion for *c.* 400/500 years in the AMS zone ('stasis'), which was an intermediate zone between typically Mediterranean and typically Balkan ('temperate') conditions. In that zone an unconscious adaptation to slightly different, harsher (especially in winter) climatic conditions happened.

3. The cultural complexes with macroblade technology/painted pottery and with Impressed ware, which originated in outskirts of the Aegean Zone, spread northward, in the final effect to the *Caput Adriae* and southern Carpathian Basin, during the time-frames of *c.* 6100/6000-5900/5600 BC. The beginning of this period overlaps with the beginning of the amelioration of climate after 8.2 ka event. It is therefore possible to accept the hypothesis of the spread of the First Neolithic beyond the Aegean Zone essentially as a result of the vanishing of conditions typical of the RCC. But again, we should not interpret this process as a kind of escape from the adverse conditions. Firstly, we have indicated that conditions during the RCC in the Aegean zone were locally favourable. Secondly, the Neolithic settlement in northern Greece did not, after all, disappear or weaken after the turn of the 7th and 6th millennia BC.

The differences between the environments of the Aegean and Balkan Zones locally were not so radical in the late RCC. One may therefore ask why there was a Neolithic 'standstill' in Greece. It seems that the adaptation that took place in the AMS zone had to be relatively long to be effective. Perhaps this is reflected in some changes in material culture and economy. When the Neolithic communities on the Greek-Macedonian-Albanian borderland had fully adapted to the new conditions (taking advantage of the transitional nature of the AMS zone), the areas north of the Aegean Zone became open to them. This opportunity was eagerly and relatively quickly seized, as there was no longer any need for long-term, complicated environmental adaptations. Consequently, the reason of movements was perhaps not so much the improvement of environmental conditions in the Balkans after the end of the RCC, although the stabilisation of climatic conditions was certainly a contributing factor, but the complete, economic, settlement, social (and ideological?) adaptation to these Balkan conditions, achieved during the almost 500-year 'standstill' of Neolithic groups in the AMS zone. Either way, the favourable conditions that became opened for Neolithic settlement were again exploited ('opportunistic colonisation').

One can repeat that the spreading of the first Neolithic in the Balkans, beyond the Aegean Zone, only from the archaeological perspective seems fast. From the point of view of a human generation or an individual person, 200/300 years is really a long time, although it is a time that can be remembered in the collective consciousness. Again, therefore, this would mean that we are not dealing with processes of the fast mitigation of disadvantageous conditions. There was enough time to recognise the resources and opportunities available regionally and locally (also through interactions with indigenous hunter-gatherers) and to generate detailed regional and local survival strategies accordingly.

4. It is difficult to say whether the Neolithic expansion stopped, due to different, central European, agro-ecological conditions. If we postulate some phenomena of this kind, it seems – in the current layout of chronological data – that there happened a 'slowing down' of this expansion, from the line reached c. 5800/5700 BC to the line reached c. 5600 BC (i.e. already including the earliest LBK and ALPC). In the course of this slowing down there would have been processes of adaptation, more intense in Transdanubia (cf. e.g., Bánffy 2019), less intense in the Tisza Basin (cf. e.g. Kozłowski, Nowak 2010; Kozłowski *et al.* 2015). Essentially, they were not environmentally driven. These were processes targeted at 'cultural change', including material culture, resulting in the formation of the ALPC and LBK.

To finally recapitulate it can be said that becoming aware of the fact that the First Neolithic spread was not uniform and regular is not a new conclusion of course. Similar ideas have already been presented (e.g., Guilaine 2003; Bocquet-Appel *et al.* 2009, 2012; Gronenborn 2010; Horejs 2019) though there are inevitable differences between individual constructs. We would like to emphasize that these steps were related to exploitation of occasions created by global and local environmental transformations (cf. Banks *et al.* 2013). Such occasions made it possible to fulfil the eternal human needs and longings, expressed in the thoughts like: 'maybe it will be better there?' or 'maybe we will find our paradise there?'. These needs and longings have always and everywhere pushed masses of people to sometimes risky and crazy migrations and wanderings.

These movements were inevitably supplemented by different adaptation processes, which, in a measure, included also changes of material culture and social realms. One has to also agree with the views that such movements and adaptations were frequently associated with mental shifts (Krauß *et al.* 2018: 25), although the question of what was the cause and what was the effect remains open. It is very difficult to determine how such adaptations looked in practice. Remarkably, in the numerous publications cited in this article, one would look in vain for attempts to make such detailed reconstructions. These adaptations had to make possible to function in new and different environmental zones. However, if these zones were similar to the old ones, adaptations did not have to be so radical.

References

Andrianov B.V., Mantellini S. 2016. *Ancient Irrigation Systems of the Aral Sea Area: The History and Development of Irrigated Agriculture.* Oxford, Oxbow.

Arponen V.P.J., Grimm S., Käppel L., Ott K., Thalheim B., Kropp Y., Kittig K., Brinkmann J., Ribeiro A. 2019. Between natural and human sciences: On the role and character of theory in socio-environmental archaeology. *The Holocene* 29, p.1671–1676.

Atici L., Pilaar Birch S.E., Erdoğu B. 2017. Spread of domestic animals across Neolithic western Anatolia: New zooarchaeological evidence from UğurluHöyük, the island of Gökçeada, Turkey. *PLoS ONE* 12(10): e0186519.

Bánffy E. 2019. *First Farmers of the Carpathian Basin. Changing patterns in subsistence, ritual and monumental figurines.* Oxford, Oxbow.

Banks W.E., Antunes N., Rigaud S., d'Errico F. 2013. Ecological constraints on the first prehistoric farmers in Europe. *Journal of Archaeological Science* 40, p.2746–2753.

Bar-Matthews M., Ayalon A. 2011. Mid-Holocene climate variations revealed by high-resolution speleothem records from Soreq Cave, Israel and their correlation with cultural changes. *The Holocene* 21, p.163–171.

Bar-Matthews M., Ayalon A., Kaufman A. 2000. Timing and hydrological conditions of Sapropel events in the Eastern Mediterranean, as evident from speleothems, Soreq cave, Israel. *Chemical Geology* 169, p.145–156.

Benjamin J., Rovere A., Fontana A., Furlani S., Vacchi M., Inglis R.H., Galili E., Antonioli F., Sivan D., Miko S., Mourtzas N., Felja I., Meredith-Williams M., Goodman-Tchernov B., Kolaiti A., Anzidei M., Gehrels R. 2017. Late Quaternary sea-level changes and early human societies in the central and eastern Mediterranean Basin: An interdisciplinary review. *Quaternary International* 449, p.29–57.

Berger J.F., Lespez L., Kuzucuoğlu C., Glais A., Hourani F., Barra A., Guilaine J. 2016. Interactions between climate change and human activities during the early to mid-Holocene in the eastern Mediterranean basins. *Climate of the Past* 12, p.1847–1877.

Blagojević T., Porčić M., Penezić K., Stefanović S., 2017. Early Neolithic population dynamics in the Eastern Balkans and the Great Hungarian Plain. *Documenta Praehistorica* 44, p.18–33.

Bocquet-Appel J.P., Naji S., Vander Linden M., Kozłowski J.K. 2009. Detection of diffusion and contact zones of early farming in Europe from the space-time distribution of 14C dates. *Journal of Archaeological Science* 36, p.807–820.

Bocquet-Appel J.P., Naji S., Vander Linden M., Kozłowski J.K. 2012. Understanding the rates of expansion of the farming system in Europe. *Journal of Archaeological Science* 39, p.531–546.

Bradtmöller M., Grimm S., Riel-Salvatore J. 2017. Resilience theory in archaeological practice – An annotated review. *Quaternary International* 446, p.3–16.

Brami M., Heyd V. 2011. The origins of Europe's first farmers: The role of Hacılar and Western Anatolia, fifty years on. *Prähistorische Zeitschrift* 86, p.165–205.

Brandt G., Szécsényi-Nagy A., Roth Ch., Alt K., Haak W. 2014. Human paleogenetics of Europe - The known knowns and the known unknowns,. *Journal of Human Evolution* 79, p.73–92

Bronk Ramsey Ch., Higham T., Whittle A., Bartosiewicz L. 2007. Radiocarbon chronology, In: A. Whittle (ed.) *The Early Neolithic on the Great Hungarian Plain. Investigations of the Körös culture site of Ecsegfalva 23, County Békés, volume 1.* Budapest, Archaeological Institute of the Hungarian Academy of Science, p. 173-188.

Bulkan Ö., Yalçın M.N., Wilkes H. 2018. Geochemistry of Marmara Lake sediments - Implications for Holocene environmental changes in Western Turkey. *Quaternary International* 486, p. 199–214.

Bunguri A. 2019. Problems of Neolithisation in Albania. *Eurasian Prehistory* 15, p.47–100.

Chapman J. 2018. Climatic and human impact on the environment?: A question of scale. *Quaternary International* 496, p.3–13.

Checa H., Margaritelli G., Pena L.D., Frigola J., Cacho I., Rettori R., Lirer F. 2020. High resolution paleo-environmental changes during the Sapropel 1 in the North Ionian Sea, central Mediterranean. *The Holocene* 30, p.1504–1515.

Ciută M.-M. 2005. *Începuturile neoliticului timpuriu în spațiul intracarpatic transilvănean* (*The Beginning of the Neolithic in the Intra-Carpathian Area*). Alba Iulia, Aeternitas.

Contreras D.A. 2017. Correlation is not enough: Building better arguments in the archaeology of human-environment interactions, In: "D.A. Contreras (ed.) *The Archaeology of Human-Environment Interactions Strategies for Investigating Anthropogenic Landscapes, Dynamic Environments, and Climate Change in the Human Past*". New York-London, Routledge, p.3-22.

Czerniak L. 1992. Kto wierzy w bociany? Uwagi w kwestii osobliwości nauki (Who believes in storks? Notes on the peculiarities of science). *Przegląd Archeologiczny* 40, p.113–116.

Çakirlar C. 2012. The evolution of animal husbandry in Neolithic central-west Anatolia: The zooarchaeological record from Ulucak Höyük (c. 7040-5660 cal. BC, Izmir, Turkey). *Anatolian Studies* 62, p.1–33.

Çevik Ö., Vuruskan O. 2020. Ulucak Höyük: the pottery emergence in Western Anatolia. *Documenta Praehistorica* 47, p.96–109.

Davies A.L., Streeter R., Lawson I.T., Roucoux K.H., Hiles W. 2018. The application of resilience concepts in palaeoecology. *The Holocene* 28, p.1523–1534.

Dearing J.A. 2006. Climate-human-environment interactions: Resolving our past. *Climate of the Past* 2, p.187–203.

Derrick J. 1977. The great West African drought. 1972-1974. *African Affairs* 76, p.537–586.

Domboróczki L. 2009. Settlement structures of the Alföld Linear Pottery culture (ALPC) in Heves County (north-eastern Hungary): Development models and historical reconstructions on micro, meso and macro levels. In: J.K. Kozłowski (ed.) *Interactions between different models of Neolithization North of the Central European Agro-Ecological Barrier.* Kraków, Polska Akademia Umiejętności. p.75-128.

Domboróczki L. 2010. Report on the excavation at Tiszaszőlős-Domaháza-puszta and a new model for the spread of the Körös culture. In: J.K. Kozłowski and P. Raczky (eds) *Neolithisation of the Carpathian Basin: Northernmost Distribution of the Starčevo/Körös Culture.* Kraków-Budapest, Polish Academy of Arts and Sciences & Institute of Archaeological Sciences of the Eőtvős Loránd University, p.137-176.

Domboróczki L., Kaczanowska M., Kozłowski J.K. 2010. The Neolithic settlement of Tiszaszőlős-Domaháza-puszta and the question of the northern spread of the Körös culture. *Atti della Società per la Preistoria e Protostoria della Regione Friuli-Venezia Giulia* 17 (2008-2009), p.101–155.

Domboróczki L., Raczky P. 2010. Excavations at Ibrány-Nagyerdő and the northernmost distribution of the Körös culture in Hungary. In: J.K. Kozłowski and P. Raczky (eds) *Neolithisation of the Carpathian Basin: Northernmost Distribution of the Starčevo/Körös Culture.* Kraków-Budapest, Polish Academy of Arts and Sciences & Institute of Archaeological Sciences of the Eőtvős Loránd University, p.191-218.

Efstratiou N. 2019. Neolithic transition in north-western Greece. *Eurasian Prehistory* 15, p.7-24.

Efstratiou N., Karetsou A., Banou E., Margomenou D. 2004. The Neolithic settlement of Knossos: new light on an old picture, In: G. Cadogan, E. Hatzakiand E. Vasilakis (eds) *Knossos: Palace, City, State.* (British School at Athens Studies 12). London, British School at Athens, p.43-49.

Fidanoski L. 2019. The beginning of the end; the story of the Neolithisation of North Macedonia. *Eurasian Prehistory* 15, p.145-162.

Finné M., Woodbridge J., Labuhn I., Roberts C.N. 2019. Holocene hydro-climatic variability in the Mediterranean: A synthetic multi-proxy reconstruction. *The Holocene* 29, p.847–863.

Flohr P., Fleitmann D., Matthews R., Matthews W., Black S. 2016. Evidence of resilience to past climate change in Southwest Asia: early farming communities and the 9.2 and 8.2 ka events. *Quaternary Science Reviews* 136, p. 23–39.

Forenbacher S. 2019. Trans-Adriatic contacts and the transition to farming. *Eurasian Prehistory* 15, p.25–46.

Forenbaher S., Kaiser T., Miracle P.T. 2013. Dating the East Adriatic Neolithic. *European Journal of Archaeology* 16, p.89–609.

Freeman J., Robinson E., Beckman N.G., Bird D., Baggio J.A., Anderies J.M. 2020. The global ecology of human population density and interpreting changes in paleo-population density. *Journal of Archaeological Science* 120, p.105-168.

Gaastra J.S., Vander Linden M. 2018. Farming data: Testing climatic and palaeoenvironmental effect on Neolithic Adriatic stockbreeding and hunting through zooarchaeological meta-analysis. *The Holocene* 28, p.1181–1196.

Gauthier N. 2016. The spatial pattern of climate change during the spread of farming into the Aegean. *Journal of Archaeological Science* 75, p.1–9.

Gkouma M., Karkanas P. 2018. The physical environment in Northern Greece at the advent of the Neolithic. *Quaternary International* 496, p.14–23.

Goslar T., Kalicki T., Kaczanowska M., Kozłowski J.K. 2016. Stratigraphic sequence in trench A: complex II, layers 2-12 – from the Early Neolithic to the Palaeolithic, In: M. Kaczanowska, J.K. Kozłowski and A. Sampson (eds) *The Sarakenos Cave at Akraephnion, Beotia, Greece, vol. II. The Early Neolithic, the Mesolithic and the Final Palaeolithic (Excavations in Trench A).* Kraków, Polish Academy of Arts and Sciences, p.18–34.

Gronenborn D. 2005. Einführung: Klimafolgenforschung und Archäologie, In: D. Gronenborn (ed.) *Klimaveränderung und Kulturwandel in neolithischen Gesellschaften Mitteleuropas, 6700-2200 v. Chr. RGZM – Tagungen 1 (Mainz 2005).* Mainz, Römisch-Germanisches Zentralmuseum, p.1–16.

Gronenborn D. 2010. Climate, Crises, and the "Neolithisation" of Central Europe between IRD-events 6 and 4. In: D. Gronenborn and J. Petrasch (eds) *Die Neolithisierung Mitteleuropas. Internationale Tagung, Mainz 24. bis 26. Juni 2005.* Mainz, Römisch-Germanisches Zentralmuseum, Forschungsinstitut für Vor- und Frühgeschichte, p.61–80.

Gronenborn D., Strien H.-Ch., Dietrich S., Sirocko F. 2014. 'Adaptive cycles' and climate fluctuations: a case study from Linear Pottery Culture in western Central Europe. *Journal of Archaeological Science* 51, p.73–83.

Gronenborn D., Strien H.-Ch., Lemmen C. 2017. Population dynamics, social resilience strategies, and Adaptive Cycles in early farming societies of SW Central Europe. *Quaternary International* 446, p.54–65.

Guilaine J. 2003. Aspects de la Néolithisation en Méditerranée et en France, In: A.J. Ammerman and P. Biagi (eds) *The Widening Harvest: the Neolithic Transition in Europe. Looking Back, Looking Forward.* Boston, Archaeological Institute of America, p.189-206.

Guilbeau D., Kayacan N., Altlnbilek-Algül C., Erdogu B., Çevik Ö. 2019. A comparative study of the Initial Neolithic chipped-stone assemblages of Ulucak and Uğurlu. *Anatolian Studies*, 69, p.1–20.

Henry D.O., Cordova C.E., Portillo M., Albert, R.M., DeWitt, R., Emery-Barbier, A. 2017. Blame it on the goats? Desertification in the Near East during the Holocene. *The Holocene* 27, p.625–637.

Hercman H., Gąsiorowski M., Pawlak J., Błaszczyk M., Gradziński M., Matoušková Š., Zawidzki P., Bella P. 2020. Atmospheric circulation and thedifferentiation of precipitation sources during the Holocene inferred from five stalagmite records from Demänová Cave System (Central Europe). *The Holocene* 30, p.834–846.

Hofmanová Z., Kreutzer S., Hellenthal G., et al. 2016. Early farmers from across Europe directly descended from Neolithic Aegeans. *Proceedings of the National Academy of Sciences* 113.25: 201523951.

Horejs B. 2019. Long and short revolutions towards the Neolithic in western Anatolia and Aegean. *Documenta Praehistorica* 46, p.68–83.

Horejs B., Milić B., Ostmann F., Thanheiser U., Weninger B., Galik A. 2015. The Aegean in the early 7th millennium BC: Maritime networks and colonization. *Journal of World Prehistory* 28, p.289–330.

Imber C. 2020. *Imperium Osmańskie 1300-1650*. Kraków, Wydawnictwo Astra (orig.: C. Imber 2002. *The Ottoman Empire, 1300-1650. The Structure of Power*. Basingstoke-New York, Palgrave Macmillan).

Ivanova M., De Cupere B., Ethierand J., Marinova E. 2018. Pioneer farming in southeast Europe during the early sixth millennium BC: Climate-related adaptations in the exploitation of plants and animals. *PLoS ONE* 13(5): e0197225.

Kaczanowska M., Kozłowski J.K. 2014. The Aegean Mesolithic: Material culture, chronology, networks of contacts. *Eurasian Prehistory* 11, p. 31–62.

Kaczanowska M., Kozłowski J.K. 2016. Taxonomy, chronology and function of the cave during the sedimentation of layer 3. In: M. Kaczanowska, J.K. Kozłowski, A. Sampson (eds) *The Sarakenos Cave at Akraephnion, Beotia, Greece, vol. II. The Early Neolithic, the Mesolithic and the Final Palaeolithic (Excavations in Trench A)*. Kraków, Polish Academy of Arts and Sciences, p.130-131.

Kalicki T. 2006. *Zapis zmian klimatu oraz działalności człowieka i ich rola w holoceńskiej ewolucji dolin środkowoeuropejskich (Reflection of climatic changes and human activity and their role in the Holocene evolution of Central European valleys)*. Warszawa, Instytut Geografii i Przestrzennego Zagospodarowania Polskiej Akademii Nauk.

Kalicz N. 2011. *Méhtelek. The First Excavated Site of the Méhtelek Group of the Early Neolithic Körös Culture in the Carpathian Basin* (British Archaeological Reports International Series 2321). Oxford, Archaeopress.

Kalicz N., Koós J. 2002. Eine Siedlung mitältest neolitischen Gräbern in Nordostungarn. *Prehistoria Alpina* 37, p.45–79.

Kalicz N., Kreiter A., Kreiter E., Tokai Z.M., Tóth M., Bajnóczi B. 2012. A neolitikum történeti és kronológiai kérdései Becsehely–Bükkaljai-dűlőlelőhelyen (The Neolithic historical and chronological questions in the Becsehely–bükkaljaidűlő). In: B. Kolozsi (ed.) *Momos 4. Őskoros Kutatók IV. Összejövetelének konferenciakötete, Debrecen, 2005. Március 22-24.* Debrecen, Déri Múzeum, p.87-170.

Kalicz N., Makkay J., 1977. *Die Linienbandkeramik in der Grossen Ungarischen Tiefebene.* Budapest, Akadémiai Kiadó.

Karamitrou-Mentessidi G., Efstratiou N., Kozłowski J.K., Kaczanowska M., Maniatis Y., Curci A., Michalopoulou S., Papathanasiou A., Valamoti S.M. 2013. New evidence on the beginning of farming in Greece: the Early Neolithic settlement of Mavropigi in western Macedonia (Greece). *Antiquity* 87(336): http://antiquity.ac.uk/projgall/mentessidi336/

Kealhofer L., Marsh B. 2019. Agricultural impact and political economy: Niche construction in the Gordion region, central Anatolia. *Quaternary International* 529, p.91–99.

Kempf M. 2020. Neolithic land-use, landscape development, and environmental dynamics in the Carpathian Basin. *Journal of Archaeological Science: Reports* 34: 102637.

Kertész R., Sümegi P. 2001. Theories, critiques and a model: Why did the expansion of the Körös-Starčevo culture stop in the centre of the Carpathian Basin? In: R. Kertész and J. Makkay (eds), *From the Mesolithic to the Neolithic. Proceedings of the International Archaeological Conference held in the Damjanich Museum of Szolnok, September 22-27, 1996.* Budapest, Archaeolingua. p.225–246.

Kintigh, K.W., Ingram2018. Was the drought really responsible? Assessing statistical relationships between climate extremes and cultural transitions. *Journal of Archaeological Science* 89, p.25–31.

Kovács Z.E., Gál E., Bartosiewicz L. 2010. *Early Neolithic animal bones from Ibrány-Nagyerdő, Hungary.* In: J.K. Kozłowski and P. Raczky (eds) *Neolithisation of the Carpathian Basin: Northernmost Distribution of the Starčevo/Körös Culture.* Kraków-Budapest, Polish Academy of Arts and Sciences & Institute of Archaeological Sciences of the Eötvös Loránd University, p.238–253.

Kozłowski J.K., Nowak M. 2010. From Körös/Criş to the early Eastern Linear Complex: multidirectional transitions in the north-eastern fringe of the Carpathian Basin, In J.K. Kozłowski and P. Raczky (eds) *Neolithisation of the Carpathian Basin: Northernmost Distribution of the Starčevo/Körös Culture.* Kraków-Budapest, Polish Academy of Arts and Sciences & Institute of Archaeological Sciences of the Eötvös Loránd University, p.65-90.

Kozłowski J.K., Nowak M., Vizdal M. (eds) 2015. *Early Farmers of the Eastern Slovak Lowland: The Settlement of the Eastern Linear Pottery Culture at Moravany.* Kraków, Polska Akademia Umiejętności.

Krauß R., Marinova E., De Brue H., Weninger B. 2018. The rapid spread of early farming from the Aegean into the Balkans via the Sub-Mediterranean-Aegean Vegetation Zone. *Quaternary International* 496, p.24–41.

Kukawka S. 1992. Czy wierzę w bociany? (Do I believe in storks?). *Przegląd Archeologiczny* 40, p.117–122.

Kyrikou S., Kouli K., Triantaphyllou M.V., Dimiza M.D., Gogou A., Panagiotopoulos I.P., Anagnostou C., Karageorgis A.P. 2020. Late Glacial and Holocene vegetation patterns of Attica: A high-resolution record from Elefsis Bay, southern Greece. *Quaternary International* 545, p.28–37.

Lemmen C., Wirtz K.W. 2014. On the sensitivity of the simulated European Neolithic transition to climate extremes. *Journal of Archaeological Science* 51, p.65–72.

Lespez L., Tsirtsoni Z., Darcque P., Koukouli-Chryssanthaki H., Malamidou D., Treuil R., Davidson R., Kourtessi-Philippakis G., Oberlin C. 2013. The lowest levels at DikiliTash, northern Greece: a missing link in the Early Neolithic of Europe. *Antiquity* 87, p.30–45.

Lipson M., Szécsényi-Nagy A., Mallick S., *et al.* 2017. Parallel palaeogenomic transects reveal complex genetic history of early European farmers. *Nature* 551, 7680, p.368–372.

Litt T., Ohlwein Ch. 2017. Pollen as palaeoclimate indicators in the Levant. In: Y. Enzel, O. Bar-Yosef (eds) *Quaternary of the Levant. Environments, Climate Change, and Humans.* Cambridge, Cambridge University Press, p.337-345.

Luca S.A., Suciu C.I., Dumitrescu-Chioar F. 2010. Starčevo-Criş culture in western part of Romania – Transylvania, Banat, Çrişana, Maramureş, Oltenia and western Muntenia: repository, distribution map, state of research and chronology. In: J.K. Kozłowski and P. Raczky (eds) *Neolithization of the Carpathian Basin: Northernmost Distribution of the Starčevo/Körös Culture.* Kraków-Budapest, Polish Academy of Arts and Sciences & Institute of Archaeological Sciences of the Eőtvős Loránd University, p.254–265.

Maniatis G. 2014. Radiocarbon dating of the major cultural changes in prehistoric Macedonia: recent developments. In: E. Stefani, N. Merousis and A. Dimoula (eds), *A Century of Research in Prehistoric Macedonia 1912-2012 (International Conference Proceedings, Archaeological Museum of Thessaloniki, 22-24 November 2012).* Thessaloniki, Archaeological Museum of Thessaloniki Publications, p.205–222.

Marinova E., Ntinou M. 2018. Neolithic woodland management and land-use in south-eastern Europe: The anthracological evidence from Northern Greece and Bulgaria. *Quaternary International* 496, p.51–67.

Mathieson I., Alpaslan-Roodenberg S., Posth C., *et al.* 2018. The genomic history of southeastern Europe. *Nature* 555, 7695, p.197–203.

Mercuri A.M., Florenzano M.A., Terenziani R., Furia E., Dallai D., Torri P. 2019. Middle to late-Holocene fire history and the impact on Mediterranean pine and oak forests according to the core RF93-30, central Adriatic Sea. *The Holocene* 29, p.1362–1376.

Migowski C., Stein M., Prasad S., Negendank J.F.W., Agnon A. 2006. Holocene climate variability and cultural evolution in the Near East from the Dead Sea sedimentary record. *Quaternary Research* 66, p.421–431.

Müller J., Kirleis W. 2019. The concept of socio-environmental transformations in prehistoric and archaic societies in the Holocene: An introduction to the special issue. *The Holocene* 29, p.1517–1530.

Nicoll K., Zerboni A. 2020. Is the past key to the present? Observations of cultural continuity and resilience reconstructed from geoarchaeological records. *Quaternary International* 545, p.119–127.

Nikitin A.G., Stadler P., Kotova N., Teschler-Nicola M., Price T.D., Hoover J., Kennett D., Lazaridis I., Rohland N., Lipson M., Reich D. 2019. Interactions between earliest Linearbandkeramik farmers and central European hunter gatherers at the dawn of European Neolithization. *Scientific Reports* 9: 19544.

Oross K., Bánffy E., Osztás A., Marton T., Nyerges É.Á., Köhler K., Szécsény-Nagy A., Alt K.W., Bronk-Ramsey Ch., Goslar T., Kromer B., Hamilton D. 2016. The early days of Neolithic Alsónyék: the Starčevo occupation. *Bericht der Römisch-Germanischen Kommission* 94 (2013), p.93-122.

Oross K., Siklosi Z., 2012. Relative and absolute chronology of the Early Neolithic in the Great Hungarian Plain. In: A. Anders and Z. Siklosi (eds) *The First Neolithic Sites in Central / South-East European Transect III. The Körös Culture in Eastern Hungary.* British Archaeological Reports International Series 2334. Oxford, Archaeopress. p. 129-159.

Palmisano A., Woodbridge J., Roberts C.N., Bevan A., Fyfe R., Shennan S., Cheddadi R., Greenberg R., Kaniewski D., Langgut D., Leroy S.A.G., Litt T., Miebach A. 2019. Holocene landscape dynamics and long-term population trends in the Levant. *The Holocene* 29, p.708–727.

Perlès C. 2001. *The Early Neolithic in Greece.* Cambridge, Cambridge University Press.

Perlès C., Quiles A., Valladas H. 2013. Early seventh-millennium AMS dates from domestic seeds in the Initial Neolithic at Franchti Cave (Argolid, Greece). *Antiquity* 87 (338), p.1001–1015.

Perşoiu A., Onac B.P., Wynn J.G., Blaauw M., Ionita M., Hansson M. 2017. Holocene winter climate variability in Central and Eastern Europe. *Scientific Reports* 7, p.1196.

Pilaar Birch S.E., Vander Linden M. 2018. A long hard road... Reviewing the evidence for environmental change and population history in the eastern Adriatic and western Balkans during the Late Pleistocene and Early Holocene. *Quaternary International* 465, p.177–191.

Porčić M., Blagojević T., Stefanović S. 2016. Demography of the Early Neolithic population in Central Balkans: Population dynamics reconstruction using summed radiocarbon probability distributions. *PLoS ONE* 11(8): e0160832.

Rambeau C.M.C. 2010. Palaeoenvironmental reconstruction in the Southern Levant: Synthesis, challenges, recent developments and perspectives. *Philosophical Transactions of the Royal Society A: Mathematical, Physical and Engineering Sciences* 368(1931), p.5225–5248.

Regenye J. 2010. What about the other side: Starčevo and LBK settlements north of Lake Balaton. In: J.K. Kozłowski and P. Raczky (eds) *Neolithization of the Carpathian Basin: Northernmost Distribution of the Starčevo/Körös Culture.* Kraków-Budapest: Polish Academy of Arts and Sciences & Institute of Archaeological Sciences of the Eőtvős Loránd University, p.53-64.

Reingruber A., Thissen L. 2017. The 14 SEA Project. A ^{14}C database for Southeast Europe and Anatolia (10,000–3000 cal. BC), http://www.14sea.org/index.html

Reingruber A., Toufexis G., Kyparissi-Apostolika N., Anetakis M., Maniatis Y., Facorellis Y. 2017. Neolithic Thessaly: radiocarbon dated periods and phases. *Documenta Praehistorica* 44, p.34–53.

Richerson P.J., Boyd R., Bettinger R.L. 2001. Was agriculture impossible during the Pleistocene but mandatory during the Holocene? A climate change hypothesis. *American Antiquity* 66, p.387–411.

Sampson A., Kozłowski J.K., Kaczanowska M., Budek A., Nadachowski A., Tomek T., Miękina B. 2009. Sarakenos Cave in Boeotia. From Palaeolithic to the Early Bronze Age. *Eurasian Prehistory* 6, p.199–231.

Sampson A., Kaczanowska M., Kozłowski J.K. 2010. *The Prehistory of the Island of Kythnos (Cyclades, Greece) and the Mesolithic Settlement at Maroulas.* Kraków, Polish Academy of Arts and Sciences & The University of the Aegean.

Sánchez Goñi M.F., Ortu E., Banks W.E., Giraudeau J., Leroyer Ch., Hanquiez V. 2016. The expansion of Central and Northern European Neolithic populations was associated with a multi-century warm winter and wetter climate. *The Holocene* 26, p.1188–1199.

Sanlaville P. 1996. Changements climatiques dans la région levantine à la fin du Pléistocène supérieur et au début de l'Holocène. Leurs relations avec l'évolution des sociétés humaines. *Paléorient* 22, p.7-30.

Stadler P., Kotova N. 2019. *Early Neolithic Settlement Brunn am Gebirge, Wolfholz, in Lower Austria, Volume 1 / part a.* Langenweissbach-Wien, Beier & Beran.

Stock F., Ehlers L., Horejs B., Knipping M., Ladstätter S., Seren S., Brückner H. 2015. Neolithic settlement sites in Western Turkey - palaeogeographic studies at çukuriçi Höyük and Arvalya Höyük. *Journal of Archaeological Science: Reports* 4, p.565–577.

Styllas M.N., Ghilardi M. 2017. Early- to mid-Holocene paleohydrology in northeast Mediterranean: The detrital record of Aliakmon River in Loudias Lake, Greece. *The Holocene* 27, p.1487–1498.

Sümegi P., Kertész R. 2001. Palaeographic characteristics of the Carpathian Basin – an ecological trap during the Early Neolithic. In: R. Kertész and J. Makkay (eds.), *From the Mesolithic to the Neolithic. Proceedings of the International Archaeological Conference held in the Damjanich Museum of Szolnok, September 22-27, 1996.* Budapest, Archaeolingua, p.405–415.

Sümegi P. 2003. Early Neolithic man and riparian environment in the Carpathian Basin. In: E. Jerem and P. Raczky (eds), *Morgenrot der Kulturen.*

Frühe Etappen der Menscheitsgeschichte in Mittel- und Südosteuropa. Festschrift für Nándor Kalicz zum 75. Geburtstag. Budapest, Archaeolingua, p.53–60.

Thienemann M., Masi A., Kusch S., Sadori L., John S., Francke A., Wagner B., Rethemeer J. 2017. Organic geochemical and palynological evidence for Holocene natural and anthropogenic environmental change at Lake Dojran (Macedonia/Greece). *The Holocene* 27, p.1103–1114.

Verheyden S., Nader F.H., Cheng H.J., Edwards L.R., Swennen R. 2008. Paleoclimate reconstruction in the Levant region from the geochemistry of a Holocene stalagmite from the Jeita cave, Lebanon. *Quaternary Research* 70, p.368–381.

Vossel H., Roeser P., Litt T., Reed J.M. 2018. Lake Kinneret (Israel): New insights into Holocene regional paleoclimate variability based on high-resolution multi-proxy analysis. *The Holocene* 28, p.1395–1410.

Walsh K., Berger J.-F., Roberts C.N., Vanniere B., Ghilardi M., Brown A.G., Woodbridge J., Lespez L., Estrany L., Glais A., Palmisano A., Finné M., Verstraeten G. 2019. Holocene demographic fluctuations, climate and erosion in the Mediterranean: A meta data-analysis. *The Holocene* 29, p.864–885.

Walther B.A. 2016. A review of recent ecological changes in the Sahel, withparticular reference to land-use change, plants, birds and mammals. *African Journal of Ecology* 54, p.268–280.

Weninger B. 2017. Niche construction and theory of agricultural origins. Case studies in punctuated equilibrium. *Documenta Praehistorica* 44, p.6-17.

Weninger B., Alram-Stern E., Bauer E., Clare L., Danzeglocke U., Jöris O., Kubatski C., Rollefson G., Todorova H., Van Andel T. 2006. Climate forcing due to the 8200 cal yr BP event observed at Early Neolithic sites in the eastern Mediterranean. *Quaternary Research* 66, p.401–420.

Weninger B., Clare L., Rohling E.J., Bar-Yosef O., Böhner U., Budja M., Bundschuh M., Feurdean A., Gebel H.G., Jöris O., Linstädter J., Mayewski P., Mühlenbruch T., Reingruber A., Rollefson G., Schyle D., Thissen L., Todorova H., Zielhofer C. 2009. The impact of Rapid Climate Change on prehistoric societies during the Holocene in the Eastern Mediterranean. *Documenta Praehistorica* 36, p.7–59.

Weninger B., Clare L., Gerritsen F., Horejs B., Krauß R., Linstädter J., Özbal R., Rohling J.H. 2014. Neolithization of the Aegean and Southeast Europe during the 6600-6000 calBC period of Rapid Climate Change. *Documenta Praehistorica* 41, p.1–31.

Weninger B. Clare L. 2017. 6600–6000 cal BC abrupt climate change and Neolithic dispersal from West Asia. In: H. Weiss (ed.), *Megadrought and Collapse: From Early Agriculture to Angkor.* New York, Oxford University Press, p.69–92.

Whittle A., Bartosiewicz L., Borić D., Pettitt P., Richards M. 2002. In the beginning: new radiocarbon dates for the Early Neolithic in northern Serbia and south-east Hungary. *Antaeus* 25, p.63–117.

Woodbridge J., Roberts C.N., Palmisano A., Bevan A., Shennan S., Fyfe R., Eastwood W.J., Izdebski A., Çakırlar C., Woldring H., Broothaerts N., Kaniewski D., Finné M., Labuhn I. 2019. Pollen-inferred regional vegetation patterns and demographic change in Southern Anatolia through the Holocene. *The Holocene* 29, p.728–741.

Hiatus et recompositions culturelles dans le néolithique méditerranéen : le climat en cause?

Jean Guilaine

Résumé

Cette contribution tente d'analyser les impacts sur les sociétés méditerranéennes des deux crises climatiques qui bornent, à l'amont comme à l'aval, le déroulement du Néolithique lato sensu : les événements datés à 8200 et à 4200 BP. Les îles sont tout particulièrement concernées dans ce débat. Les effets de ces péjorations ont été différents. Le « 8200 BP event » a entrainé une déstabilisation des premières communautés agricoles en Orient et des ultimes sociétés de chasseurs-cueilleurs en Méditerranée de l'Ouest. Il en a souvent résulté des ruptures stratigraphiques en liaison avec des processus d'exodes, de recompositions culturelles ou, souvent de hiatus de durée diverse.

En revanche le « 4200 BP event » survient dans un contexte totalement différent : systèmes étatiques en Egypte ou systèmes urbanisés à l'Est, brillantes cultures chalcolithiques à l'Ouest. Les perturbations qui interviennent alors, sans mésestimer le facteur climatique, ont souvent, semble-t-il, des causes systémiques, impliquant parallèlement dans leur motivation des situations politiques, économiques, sociales. L'impact produit se soldera souvent par des réorganisations au plan des structures sociales et des réseaux de circulation.

Abstract

This contribution attempts to analyze the impacts on Mediterranean societies of the two climate crises that limit, upstream and downstream, the course of the Neolithic lato sensu: the events dated to 8200 BP and 4200 BP. Islands are particularly affected in this debate. The effects of such a pejoration have been different. The "8200 BP event" led to a destabilization of the first agricultural communities in the East and the ultimate hunters-gatherer societies in the Western Mediterranean. This has often resulted in stratigraphic ruptures in connection with processes of exodus, cultural changes or, often, hiatuses of various durations.

On the other hand, the "4200 BP event" occurs in a totally different context: state systems in Egypt or urbanized systems in the East, brilliant Chalcolithic cultures in the West. The disturbances that occur then, without underestimating the climate factor, often have, it seems, systemic causes, implying in parallel in their motivation political, economic and social situations. The impact will often result in reorganizations of social structures and exchange networks.

Introduction

Le déroulement du Néolithique méditerranéen est « encadré » par deux crises climatiques de l'Holocène : les « 8200 et 4200 Cal BP Events ». On se propose de discuter les corrélations possibles entre ces phénomènes et les perturbations survenues au sein des sociétés néolithiques contemporaines de ces processus. Rappelons rapidement que le « 8200 Cal BP Event » (environ 6200 avant notre ère), observé dans les carottes du Groenland, serait dû au déversement dans l'Océan atlantique d'eaux provenant de la calotte glaciaire des Laurentides en Amérique du Nord ; les barrages des lacs nés de la fonte des glaces auraient cédé et leurs eaux très froides, lâchées dans l'Atlantique, auraient entraîné un refroidissement général et une baisse des températures. Pour autant les effets de cette détérioration n'auraient pas été identiques partout : froids et arides dans le Nord de l'Eurasie, frais et humides en Europe (et marqués notamment par des transgressions lacustres), plus secs dans certaines zones sud-méditerranéennes (avec, dans ces cas, des phases de bas niveaux des lacs) (Berger 2009). L'autre phénomène climatique, le « 4200 Cal BP Event » (environ 2300/2200 avant notre ère), survenu vers la fin du Néolithique européen, se caractérise par une augmentation des conditions sèches ayant entrainé une aridité plus ou moins marquée selon les aires géographiques et une pénurie en eau ayant affecté, en plusieurs points du monde, les organisations sociales, entrainant des crises des systèmes politiques en place (Meller *et al.* 2015). Envisageant les incidences de ces phénomènes sur les sociétés néolithiques méditerranéennes, on livrera ici quelques réflexions limitées au seul point de vue archéologique.

Le « 8200 Cal BP Event » et les hiatus archéologiques insulaires

Il nous a été donné, à plusieurs reprises, d'insister sur certains « silences archéologiques » qui caractérisent les régions méditerranéennes vers la fin du VIIe millénaire avant notre ère et, plus particulièrement, les milieux insulaires (Guilaine 1996, Berger, Guilaine 2009). Rappelons qu'à cette époque des pointes de sècheresse sont signalées au Levant entrainant l'abandon ou la reconfiguration de sites d'envergure avec des discontinuités stratigraphiques. Ainsi O. Bar Yosef considère que l'impact climatique du 8200 Cal BP Event a mis un terme définitif aux dernières expressions du PPNB, rupture suivie par la constitution dans le Levant-Sud du Néolithique à poterie (Yarmoukien) (Bar Yosef 2006). Il n'est pas certain toutefois que cet impact ait eu les mêmes effets de façon systématique et avec les mêmes conséquences pour le devenir des sites, la variété des situations écologiques entrainant des réponses différentes. Çatal Huyuk (Turquie) serait alors abandonné au profit d'un autre établissement (Çatal Huyuk-West) (Weninger *et al.* 2006). A Jericho et Byblos, de simples fosses yarmoukiennes se surimposent aux maisons rectangulaires plâtrées du PPNB (les auteurs mêlant à Byblos ces deux phases au sein d'un même « Néolithique ancien », mélange en fait des deux occupations) (Garfinkel 2004). A Ain Ghazal (Jordanie), le grand centre PPNB avait commencé de se fragiliser au cours du PPNC, vulnérabilité que la crise accentuera en entrainant au Néolithique céramique une reconversion vers des unités plus restreintes (Simmons 2000). Ces rapides exemples montrent la nécessité d'examiner les situations au cas par cas avant de tirer des conclusions à un niveau plus général.

C'est ainsi que l'observation de hiatus archéologiques dans le déroulement des néolithiques insulaires peut être interprétée comme une rétraction des peuplements, voire une désertion consécutive à un changement des conditions environnementales déterminées par les effets climatiques. La question qui se pose alors d'un strict point de vue archéologique est : ces hiatus correspondent-ils à une réalité ou ne sont-ils que des moments provisoires dans l'histoire de la recherche dus à l'absence de documentation ? L'avenir répondra à ces dilemmes. C'est pourquoi les présentes considérations doivent-elles être perçues dans le contexte de l'état actuel des données. On examinera les cas suivants : Chypre, la Crète, Corfou, Malte, la Sicile, la Sardaigne et la Corse (les Baléares s'excluant en raison de leur peuplement plus récent que la période considérée).

Chypre

Le long déroulement de la séquence PPNB de Shillourokambos, montre, à son stade terminal (vers 7000/6900 BC), une progressive transition vers une culture acéramique tardive, illustrée par le site de Khirokitia qui couvrira essentiellement le VIIe millénaire alors qu'une large partie du VIe millénaire restera à Chypre non documentée. Y eut-il alors une chute démographique, voire un dépeuplement de l'ile ? La documentation ne reprendra réellement qu'au Ve millénaire avec le rapide développement d'une culture néolithique à poterie (le Sotira), très dynamique et conquérante. L'une des questions posées est donc celle de l'éventuel impact de l'épisode 8200 BP sur la culture de Khirokitia. Deux positions s'affichent, fondées sur une discussion autour de datations [14]C. Pour certains auteurs, les datations du site–éponyme qui montrent un pic autour de 8 300 BP supposent un abandon des lieux lors de l'évènement aride (Weninger *et al.* 2006). Pour les fouilleurs actuels, certaines datations indiqueraient une perduration de l'occupation lors des premiers siècles du VIe millénaire, le site ayant été moins affecté que d'autres établissements levantins (Daune- Le Brun *et al.* 2017). Les datations du site de Tenta, dont plusieurs se situaient au VIe millénaire, poussaient donc les climatologues à considérer son occupation postérieurement à celle de Khirokitia, après la « crise » (Weninger *et al.* 2006). Toutefois une reconsidération des divers épisodes d'occupation de Tenta, fondée sur les comparaisons avec les séquences de Shillourokambos et de Mylouthkia, ont montré que le déroulement de fréquentation du site se positionnait totalement avant 8200 BP et que les datations anciennes devaient être abandonnées car trop hasardeuses (Todd 2003). Le problème reste donc ouvert sur les causes ayant entrainé le « silence archéologique » du VIe millénaire, la datation de ses débuts et ses causes.

La Crète

La mer Egée semble assez largement fréquentée dès le Mésolithique et des réseaux de circulation de l'obsidienne de Mélos dès lors attestés. L'implantation en Crète des premiers agro-pasteurs est essentiellement documentée sur le site de Cnossos, dont les niveaux néolithiques ont été révélés par les fouilles de J.Evans, puis revus plus récemment par N.Efstratiou (Evans, 1964, 1968 ; Efstratiou *et al.* 2004, Efstratiou 2013), la séquence stratigraphique relevée et les datations obtenues par le premier ayant été confirmées et validées par les travaux du second. Les néolithiques acéramiques, probablement originaires des côtes sud-anatoliennes, où existe un faciès néolithique sub-contemporain de souche proche-orientale (Cukuriçi) (Horejs *et al.* 2015), introduisent sur l'ile la culture des céréales, l'élevage, les maisons en « dur ». Mais cette première occupation, datée vers 6800/6700 BC, n'aura pas de postérité. En effet dans la seconde moitié du VIIe millénaire et les débuts du VIe, le site ne connait que des fréquentations clairsemées, sans consistance, peu identifiables au plan culturel. Une fréquentation plus pérenne redémarrera vers 5500 BC dans un contexte de type Néolithique moyen. L'expression « Early Neolithic I and II » utilisée par J. Evans pour désigner dès lors ces premières installations avec céramique a été à plusieurs reprises critiquée par nous-même car ces fréquentations s'inscrivent en fait dans le Néolithique moyen de la séquence grecque, le « vrai » Néolithique ancien avec céramique de Crète restant à identifier (Guilaine 2013). Le 8200 Cal BP Event intervient donc au cours de ce hiatus culturel.

Corfou

Le site de Sidari a livré une séquence comportant un Mésolithique local daté vers 8000 BC, un Néolithique initial qui se développe vers 6400-6200 BC, et un Néolithique ancien daté à 6000 BC. (Berger *et al.* 2014). Un hiatus existe donc ici entre les deux versions du Néolithique, l'un s'inscrivant dans la sphère culturelle égéenne, l'autre relevant de l'aire italo-adriatique *a ceramica impressa*. Le 8200 BP Event pourrait donc s'intercaler entre ces deux moments.

Malte

En l'état des données, le premier néolithique, se rattachant vraisemblablement à la vague précoce à céramique imprimée, est daté à 5900 BC c'est-à-dire en concordance avec les plus anciennes manifestations néolithiques du Sud de la péninsule italienne. La datation est obtenue à partir d'un horizon sédimentaire à marqueurs d'anthropisation (pollens de céréales) et non d'un site archéologique (travaux du Fragsus Project, C. Malone et R. McLaughlin). A cette étape succèdent vers 5700 BC des occupations Stentinello attestées sur plusieurs sites (Skorba, Ghar Dalam). En l'état de la recherche, l'absence de mésolithique (provisoire ?) ne permet pas de poser la question d'un hiatus entre chasseurs et néolithiques.

La Sicile

Comme l'Italie péninsulaire, la Sicile fait partie de la sphère ouest-méditerranéenne à poterie imprimée dont les premières manifestations se placent vers 6000-5900 BC. Le site de la grotte de l'Uzzo à Trapani a livré, sur le talus extérieur de la cavité, une séquence qui voit se succéder des niveaux du Mésolithique et du Néolithique ancien. Les industries à trapèzes y apparaissent dès 6700 BC. Entre les phases récentes du Mésolithique (tranchées F et M, niveaux 22-15) et le Néolithique ancien *a ceramica impressa*, (niveaux 10-7) se développent des niveaux (14-11), dits « de transition », mais en fait toujours mésolithiques, caractérisés par une économie désormais tournée vers les ressources marines (cétacés, mollusques) (Collina 2015, Natali, Forgia 2018). Deux hypothèses sont possibles : ces niveaux sont antérieurs au 8200 Cal BP Event et s'inscrivent dans un Castelnovien terminal, antérieur à l'arrivée des Néolithiques, ou bien ils sont contemporains de l'extrême fin du VIIe millénaire avant notre ère et le changement économique drastique observé en direction des ressources de la mer pourrait être une réponse à la péjoration climatique. Ce point chronologique est à préciser. Notons que dans la grotte de Latronico 3, dans le Sud péninsulaire (Basilicate), un hiatus existe bien entre le Mésolithique castelnovien, daté entre 6600 et 6200 BC, et le premier néolithique qui se manifeste dans le courant du VIe millénaire (Guilaine *et al.* 2019).

Corse et Sardaigne

Principalement attesté en Corse, est connu un Mésolithique aux caractères « indifférenciés » : les datations ^{14}C le placent essentiellement dans le courant des IXe et VIIIe millénaires. Mais une présence « mésolithique » plus récente est attestée sur le site de Campu Stefanu (Sollacaro) par une tombe plurielle datée entre 7000 et 6600 BC (Cesari *et al.* 2014). Quelques autres sites témoignent d'occupations vers le milieu du VIIe millénaire. Les premières manifestations du Néolithique relèvent du courant impressa de souche italique identifié en Corse (Abri Albertini, Campu Stefanu) et dans l'archipel toscan (Le Secche dans l'ile de Giglio), cette phase n'ayant pas encore été repérée en Sardaigne. Sur les deux iles, le relais est pris ensuite par le Cardial tyrrhénien (Filiestru, Santa Chiara-Terralba, Basi) autour de 5600

BC, puis par des faciès « épicardiaux » locaux (Curaghiaghiu) ou des influx centro-italiques à céramique cannelée (Sasso) (Luglié 2018). La question du hiatus et de ses causes, observable entre Mésolithique et Néolithique ancien à la charnière VIIe/VIe millénaires reste donc à éclaircir.

Bilan

Des hiatus sont bien observables, plus ou moins prononcés, en Méditerranée vers la fin du VIIe millénaire avant notre ère. Nous ne pensons pas que ces « absences archéologiques » aient automatiquement entrainé des désertifications humaines. Mais on peut penser que les changements climato-environnementaux, induits par le « 8200 Cal BP Event », ont pu déstabiliser les communautés en place et les contraindre à repenser leur style de vie et leurs processus d'acquisition de nourriture. Il convient toutefois de prendre en compte dans cette réflexion les disparités dans les degrés socio-économiques atteints. Au Levant, dans un contexte où l'économie de production est en place depuis près de deux millénaires, c'est, avec le déclin des méga-sites, une reconversion vers le pastoralisme, des genres de vie plus mobiles, une segmentation en hameaux, voire à des exodes que l'on assiste.En Crète, à Cnossos, ce sont des passages fugaces, diffus, qui font suite à l'abandon des lieux par les Néolithiques acéramiques avant que le site ne soit réinvesti vers 5500 BC. Á Corfou, un hiatus semble s'introduire entre deux moments du Néolithique, initial et *impresso*. En Méditerranée centrale et occidentale (Sicile, Sardaigne, Corse), le « hiatus » entre Mésolithique et Néolithique ancien est sensible et reste un thème de recherche à approfondir. Analyser ces « absences », ainsi que les comportements économiques et autres processus d'adaptation des groupes humains, demeure un challenge motivant.

Le « 4200 Cal BP Event » et l' « effondrement » des apogées du IIIe millénaire : déstructurations et repliements

Un autre épisode climatique sévère, noté par les climatologues vers 4200 BP (2300-2200 BC), marque la transition de l'Holocène moyen à l'Holocène récent (Magny *et al.* 2009, Bini *et al.* 2019). Dans la littérature cet événement est fréquemment mis en avant pour expliquer les perturbations sociales qui affectent les sociétés du IIIe millénaire et, notamment, les organisations jugées comme le plus « robustes » : effondrement de l'Ancien Empire égyptien, chute de l'empire akkadien en Mésopotamie. Il se caractérise, sur le temps court ou long selon les auteurs, par des pics d'assèchement et d'aridité qui, réduisant la pluviosité, ont pu entrainer une raréfaction des nappes phréatiques. Les conséquences économiques et sociales induites révèleraient l'une des premières grandes crises imputables au manque d'eau dans la gestion des villes ou des communautés rurales. On tâchera d'aborder ces questions en les confrontant aux données archéologiques de quelques régions méditerranéennes et en les situant dans leur contexte.

Du Levant à l'Asie mineure

Les agglomérations fortifiées de Palestine du Bronze ancien II-III disparaissent dans la seconde moitié du IIIe millénaire. Cette dilution, qui inaugure le Bronze ancien IV, est parfois attribuée à une dessiccation consécutive à la raréfaction des pluies (Rosen 1989). D'autres chercheurs, sans renier la dégradation environnementale, avancent plutôt une interaction de facteurs : raids et invasions périodiques des nomades du désert (les Amorites de la littérature), incapacité à nourrir une population urbaine croissante et désagrégation de la structure sociale, chute des circuits d'échange avec l'Egypte, etc. (Dever 1989). La désurbanisation aurait engendré une reconversion vers de multiples petits sites à même de mieux gérer leurs besoins alimentaires.

Plus au Nord, Sargon (c. 2300-2240) organise l'état d'Akkad dont il n'aura de cesse d'agrandir le territoire. Vers l'Ouest, ses conquêtes aboutiront à l'annexion du Nord de la Syrie. Mais Akkad s'effondrera à son tour dans le courant du XXIIe siècle. Des démolitions, dont les raisons sont controversées et les dates pas forcément concordantes, concernent divers établissements et des discontinuités stratigraphiques étayent ces épisodes autant dans le Levant qu'en Anatolie. En Anatolie occidentale, la « deuxième ville » de Troie subit des dommages de même que sa voisine Poliochni (Lemnos). Ces temps troublés pourraient expliquer l'enfouissement prudent des « trésors » des deux cités (Huot 2004). Si donc, en Orient, des motifs climatiques ont pu entrainer des réorganisations sociales contraintes par des impératifs économiques, la guerre, devenue endémique, a pu être très souvent la cause de divers troubles.

Les Cyclades et la Crète

Le IIIe millénaire a été l'un des âges d'or de la sphère cycladique et crétoise. Cet apogée, concrétisé notamment par la culture de Keros-Syros, se caractérise par un large fonctionnement de réseaux maritimes unissant la Grèce continentale, les iles et les littoraux ouest-anatoliens. Divers produits circulent en mer Egée : céramiques, armes, parures, figurines de marbre. Or cet « esprit international » (Renfrew 1972) entre vite en difficulté vers 2200 BC par suite d'une rétraction des échanges maritimes. Si les relations entre l'Anatolie, le continent grec et les archipels ne sont pas totalement rompus, elles chutent en intensité. En Crète, des sites prospères sont abandonnés (Myrtos). Des destructions sont, comme déjà notées en Anatolie, observables en Grèce (incendie de la Maison des Tuiles, une sorte de proto-palais, à Lerne). L'insécurité a pu déstructurer les réseaux maritimes tandis que des crises politiques et sociales sont généralement avancées pour expliquer abandons ou destructions.

Malte

Le cas de Malte est emblématique. L'archipel connait un premier hiatus archéologique entre le milieu du Ve millénaire et le début du IVe (4600-3800 BC) (travaux de Fragsus Project, C. Malone et R. McLaughlin dir.). Ensuite s'amorce un apogée de plus d'un millénaire (3600-2350 BC) marqué par une architecture mégalithique originale (temples) au service d'un culte placé sous le contrôle d'élites lesquelles, à travers la maitrise de manifestations liturgiques, assurent la cohésion sociale. Cette « Culture des Temples » semble ensuite péricliter assez rapidement pour disparaitre autour de 2300 BC. Une rupture intervient alors, entrainant une possible désertification de l'archipel, avant que ne survienne une nouvelle culture (le « Cimetière Tarxien ») apportant sur les iles la métallurgie et le rite de la crémation des défunts.

On a tenté de corréler l'effondrement de la culture des Temples avec le 4200 Cal BP Event, sous l'effet d'un double impact, climatique et social. Déterministe dans la mesure où l'aridité croissante a pu entrainer des difficultés auprès d'une population en progressive augmentation sur un espace restreint. Social car les élites n'ont plus été à même de satisfaire les besoins des habitants et ont ainsi perdu toute possibilité de maintenir leur autorité et leur reproduction.

Le Sud de la France

A partir de 2500 BC survient le déclin des grandes cultures chalcolithiques de Méditerranée occidentale, caractérisées par la pratique des tombes collectives (hypogées, mégalithes) : Gaudo, Rinaldone, Fontbouisse, Véraza, Los Millares. Un premier élément perturbateur en a été la rapide diffusion du gobelet campaniforme international qui, associé à des marqueurs singuliers, a introduit une idéologie individualiste contestant l'autorité des cercles dominants, fondés sur des relations de parenté et l'usage de tombes communautaires ou familiales. Après ces remises en question sociales, des pics d'aridité, notés par les climatologues (Berger et al.2019), ont-ils, vers 2200 BC, fait décliner les productions céréalières et contribué à la désagrégation définitive des structures chalcolithiques ? Les premiers temps du Bronze ancien sont ambigus. D'une part, les sépultures collectives ou individuelles (grottes sépulcrales, usage prolongé des mégalithes, caissons) ne semblent pas indiquer un déficit de population (mais ce point reste à préciser). En revanche les habitats montrent, comparativement au Fontbouisse ou au Vérazien, une chute de l'effectif des localités, un moindre investissement architectural, avec, souvent, une simple « squaterisation » de sites antérieurs. Retour à un style de vie plus mobile orienté vers le pastoralisme ? Vers un cycle de mise en jachères ?

Le Sud de la Péninsule Ibérique

Dans le contexte de la Méditerranée occidentale, le Sud de la péninsule Ibérique, d'Almeria à la baie de Lisbonne, occupe une place à part. Dans la première moitié du IIIe millénaire y cohabitent des sites « surdimensionnés » (tels Valencina de la Concepción, plus de 400 hectares, Marroquies Bajos, 113 hectares, Porto Torrao, 100 hectares, La Pijotilla, 80 hectares), des localités fortifiées par plusieurs lignes de murailles (Los Millares, Zambujal, Leceia), enfin des sites secondaires, le tout selon un système hiérarchisé. Ces processus de centralisation ont généré des élites dont la domination s'affiche à travers des marqueurs sociaux en matériaux importés comme l'ivoire d'éléphant d'Asie ou d'Afrique. Or un tel système va s'effondrer vers 2300/ 2100 BC. Des causes climatiques ont-elles joué dans un tel déclin ? Une corrélation entre la montée en puissance de l'aridité et la chute du nombre des établissements semble effective (Hinz *et al.* 2019). Les recompositions culturelles qui engendreront la culture d'El Argar verront un déplacement du centre de gravité du dynamisme moteur vers le Sud-Est de la péninsule.

Bilan

On ne saurait écarter le rôle des processus climatiques dans les repliements ou les réorganisations qui caractérisent le dernier quart du IIIe millénaire : ainsi de la désintégration urbaine en Palestine avec retour vers un peuplement plus dispersé, déclin de la société des temples de Malte, désagrégation des chalcolithiques méridionaux. Mais en général ces événements à portée économique sont à tel point imbriqués avec des mouvements socio-politiques qu'il semble difficile d'en attribuer la cause à un facteur unique : confrontations menant à la destruction des cités anatoliennes, dilution des systèmes centralisés (Egypte) ou en marche vers une forme de centralisation (Sud Ibérique), ruptures dans les réseaux d'échanges (Egée). On envisage donc plutôt dans la majorité des cas des causes multifactorielles, interactives, les difficultés économiques engendrées par les perturbations climatiques se combinant avec des situations sociales pour aboutir à l'effondrement d'entités jusque là prospères et, en apparence, solides.

Conclusions

S'agissant des effets respectifs dus aux 8200 Cal BP et 4 200 cal BP Events, on ne saurait mettre en parallèle le contexte social de la Méditerranée caractérisant chacun de ces deux épisodes. Au VIIe millénaire avant notre ère, le Proche-Orient est une aire de villages, parfois de forte taille (Ain Ghazal : 15 hectares, Çatal Huyuk : 12 hectares), le Sud-est européen est le siège de petites localités villageoises tandis que toute l'autre moitié de la Méditerranée, à l'Ouest, est parcourue par des bandes réduites de chasseurs-cueilleurs. L'événement 8200 BP entrainera la dilution des plus grands établissements d'Orient et une reconfiguration vers des communautés de taille plus restreinte. A l'Ouest, la déstabilisation des chasseurs les rendra souvent peu « lisibles » pour l'archéologue parti sur leurs traces (les « hiatus »). Les mésolithiques ne disparaitront certes pas puisque certains de leurs marqueurs génétiques viendront, quelques siècles après, se mêler à ceux des nouveaux venus néolithiques. Il est donc certain que, selon des amplitudes variées, les sociétés de l'ensemble du champ méditerranéen ont été affectées par l'épisode climatique.

L'impact de l'événement survenu quatre millénaires après, orienté cette fois vers des tendances arides, n'est pas comparable. Il se manifeste en effet au sein de sociétés que le degré de complexité et de hiérarchisation interne, en liaison avec un niveau technique sophistiqué, rendent a priori moins vulnérables : centralisme étatique en Egypte, urbanisation ou sub-urbanisation du Levant à l'Egée, brillantes cultures chalcolithiques de Malte au Portugal. Par ailleurs le contexte démographique, des capitales orientales jusqu'aux méga-sites sud-ibériques, n'a plus rien à voir avec les faibles densités du VIIe millénaire. Partout la production agricole est fortement sollicitée pour satisfaire les besoins en pleine croissance des effectifs de population. Le pic d'aridité de la fin du IIIe millénaire a donc eu des incidences certaines sur l'économie. Mais les déstabilisations alors survenues ont pu avoir aussi d'autres facteurs : guerres et mouvements de contestation socio-politiques des élites en place notamment. Déjà, en Occident, les cultures chalcolithiques étaient secouées dans leurs structures mêmes par l'idéologie campaniforme. De sorte que ces divers facteurs, climatiques, économiques, politiques, sociaux ont pu se combiner, interagir pour, par effet domino, entrainer des remises en question sensibles à l'échelle de la Méditerranée entière. Pourtant une forme de résilience n'entrainera pas la disparition des populations elles-mêmes et les hiatus sont ici moins lisibles que pour l'événement 8200 BP. Si les élites font les frais de tels bouleversements, les héritages dans la culture matérielle au sein du Bronze ancien signent souvent des formes de continuité. Ainsi les traditions campaniformes sont-elles manifestes dans le Bonnanaro sarde, les épicampaniformes du Midi ou dans l'Argarique. Les redéfinitions qui interviennent dans les réseaux de circulation, les réorientations économiques suggèrent des initiatives dont certaines impliquent des mouvements d'individus (par exemple la technique métallurgique des premiers alliages). Au fond l'un des effets majeurs des perturbations de la fin du IIIe millénaire réside dans la contestation des centralismes qu'il soit étatique comme en Egypte ou à un stade inférieur comme dans les sociétés chalcolithiques du Sud ibérique. Mais la « crise » passée, ces centralismes et le pouvoir de dominants ne tarderont pas à refaire surface.

Tableau 1. Evolution chrono-culturelle du Néolithique au sein des principales îles de la Méditerranée avec indication des crises climatiques. On constate l'existence de hiatus archéologiques paraissant en liaison avec l'épisode 8200 BP ainsi que les recompositions culturelles intervenant après l'épisode 4200 BP.

BC	CYPRUS	CRETE	MALTA	SARDINIA/CORSICA	BALEARIC ISLANDS	
2000	MIDDLE BRONZE AGE	MIDDLE MINOAN	TARXIEN CEMETERY	BONNANARO	CAMPANIFORME	
	EARLY BRONZE AGE					4200 BP EVENT
3000	CHALCOLITHIC	EARLY MINOAN	TEMPLE CULTURE	BEAKERS / CHALCOLITHIC / MONTE CLARO / ABEALZU-FILIGOSA	NEOLITHIC	
4000	NEOLITHIC (SOTIRA)	LATE / NEOLITHIC	ZEBBUG / ?	OZIERI / S. CIRIACO / B. IGHINU / NEOLITHIC / FILIESTRU / EL GIGLIO	?	
5000		MIDDLE	SKORBA / NEOLITHIC			
	?		G. DALAM		?	
6000		?		?		8200 BP EVENT
	ACERAMIC NEOLITHIC (KHIROKITIA)	ACERAMIC NEOLITHIC	?			
7000				MESOLITHIC	?	
8000	PRE-POTTERY NEOLITHIC / PPNB	MESOLITHIC ?	?		?	
9000	PPNA					

Chrono-cultural evolution of the Neolithic within the main islands of the Mediterranean with indication of the climatic crises. We note the existence of archaeological hiatuses appearing in connection with the 8200 BP event as well as the cultural changes occurring after the 4200 BP event.

Bibliographie

Bar Yosef O. 2006. L'impact des changements climatiques du Dryas récent et de l'Holocène inférieur sur les sociétés de chasseurs-cueilleurs et d'agriculteurs du Proche-Orient. *In*: E. Bard (dir) : *L'Homme face au climat*, Paris, O. Jacob, p.283-301.

Berger J.-F. 2009. Les changements climato-environnementaux de l'Holocène ancien et la Néolithisation du bassin méditerranéen. *In*: J.-P. Demoule (dir) : *La Révolution néolithique dans le monde*, Paris, CNRS Editions, p.121-144.

Berger J.-F., Guilaine J. 2009. The 8200 cal BP abrupt environmental change and the Neolithic Transition: a Mediterranean Perspective, *Quaternary International*, 200, p.33-49.

Berger J.-F., Metallinou G., Guilaine J. 2014. Vers une révision de la transition méso-néolithique sur le site de Sidari (Corfou, Grèce). Nouvelles données géoarchéologiques et radiocarbone, évaluation des processus post-dépositionnels. *In:* C. Manen, T. Perrin, J. Guilaine (dirs) : *La Transition Néolithique en Méditerranée*, Errance/Archives d'Ecologie Préhistorique, Arles et Toulouse, p.213-232.

Berger J.-F., Shennan S., Woodbridge J., Pamisaro A., Mazier F., Nuninger C., Guillon S., Doyen E., Begeot C., Andrieu-Poncel V., Azuara J., Bevan A., Fyfe R., Neil Roberts C., 2019. Holocene land cover and population dynamics in Southern France, *The Holocene*, doi : 10. 1177/ 0959683619826698 journals.sagpub.com/home/hol

Bini M., Zanchetta G. *et al.* 2019. The 4.2 ka BP Event in the Mediterranean Region: an overview, *Climate of the Past*, 15 (2), p.555-577.

Cesari J., Courtaud P., Leandri F., Perrin T., Manen C. 2014. Le site de CampuStefanu (Sollacaro, Corse-du-Sud). Une occupation du Mésolithique et du Néolithique ancien dans le contexte corso-sarde. *In:* C. Manen, T. Perrin, J. Guilaine (dirs) : *La Transition Néolithique en Méditerranée*. Errance/Archives d'Ecologie Préhistorique, Arles et Toulouse, p.283-305.

Collina C. 2015. *Le Néolithique ancien en Italie du Sud*, Archaeopress Archaeology, Oxford.

Daune-Le Brun O., Hourani F., Le Brun A. 2017. Khirokitia (Chypre, VII, VIe millénaires avant J.C.), la séquence stratigraphique dans son contexte. *In:* J.-D. Vigne, F. Briois, M. Tengberg (dirs) : *Nouvelles données sur les débuts du Néolithique à Chypre*, Actes de la séance de la Société Préhistorique Française (2015), p.217-228.

Dever W.G. 1989. The collapse of the Urban Early Bronze Age in Palestine: towards a systemic analysis. *In:* P. de Miroschedji (dir): *L'Urbanisation de la Palestine à l'âge du Bronze ancient*, BAR International Series 527, Oxford, p.225-246.

Efstratiou N. 2013 Climate changes and social responses in the early prehistory of Greece: a complex and dynamic Interaction. *In:* E. Starnini (dir): *Unconformist Archaeology. Papers in honour of Paolo Biagi*, BAR International series 2528, p.27-34.

Efstratiou N., Karetsou A., Banou E.S., Margomenou D., 2004. Neolithic Settlement at Knossos : New Light on an Old Picture. *In:* G.Cadogan, E.Hatzaki, A.Vasilakis (dirs) : *Knossos : Palace, City, State*, British School at Athens, London, 12, p.39-49.

Evans J. 1964. Excavations at the Neolithic Settlement of Knossos 1957-1960. *Annual of the British School at Athens*, 59, p.132-230.

Evans J. 1968. Knossos Neolithic: II. Summary and Conclusions, *Annual of the British School at Athens,* 63, p.239-276.

Garfinkel J. 2004. « Néolithique » et « Enéolithique » Byblos in Southern Levantine Context. *In:* E. Peltenburg and A. Wasse (dirs) : *Neolithic Revolution*, Oxbow Books, Oxford, p.175-188.

Guilaine J. 1996. La Néolithisation de la Méditerranée occidentale. *In: The Neolithic in the Near East and Europe*, Colloquium XVII, XIII International Congress of Prehistoric and Protohistoric Sciences, abaco edizioni, Forli, p.53-68.

Guilaine J. 2018. A Personal view of the Neolithization of the Western Mediterranean, *Quaternary International*, 470, mars, p.211-225.

Guilaine J. 2013. The Neolithic Transition in Europe: some comments on gaps, contacts, arrythmic model, genetics. *In:* E. Starnini (dir) : *Unconformist Archaeology. Papers in honour of Paolo Biagi*, International series 2528, Oxford, p.55-64.

Guilaine J., Radi G., Angeli L. 2019. La néolithisation de l'Italie du Sud-Est, *Eurasian Prehistory*, 15 (1-2), p.101-144.

Hinz M., Schirrmacher J., Kneisel J., Rinne C., Weinelt 2019. The Chalcolithic-Bronze Age transition in Southern Iberia under the influence of the 4.2 Ka BP event? *Journal of Neolithic Archaeology,* 6 dec 2019. doi : 10.12766/jna.2019.1

Horejs B., Milíc B., Ostmann F., Thaneiser U., Weninger B., Galik A. 2015. The Aegean in the Early 7e millenium BC: Maritime Networks and Colonization, *Journal of World Prehistory*, 28, p.289-330.

Huot J.-L. 2004. *Une archéologie des peuples du Proche-Orient. I. Des premiers villageois aux peuples des cités-Etats (Xe-IIIe millénaire av. J.C.), Paris,* Errance.

Luglié C. 2018. Your path led through the Sea... The Emergence of Neolithic in Sardinia and Corsica, *Quaternary International*, 470, mars, p.285-300.

Magny M., Vanniere B., Zanchetta G., Fouache E., Touchais G., Petrika L., Coussot C., Walter-Somonnet A.V., Arnaud F. 2009. Possible complexity of the climatic event around 4300-3800 cal BP in the Central and Western Mediterranean*, The Holocene*, 19, 6, p.823-833.

Meller H., Arz H.W., Jung R., Risch R. (dirs) 2015. *A Climatic Breakdown as a cause for the collapse of Old World ?* , Landesmuseum fur Vorgeschichte, Halle, 12/1.

Natali E., Forgia V. 2018. The Beginning of the Neolithic in Southern Italy and Sicily. *Quaternary International*, 470, mars, p.253-269.

Perlès C. 2001. *The Early Neolithic in Greece*, Cambridge University Press.

Renfrew C. 1972. *The Emergence of Civilisation. The Cyclades and the Aegean in the Third Millenium BC*, Methuen, London.

Rosen A.M. 1989. Environmental Change at the end of Early Bronze Age Palestine. *In:* P. de Miroschedji (dir): *L'Urbanisation de la Palestine à l'âge du Bronze ancien*, BAR international Series 527, Oxford, p.247-255.

Simmons A. 2000. Villages on the Edge. Regional Settlement Change and the End of the Levantine Pre-Pottery Neolithic. *In:* I. Kuijt (dir) : *Life in Neolithic Farming Communities*, New-York, Kluwer Academy/ Plenum Publishers, p.211-230.

Todd I. 2003. Kalavasos-Tenta : a Reappraisal. *In:* J. Guilaine et A. Le Brun (dirs) : *Le Néolithique de Chypre,* Supplément 43 au Bulletin de Correspondance Hellénique, Ecole française d'Athènes, p.35-44.

Weninger B., Alram-Stern E., Bauer E., Clare L., Danzeglocke U., Jöris O., Kubatzki C., Rollefson G., Todorova H. 2006. Climate forcing due to the 8200 cal yr BP event observed at Early Neolithic sites in the Eastern Mediterranean , *Quaternary Research*, 66, p.401-420.

Cultural adaptations in Libya From Upper Pleistocene to Early Holocene – Chronology and Stratigraphy from littoral to desert

Barbara E. Barich[1]

Abstract

Within the huge Saharan territory, Libya represents a privileged observatory of the environmental changes that have affected human occupation over millennia, both in the Pleistocene and in the Holocene. The available data also show the presence of different climatic trends between the southern Libyan territory in the Saharan area, and the coast overlooking the Mediterranean Sea. This paper considers the most recent data coming from the three main geographical and cultural contexts of Libya. First it describes the sequence reconstructed for the Tripolitanian plateau – the Jebel Gharbi – by the Sapienza University of Rome, and then compares it with the sequence from Haua Fteah, in the Jebel Akhdar, Cyrenaica, on the basis of the revision recently made by the Cambridge University's investigation.

The comparison of these two sequences, respectively at the western and eastern ends of the Libyan coastal belt and separated from each other by the Gulf of Sirte, revealed significant concordances in climatic alternations, and in the dating of human presence here starting from MIS 5 with generalized MSA Levallois technology. Differences between the two regions are represented by the presence/absence of Aterian and Dabban industries, which can generally be associated with the MIS 4 and MIS 3 phases of climatic instability and variability. Evidence of the Aterian, found in the Jebel Gharbi, seems lacking from Haua Fteah, while for the Dabban we have an inverse situation with significant presence at Haua and absence in the western territory. The concordances between the two regions become most evident in the final Pleistocene and early Holocene with the Later Stone Age occupation, characterized by microlithic industry. This last phase allows us to extend the observation to the Saharan region, the Tadrart Akakus massif in particular, which has also been the subject of a long investigation by the Sapienza University of Rome. This region, which had experienced a long phase of abandonment due to the recurrent arid phases of the Pleistocene (MIS 4 and MIS 2), was reoccupied at the beginning of the Holocene (ca. 11,500 cal. BP) with the re-establishment of the monsoon circulation, and witnessed an important cultural flowering. It is evident that the reoccupation of the Saharan area is due to successive arrivals of groups of different provenance. However, two main paths can be identified: a movement of

1   International Association for Mediterranean and Oriental Studies (ISMEO) and University of Rome Foundation

populations from the north towards the previously deserted areas can be detected; an alternative idea is to focus on the movement of groups from the Saharo-Sudanese belt from the east (or southeast). Since Neolithic times southern Libya has turned to the Sahara and sub-Sahara, thus prefiguring the characteristics of modern Libya.

Résumé

Au sein de l'immense territoire saharien, la Libye représente un observatoire privilégié des changements environnementaux qui ont affecté l'occupation humaine au cours des millénaires tant au Pléistocène qu'à l'Holocène. Les données disponibles montrent également des tendances climatiques différentes entre le sud du territoire libyen dans la zone saharienne, et la côte bordant la mer Méditerranée. Le présent article examine les données les plus récentes provenant des trois principaux contextes géographiques et culturels de la Libye. Tout d'abord, il décrit la séquence reconstruite pour le plateau tripolitain - le Jebel Gharbi - par les recherches de l'Université Sapienza de Rome et, ensuite, il la compare avec la séquence de Haua Fteah, dans le Jebel Akhdar en Cyrénaïque, sur la base de la révision faite ces dernières années par les recherches de l'Université de Cambridge.

La comparaison de ces deux séquences, aux extrémités occidentale et orientale de la ceinture côtière libyenne, séparées l'une de l'autre par le golfe de Syrte, a révélé des concordances significatives dans les alternances climatiques et dans la datation de la présence humaine qui se développe à partir du Stade Isotopique Marin (SIM) 5 avec une technologie Levallois MSA généralisée. Les différences entre les deux régions concernent la présence/absence d'industries atériennes et dabéennes, qui peuvent généralement être associées aux SIM 4 et SIM 3 d'instabilité et de variabilité climatiques. Les preuves de la présence de l'Atérien dans le Jebel Gharbi semblent être absentes de Haua Fteah, tandis que pour le Dabéen nous avons une situation inverse, avec une présence significative à Haua Fteah et une absence dans le territoire occidental. Les concordances entre les deux régions sont plus évidentes en ce qui concerne l'occupation de la Later Stone Age, à la fin du Pléistocène et au début de l'Holocène, caractérisée par une industrie microlithique. Cette dernière phase permet notamment d'étendre l'observation à la région saharienne, le massif de la Tadrart Akakus en particulier, qui a fait l'objet d'une longue investigation par l'Université Sapienza de Rome. Cette région, qui a connu une longue phase d'abandon due aux phases arides récurrentes du Pléistocène (SIM 4 et SIM 2), est réoccupée au début de l'Holocène (ca. 11 500 cal. BP) avec le rétablissement de la mousson et connaît une importante floraison culturelle. Il est évident que la réoccupation de la zone saharienne est due à des arrivées successives de groupes d'origines différentes. Cependant, deux voies principales peuvent être identifiées. D'une part, on ne peut exclure qu'il y ait eu un mouvement de populations du Nord vers les zones précédemment désertées. Une autre hypothèse, en revanche, se tourne vers l'Est (ou le Sud-est), en mettant l'accent sur le déplacement de groupes en provenance de la ceinture saharo-soudanaise.

Introduction

During the Pleistocene, changes in the degree of terrestrial insolation linked to astronomical phenomena had a strong impact on the African climate, particularly in the Sahara, determining the recurrent alternations of arid and wet cycles (Larrasoaña et al. 2013; de Menocal 2015; de Menocal et al. 2000).

The glacial phases in the northern hemisphere caused several alterations in the dynamics that regulated the climate of North Africa, forcing a southward shift of the ITCZ (Intertropical Convergence Zone) and blocking the normal activity of the monsoon (Crucifix 2019: 94). The Sahara experienced repeated stages of severe aridity with higher dust production, particularly drastic in the Upper and Final Pleistocene (during MIS 4: ca. 74-60 ka BP and MIS 2: ca. 24-18 ka BP)[2].The increase in summer insolation at the end of the Pleistocene favored the retreat of the northern hemisphere ice and the rise in global sea level, and the increased intensity of Nile River flooding (Revel et al. 2010).

Due to the size of the geographical area and the general lack of information from modern surveys in various regions, the paleoenvironmental picture of North Africa is uneven. Nonetheless, the available data enable us to recognize a difference in the climatic trend between the internal Sahara and the North African belt not far from the Mediterranean where the impact of the northern ice was generally less drastic, and does not seem to have interrupted human occupation (Giraudi 2005; Douka et al. 2014).Likewise, the transition to the Holocene shows different characteristics along the Mediterranean coast and in the internal territories of the Sahara.

This paper will summarize these different aspects focusing on some important and significant case studies from present-day Libya, in the littoral and pre-desert belt and to the southern region in the Sahara, between the Upper Pleistocene and the Early Holocene.

Saharan climate during the Upper and Final Pleistocene

Marine Isotopic Stage 5 (MIS 5, ca. 130-74 ka BP)[3] is known as a stage of humid climate favorable to the formation of large water basins. Lake deposits are reported from the tropical belt of the Sahara, corresponding to the southern part of present-day Egypt and Libya. At Bir Tarfawi / Bir Sahara in the Egyptian Western Desert (Wendorf et al. 1993) a succession of paleo-lakes has been recorded, dating to MIS 5, with later manifestations in

---

2 In the text, the dates that are part of the consolidated chronology on climatic cycles, cultural phases, etc., are cited as calibrated dates from the present (BP), indicating with the acronym ka (kilo years) the thousands. The dates obtained with the conventional radiocarbon method are instead indicated in full, associated with the Laboratory code.
3 Marine Isotopic Stages (MIS) datings are indicated following (Klein 1989: 354, Fig.7.4) and (Farr et al. 2014).

MIS 3 (Nicoll 2018). In the same Western Desert, there is also clear evidence of wet phases in the territory of Kharga Oasis, just over 200 km north of Bir Tarfawi. Here, deposits of tufas (carbonate spring formations) are reported (Smith *et al.* 2004, 2007). Tufas were created during the wet phases that led to the formation of swamps, marshes, ponds. Plant casts attest to the presence of *Ficus ingens, Ficus salicifolia*, and *Ficus sycomorus*, reeds (*Arundo* sp.), ferns (*Pteris vittata*), African hackberry (*Celtis integrifolia*). Uranium-series dates at 127.9 ka BP, 114 ka BP, 103 ka BP provide a chronological reference for both the moist phases and the associated MSA complexes (Smith *et al.* 2007: 692).

In Libya the Megafezzan Lake, in the central southern area of the Fezzan, reached a maximum extension of about 135 km² during the wet phases (here dated to 100 ka BP, 74 ka BP, 47 ka BP and 30 ka BP) (Armitage *et al.* 2007; Drake *et al.*2011). Remains of other smaller paleo-lakes have also been recorded in the same region, which appears to have been one of the richest lake areas in the central Sahara (Smith 2010). It is believed that the presence of these wetlands and the relative abundance of vegetation and fauna played an important role in the northward migrations of the modern human groups, favoring their dispersal 'Out of Africa' (Farr *et al.* 2014; Smith 2010; Van Peer, Vermeersch 2000).

The Saharan climate was increasingly arid from MIS 4 (ca. 74-60/59 ka BP) with partial recovery during MIS 3 (ca. 60/59-24 ka BP). Various authors believe that these difficult environmental conditions led to an abandonment of the Saharan territories leading to a true depopulation (Carto *et al.* 2009). However, in the absence of comparable data archives, the construction of a unitary palaeoclimatic context seems largely hypothetical. It seems more useful to examine regional differences that may have created favorable local conditions, contradicting the notions of a general trend towards aridity (Nicoll 2004). In Libya, Tadrart Akakus, the date 60 ka BP from Uan Tabu (Unit IV) seems to indicate the last possible date for human presence in the region, before its reoccupation in the Holocene. However, we know that elsewhere the Sahara experienced other favorable episodes even later, as we know for the Western Desert which continued to be inhabited longer.

At the end of the Last Glacial Maximum (LGM, MIS 2: ca. 24-18 ka BP) the North American and European glaciers reached their maximum. Due to the cold temperatures, the monsoon could not rise north every summer, and was weakened by the cooling of the Atlantic waters. Added to this is the fact that the mobile polar anticyclones blocked the Atlantic monsoons on the western side of North Africa. Recent palaeoclimatic reconstructions have shown that particularly arid climatic conditions reached the coastal strip of Tripolitania and Cyrenaica (Jebel Gharbi and Haua Fteah), unlike anything that had occurred there up to that moment (Giraudi 2005, 2009; Douka *et al.* 2014). However, as shown below, in these latter territories the archaeological sequences do not provide evidence for interruptions in human occupation, instead they show a continuity of local development.

The aridity of the Sahara during the LGM was therefore caused by a combination of interacting phenomena which created a dynamic (anticyclone) and thermal barrier on the oceans. Some areas (called 'refuge areas') such as the central Saharan massifs and the mountain slopes of the Atlas, may have experienced river activity that lasted longer. Finally, in the southern extremity of the Sahara, the desert would have expanded considerably, with a shift of the dunes 300-500 kilometers further south.A resumption of more favorable conditions in the Sahara is reported at the end of MIS 2 (ca.18 ka BP) with the onset of deglaciation in the northern hemisphere that halted abruptly around 13 ka BP due to the advent of the short Younger Dryas cold interstadial. With 11.5 ka BP, the formal date for the beginning of the Holocene, there was a definitive recovery of favorable environmental conditions.

Upper Pleistocene and Early Holocene in northern Libya

The Upper Pleistocene and Early Holocene prehistory of northern Libya has remained little known for a long time, especially the westernmost region (Figure 1). On the contrary, important data have been available since the 1950s for the northeastern Cyrenaican littoral with the Jebel Akhdar sites and the pre-desert. Haua Fteah in particular has stratigraphy that has allowed us to frame major human events, starting from MIS 5.Remains of anatomically modern humans are recorded at Haua Fteah presumably during MIS 4, at a date that the most recent research estimates between 73 and 65 ka BP (Hublin 2001; Douka *et al.* 2014: 59). The pioneering 1950s investigations at Haua Fteah, published in McBurney's monograph (McBurney 1967), were unfortunately followed by a long halt due to problems in the region and an inability to return to the site for military reasons. Only since the first half of the 2000s has research in Cyrenaica been resumed thanks to the Cyrenaican Prehistoric Project (CPP) of the University of Cambridge. Through a connected series of studies this multidisciplinary project produced an important chronostratigraphic revision of the McBurney sequence (Barker *et al.* 2008, 2009, 2010).

Figure 1. Map of Libya showing the regions cited in the text.

On the other hand, apart from brief reports by McBurney himself (McBurney, Hey 1955), the northwestern part of Libya, west of the Gulf of Sirte up to the Tunisian border, remained practically unknown until the implementation of the Italian-Libyan Joint Project in the Jebel Gharbi. This project was initiated in 1990 by the Sapienza University of Rome (P.I. B.E. Barich), at the invitation of the Department of Antiquities of Libya (DoA). It is currently suspended due to the difficult situation that Libya country has been experiencing since 2011.

New data from Jebel Akhdar, in the east of the Libyan region, provides an opportunity to compare the two occupation sequences, which show remarkable similarities particularly for the Upper Later Stone Age aspect and the subsequent appearance of the first Neolithic sites.

Jebel Gharbi – Tripolitania

The Jebel Gharbi includes the highest part of the Tripolitanian Plateau and marks the transition between the plateau and the southern part of the coastal plain, the Jifarāh (Figure 2). Cave sites are very rare in this region, compared to open air sites which are very numerous and indicate the intensity of human occupation in the past. These open air sites are the remains of temporary camps almost always near springs or wadi courses. The regional survey model applied by the Italian-Libyan Joint Project initially had the two Ghan and Ain Zargha wadis as the main reference points, to cover later the entire region of the concession, from Gharian to Nalut. Over 100 sites were investigated and inventoried until the project was suspended. Regarding survey methods, excavation tests, technological and bio-archaeological studies, I refer to the numerous publications of the

Figure 2. Panoramic view of the Jebel Gharbi and Jifarāh plain (photo Italian-Libyan Joint Project in the Jebel Gharbi).

multidisciplinary research team (Barich 1995, 2009, 2014; Barich *et al.* 1995, 2003a, 2003b, 2006, 2010, 2015; Barich, Conati Barbaro 2003; Barich, Garcea 2008; Garcea 2009, 2010a and b; Garcea, Giraudi 2006; Giraudi 2005, 2009; Giraudi *et al.* 2012; Lucarini 2009; Lucarini, Mutri 2010a; Mutri 2008-2009; Mutri, Lucarini 2008; Barca, Mutri 2009; Alhaique, Marshall 2009).

The geomorphological and paleoenvironmental investigation (conducted by Carlo Giraudi, Enea, Rome) led to the recognition of five main landscape units (hereafter LU) (Giraudi in Barich *et al.* 2006), to which reference was made during surveys. The most densely occupied area appeared to be LU 2, between the mountain and the Jifarāh plain, where continuity of occupation was observed even through phases generally unfavorable for human settlement in North Africa. The application of radiometric dating techniques (^{14}C, U/Th, OSL) allowed the establishment of a well-founded absolute chronology (Tables 1 and 2), within which we can place the phases of occupation highlighted during the fieldwork.

The Jebel's occupation sequence was reconstructed by comparing and integrating key geo-stratigraphies identified in the three main research areas: Wadi Ghan (in the Gharian territory, to south-east), Jado area (inside the mountain at approximately 650-700 m asl) and Jifarāh's southern territory.

OSL dates cover the earliest phases of occupation even prior to MIS 5 (>150-74 ka BP) and MIS 4 (74 - 60/59 ka BP) (Table 1), providing a chronological framework for the earliest records of human occupation in the Jebel: the Early Stone Age (Acheulian) materials - bifaces and spheroids - found on the alluvial terraces of the Wadi Ginnaun, an Ain Zargha tributary (Hosha Ginnaun SJ-90-16 site) (Barich *et al.* 1995). In addition, there are numerous Early Middle Stone Age (MSA) / Levallois assemblages, collected at Wadi Ghan (Sites SJ-90-18, SJ-99-40), at Hosha Ginnaun (Sites SJ-90-15, SJ-98-30), in

Table 1. Jebel Gharbi, Libya. U/Th and OSL datings of Early Middle Stone Age.

	LOCALITY	SITE	CAL. YEARS BP	METHOD	DATED SAMPLE	LAB NUMBER
Cultural Unit						
Early MSA	Ain Zargha	27 N-trench	64,000±21,000	U/Th	Calcrete	Libia 27E
	Ain Zargha	27 N-trench	<60,000	U/Th	Calcrete	Libia 27E
	Ras El Wadi	LIB08-02 (SJ-90-12)	86,000±14,000	OSL	Sand	X3411
	Ras El Wadi	LIB08-01 (below the Aterian)	110,000±22,000	0SL	Sand	X3410
	Ras El Wadi	SJ-98-27 (near)	114,700±7400	OSL	Soil	X1512
	Ras El Wadi	SJ-98-27B (below the Aterian)	146,800±11,000	OSL	Sand	X1514
	Jado Sud		>150,000	OSL	Sand	X1515

Table 2. Jebel Gharbi, Libya. Chronology of the main phases of occupation from Aterian to Final Later Stone Age: 2-sigma calibration by run with the IntCal20 (Reimer et al. 2020, The IntCal20 Northern Hemisphere radiocarbon age calibration curve (0-55 cal kBP), *Radiocarbon* 62, p.725-757). doi: 10.1017/RDC.2020.41.

Cultural Unit	LOCALITY	SITE	YEARS bp	CAL. YEARS BP	METHOD	DATED SAMPLES	LAB NUMBER
Final LSA	Jawsh	SJ-06-87 (LIB08-03)		8200±1200	OSL	sand	X3412
	Jawsh, Auenat Dagher	SJ-03-75	8210±110	9350-9020	AMS	Charcoal	GdA-1501
	Wadi Ghan	SG-99-41	11,110±40	13,104-12,917	AMS	Organic sediment	Beta-157690
	Shakshuk East	SJ-00-55 East Hearth	11,360±55	13,338-13,158	Standard C14	Charcoal	Poz-215
	Shakshuk West	SJ-00-55 West (Test 1)	11,620±70	13,850-13,410	AMS	Charcoal	Beta-167096
	Shakshuk East	SJ-00-55 East	11,570±40	13,820-13,630	AMS	Charred roots	Beta-185498
	Shakshuk East	SJ-00-55 East	11,690±40	13,850-13,460	AMS	Charred roots	Beta-185499
	Ras El Wadi	Ras El Wadi Sez.27 LIB08-07		12,400±1400	OSL	Sand	X3416
	El Bedr- El Batn locality	SJ-03-83 Concentration 2	12,490±70	14,965-14,175	Standard C14	Ostrich egg shell	Gd-11988
Upper LSA	Wadi Ghan	SG-99-41	14,820±60	18,500-17,700	Standard C14	Charcoal	Poz-214
	Shakshuk	SJ-00-56	16,750±60	20,140-19,510	AMS	Charred material	Beta-157689
	Ras El Wadi (Jado)	SJ-90-12	18,020±190	21,610-20,090	Standard C14	Carbonate sediment	Beta-154575
	Wadi Bazina	Lacustrine series	18,760±50	22,740-21,820	AMS	Organic sediment	Beta-154554

	Site				Method	Material	Lab code
Lower LSA	Shakshuk East	Ain Soda area	24,620±400	29,780-27,894	AMS	Charcoal	Beta-167094
	Shakshuk West	SJ-00-55 West (Test 2, -106 cm)	24,740±140	29,200-28,724	AMS	Charred material	Beta-157687
	Shakshuk	SJ-00-56 Extension 2	25,410±150	30,000-29,257	AMS	Organic sediment	Beta-185497
	Shakshuk West	SJ-00-55 West (Test 2, -17 cm)	25,500±400	30,832-28,775	AMS	Charred material	Beta-167099
	Wadi Bazina	Lacustrine series	26,330±80	30,871-30,323	AMS	Organic sediment	Beta-154555
	Jado	SJ-98-12 (formerly SJ-90-12)	27,310±320	31,908-30,936	Standard C14	Carbonate sediment	Beta-154576
	Shakshuk	SJ-00-56 Excavation (hearth M/0)	27,800±430	33,043-31,116	Standard C14	Charcoal	GdA-196
	Ain Zargha	27 N-Trench		30,000±9000	U/Th	Calcrete	Libia 27B1
	Shakshuk West	SJ-00-55 West (Test 2, -64 cm)	30,870±200	35,674-34,675	AMS	Organic sediment	Beta-157688
MSA-Aterian	Shakshuk West	near Site SJ-00-56		49,200±3500	OSL	Sand	X1511
	Shakshuk (East)	Geological profile	43,530±2110	50,570-42,635	AMS	Charred material	Beta-167098
	Wadi Sel	SJ-02-68	44,600±2430	52,488-42,917	AMS	Organic sediment	Beta-167097

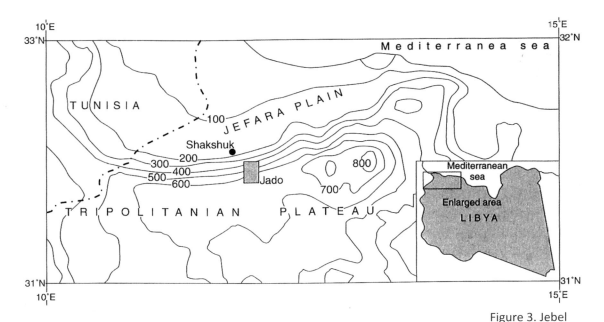

Figure 3. Jebel Gharbi, Libya. Location map with the Ras el Wadi (Jado) and Shakshuk areas.

the territory of Jado (Sites SJ-98-25, SJ-98-27B), and, finally, at El Jawsh (Site SJ-06-86). In the Wadi Ghan geo-stratigraphy these industries are recorded in the 1st terrace, 20 m above the wadi bottom, within the accumulation of boulders and pebbles (Giraudi 2005: 164, Fig.2). In the territory of Jado, they were collected at the base of the Ras el Wadi-Ain Zargha sedimentary series illustrated below, within alluvial deposits, soils and calcareous crusts, indicating moderate activity of the wadi in presence of the variable, mainly hot and humid, MIS 5 climate (Barich *et al.* 2003a : 259-260).

For the periods following MIS 5, the reconstruction of the population sequence in Jebel Gharbi was based on the wet phases, recognized by the presence of lake sediments, soils, fine silty floods and, conversely, absence of wind deflation (Giraudi 2009: 408). Of particular importance were stable ^{13}C and ^{18}O isotopes, and mineralogical (X-ray diffraction) analyses, on calcareous and pedogenetic carbonate crusts formed from rising groundwater under sub-humid to semi-arid conditions in relatively arid soils, with herbaceous vegetation or mixed herbs-shrubs (Bodrato in Barich *et al.* 2003a: 264-265).

Five wet phases have been suggested by Giraudi (Giraudi 2009: Fig.3)[4] as a paleoenvironmental reference for the human occupation of the Jebel, starting from 60-50 ka BP. From the oldest to the most recent they are:
-    Phase 5, between ca. 50 ka BP and 46 ka BP (MIS 3) with a moderately humid climate, is associated with the Aterian industry whose presence in the Jebel is estimated from 60 to 40 ka BP;

---

4   The chronology of the wet phases, reported here with calibrated dates BP, in Giraudi 2009 is instead reported as uncalibrated bp.

- Phase 4, between ca. 31.8 ka BP and 28.8 ka BP with cold-humid climate that precedes the beginnings of the Last Glacial Maximum (LGM, MIS 2), is associated with Lower Later Stone Age (LSA) industry;
- Phase 3, between ca. 22.5 ka BP and 20 ka BP with cold-humid climate of the LGM, corresponds to the first appearance of the Upper Later Stone Age;
- Phase 2, at approximately 13.5-12.5 ka BP, is present during the start of the deglaciation prior to the Younger Dryas interstadial;
- Phase 1, between ca. 8 ka BP and 5.6 ka BP represents the mid-Holocene wet phase.

Aterian

This series of climatic phases, and the human occupations associated with them, are identified in two sequences reported here in detail (Figure 3). The first, within the plateau, is present at Ras el Wadi, Jado, an area near the headwater of the Wadi Ain Zargha whose valley represents an important communication route between the mountain and the Jifarāh plain (Barich *et al.* 2003a) (Figure 4a). In this sequence the unit with Early Middle Stone Age materials, already mentioned above, is covered with red aeolian sands, devoid of materials, attesting to a period of extreme dryness (probably related to MIS 4) creating a break in the sequence (Figure 4b). These sands are in turn covered by colluvial mud interbedded with calcareous crusts within which Aterian artifacts are contained. The absence of erosion surfaces attests that the sediments were deposited during a period of weak geomorphological activity and presence of rain (i.e. Wet Phase 5). The formation containing the Aterian phase is covered by a crust indicating increased humidity and evaporation, dated 27,310 ± 310/31,908-30,936 cal. BP (Beta-154576), confirmed by another of approx. 30 ka BP, from the same area (27N - trench) (U/Th date on calcareous crust, Sample Libia 27B1).Both of these dates are an important reference for the chronology of the Aterian. Above the Aterian, a palaeosol containing Lower Later Stone Age artifacts formed on inclined slopes, attests to the presence of a humid climate (Wet Phase 4). It is accompanied by widespread vegetation that allowed its stabilization (Giraudi 2009). At the top there is a final palaeosol, deflated, from which the Upper Later Stone Age (or Epipalaeolithic) assemblages were collected.

The second sequence at the modern center of Shakshuk, at the intersection between the Jebel and the Jifarāh plain (LU 2), illustrates all the periods of occupation that we have discussed, from the Aterian, to the Lower and Upper LSA, and up to the development of the first Neolithic centers in the Holocene (Barich *et al.* 2006; Barich, Garcea 2008). In a strip of land at the foothills of Jebel Gharbi, active faults resulted in ground displacement during earthquakes of great magnitude, opening outlets for the underground drainage network (Barich, Garcea 2008, p.93). The poor coverage of alluvial sediments then facilitated the emergence of ground waters through lines of small springs. The presence of water, associated with vegetation cover (as evidenced by the geomorphological stability of the slopes), made this area desirable to hunting societies who could find their prey here in several periods. In the geo-stratigraphic section identified at Shakshuk there are three units made up of silty sands interleaved with alluvial gravels (Figure 5). The oldest of these units, referred to as E1 (Giraudi 2005), contains Aterian artifacts at the bottom, along with a dark-gray soil

Figure 4. Jebel Gharbi, Libya. a: view of the Wadi Ain Zargha; b: the Ras el Wadi stratigraphic section (Italian-Libyan Joint Project in the Jebel Gharbi).

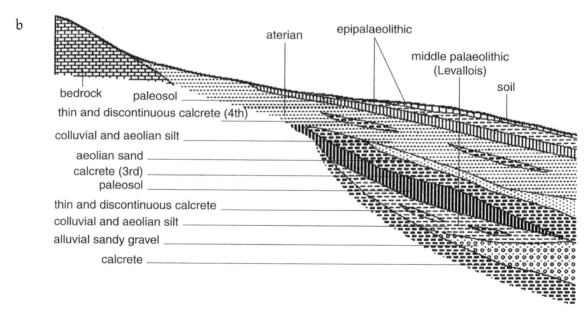

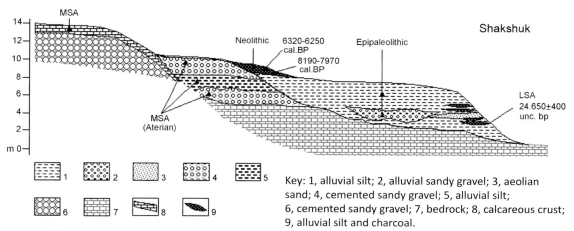

Key: 1, alluvial silt; 2, alluvial sandy gravel; 3, aeolian sand; 4, cemented sandy gravel; 5, alluvial silt; 6, cemented sandy gravel; 7, bedrock; 8, calcareous crust; 9, alluvial silt and charcoal.

Figure 5. Jebel Gharbi-Jifarāh, Libya. The Shakshuk stratigraphic section (redrawn from Barich 2014).

dated by conventional ^{14}C to 43,530 ± 2110 / 50,570-42,635 cal BP (Beta-167098). Another comparable date (also with a high standard deviation because it is close to the coverage limit of the ^{14}C method) comes from site SJ-02-68 (44,600 ± 2430/52,488-42,917 cal.BP, median 47,700, Beta-167097). Finally an OSL date, 49,200 ± 3500 BP (X1511), comes from the sands below the Upper LSA levels, near Site SJ-00-56 (Table 2). All these dates for the Aterian industry are all related to Giraudi's Wet Phase 5.

In Jebel Gharbi, the Aterian falls entirely within MIS 3, and while the beginning can be pushed back to >60 ka BP (Giraudi 2005), it is difficult to establish its end with certainty. This could be set in relation to a cold interval (perhaps around 44-43 ka BP), but the occupation could also have lasted longer, giving rise to technological innovations, up to ca. 30 ka BP. During the Aterian occupation the climate was quite moderate with a certain increase in humidity and precipitation. The basic economy was hunting, and complex tools (truncations, endscrapers, burins, notches and denticulates) (Figure 6), apart from the marker elements represented by tanged points and foliates, were used. The technological study of these artifacts (Spinapolice, Garcea 2013) has highlighted the presence of the classical Levallois core reduction sequence which is frequently associated with the Taramsa method, known from complexes of the Nile Valley (Van Peer *et al.* 2010). In various cases the presence of an alternative method was also observed, apparently aimed at obtaining blades using single or double platforms (Spinapolice, Garcea cit.) which in some ways anticipates the LSA core reduction method.

The Aterian industry appears to be widespread from the Jifarāh (LU 2) to the interior of the mountain, where surface assemblages at Ras el Wadi-Ain Zargha have been collected from open spaces and hill slopes between 650 and 700 m. asl. Here, three Aterian sites (SJ-90-12, SJ-98-27A, SJ-98-28) have yielded important collections with a high density of artifacts per square meter. As these are surface assemblages, they are probably the product of multiple occupations and visits to this locality which was particularly

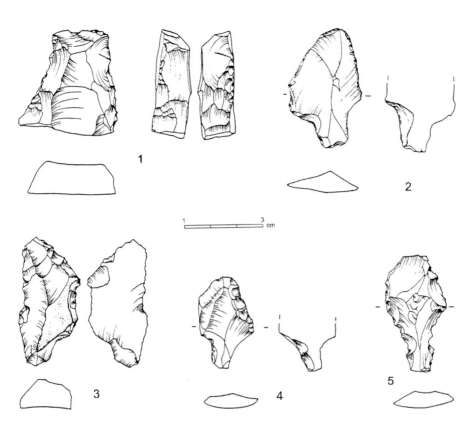

Figure 6. Jebel Gharbi, Libya. Aterian stone industry (Italian-Libyan Joint Project in the Jebel Gharbi).

favorable for hunting activities (Barich *et al.* 2003a). During the Aterian occupation, the groups' movements/exchanges seem to be oriented within the mountain. The lithic products are obtained exclusively from chert and flint that can be extracted directly from the limestone rocks on both banks of the river (Barca, Mutri 2009). Even quartzite, which was used in some assemblages, can be found locally. Sites are preferably located near water sources where the fauna used to gather. Unfortunately, no faunal remains have been preserved in association with the artifact collections. Based on comparisons with the layers of the Haua Fteah cave, specifically the Levalloiso-Mousterian ones (XXXIV-XXV), according to McBurney's (1967) definition, we can assume the presence of *Ammotragus lervia*, *Gazella dorcas*, Alcelafo, Equidae and, perhaps, also of *Rhinoceros* and wild ox (Higgs 1967; Klein, Scott 1986).

Later Stone Age

Near the top of the Ras el Wadi and Shakshuk stratigraphic sections there are soils and some organic levels that can be used as reference for a Lower Later Stone Age industry, with large blades but no microlithic elements. Conventional ^{14}C and U/Th dates place the beginnings of this LSA phase around 35-30 ka BP. In the Ras el Wadi section, Lower LSA artifacts are present in the paleosol that overlies the Aterian and is enclosed between two calcareous crusts dated respectively ca. 31.3 ka BP, the lower one, and ca. 20.5 ka BP the higher one. The lower limit is in relation to the beginning of the Wet Phase 4.

At Shakshuk West, thin charcoal-bearing beds are also associated with Lower Later Stone Age artifacts. Occupation is indicated by the organic soils of the SJ-00-55, Test 2 site [dated between 24,740 ± 140/29, 200-28,724 cal. BP (Beta-157687) and 30,870 ± 200/35,674-34,675 cal. BP (Beta-157688)]; from the base level, below the Upper LSA layers, of the site SJ-00-56 dated 25,410 ± 150/30,000-29,257 cal. BP (Beta-185497) and 27,800 ± 430/33,043-31,116 cal. BP (GdA-196); at Shakskuk East- Ain Soda, with a date 24,620 ± 400/29,780-27,894 cal.BP (Beta-167094). Therefore all artifacts are placed (by calibrated dating) between 35 ka BP and 29 ka BP (Table 2). Here there is also a situation of geomorphological stability connected to widespread vegetation due to a humid climate (Wet Phase 4).

The scarcity of the materials found in the aforementioned layers has prevented analytical definition of the technological characteristics and typology of this new Later Stone Age industry, with which the on-blade technique, first appears in the Jebel Gharbi, paving the way for the Upper LSA with bladelet microlithic production. In the absence of elements that allow it to be associated with the Dabban of Jebel Akhdar, this technique has been indicated as being Lower LSA (Barich, Garcea 2008: 90). It could be the product of on-site development at the end of the long Aterian occupation (> 60-40 ka BP), through the increase of blade technology, already present in the core reduction methods used by Aterian groups.

In the same Ras el Wadi (Site SJ-90-12) section, the Wet Phase 3, associated with the cold climate typical of the Last Glacial Maximum, is recognizable in the thin carbonate crust, overlying fine aeolian sediments, dated 18,020 ± 190 / 21,610-20,090 cal. BP (Beta-154575). Another date connected to the same cold episode was 18,760 ± 50 / 22,740-21,820 cal. BP (Beta-154554) obtained at Wadi Bazina, in a lake environment. Although the lake sediment itself represents a typical humid environment, it is the low average temperatures of the LGM and the low evaporation that produced a positive water balance (Giraudi 2009: 410). Both of these dates correlate to the cold climatic 'Heinrich events' recognized in the cores of the North Atlantic between 20,090-21,610 cal. BP (Barton et al. 2005).

Partly related to these events [(with its dating 16,750 ± 60 / 20,140-19,510 cal. BP (Beta-157689)], Shakshuk West's Site SJ-00-56 illustrates the Early Upper Later Stone Age aspect of Jebel Gharbi (Barich 2009; Mutri 2008-2009; Mutri, Lucarini 2008; Lucarini, Mutri 2010a). This site, on a bank of the Wadi Sel a few meters away from the Shakshuk spring (Figures 7a, b; 8), is an example of a hunting station of about 20 ka year ago, at which lithic artifacts were manufactured and used on the spot to dismember wildlife, in particular wild ass (*Equus africanus*), identified by archaeofaunal analysis (Alhaique, Marshall 2009).

All the materials collected belong to a clear Upper Later Stone Age horizon; the blade index is very high and the microliths are numerous (little cores, blades, bladelets and microbladelets), although bigger artifacts are also present. It is important to emphasize that no real backed products were recognized in the lithic assemblage (Mutri 2008-2009). The most commonly occurring raw material is a kind of purple flint coming from the alluvial sediments present on the bed of the wadi, although quartzite and quartz are also present. Among the faunal remains there are plentiful fragments often burned belonging to bovids and caprids, and numerous large teeth almost exclusively belonging to wild ass (*Equus africanus*). Faunal remains have provided concrete evidence for hunting practices, territorial awareness,

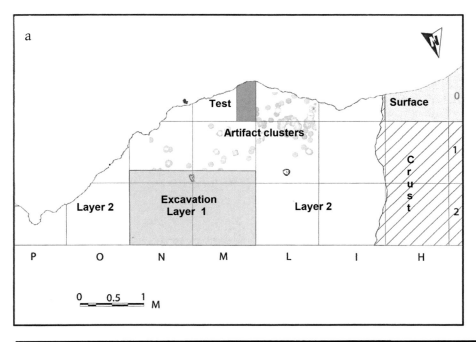

Figure 7. Jebel Gharbi- Jifarāh, Libya. a-b: plan and section of the Early Upper LSA Site SJ-00-56, Shakshuk (adapted from Mutri 2008-2009).

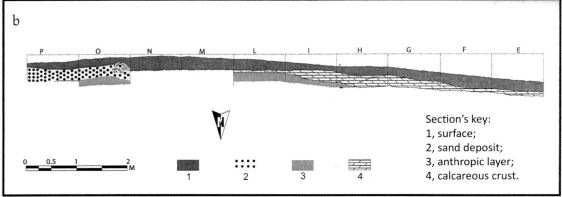

Section's key:
1, surface;
2, sand deposit;
3, anthropic layer;
4, calcareous crust.

Figure 8.
Excavation in the
Early Upper LSA
Site SJ-00-56,
Shakshuk (photo
Italian-Libyan
Joint Project
in the Jebel
Gharbi).

and the relationship between animals and a human group roughly 20 ka years ago.

This Upper LSA is widely attested throughout the Jebel and testifies to a phase with considerable intensity of occupation. In addition to the Shakshuk sites (SJ-00-56; SJ-00-55 East; SJ-00-55 West), this is also recognized in Wadi Ghan (SG-99-41; SG-02-63); in the Ras el Wadi area (SJ-98-26, SJ-98-26A, SJ-90-13); and in the El Batn area in the Jifarāh (SJ-03-83 Concentration 2). These sites included better-defined lithic typologies than Site SJ-00-56 and allowed comparison with Iberomaurusian and Eastern Oranian traditions. The sites in the Ras El Wadi area are located on the terraced alluvial fan belt (LU 2) immediately above the upper course of the Ain Zargha, from which they are separated by limestone cliffs. Bedded and pebble cherts used by groups for manufacturing their artifacts are contained within the limestone. The high artifact density indicates that the area was repeatedly inhabited by hunter-gatherer groups, attracted by the wide range of resources such as chert, game, perhaps even wild plants, and water.

Cores and debitage form 80% of the whole Upper LSA lithic assemblage. Among tools, notches, denticulates, Outchtata and backed bladelets show the highest frequencies. Backed products show standardization towards arch-backed types, while true segments are rare (Figure 9). These sites may represent a form of more 'settled' occupation in the area, similar to what seems to have happened in the upper levels at Taforalt (the *Grey Series*, GS:

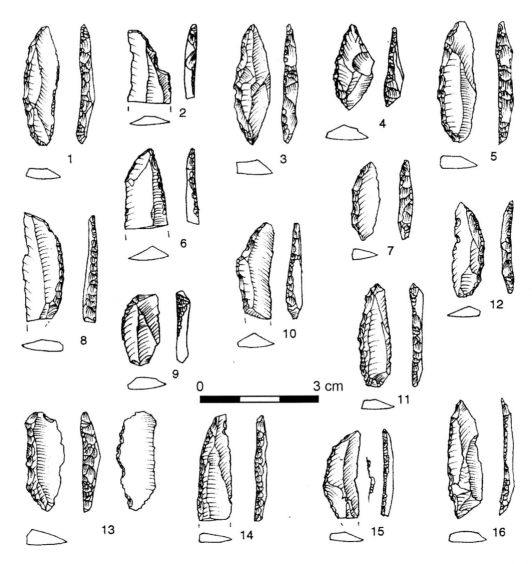

Figure 9. Jebel Gharbi, Libya. Upper LSA lithic industry from Site SJ-98-26, Ras el Wadi (from Barich, Conati Barbaro 2003).

10,990±45, OxA13517) (Barton *et al.* 2007).The study of the raw materials has shown that only local materials were used, proving that these groups of hunter-gatherers had chosen a closed artifact production and circulation system, even developing a discrete stylistic individuality, compared to that of the groups settled in the Jebel Akhdar of Cyrenaica.

The penultimate wet phase (Wet Phase 2), recognized at Shakshuk East and dated 11,690±40/13,850-13,460 cal. BP (Beta-185499), indicates a recovery of humidity at the end of the LGM, before the cold Younger Dryas interstadial (Giraudi 2009: 410). The fine sediments of this phase imply low transport due to widespread vegetation cover and regular rainfall. The date from Shakshuk East was obtained on charred roots. Another eight dates related to this same phase at the very end of Pleistocene, come not only from Shakshuk but also from El Batn (Jifarāh) and Ain Zargha-Ras el Wadi (Table 2).The latter, an OSL date (12,400 ± 1400 BP, X3416), together with the one from SJ-00-55 Hearth (11,360 ± 55 / 13,338-13,158 cal. BP,Poz-215) can represent a useful chronological reference for the Upper LSA / Epipalaeolithic surface sites from Ain Zargha.

In the Jifarāh there is a break in the sequence in relation to the harsher phase of the Younger Dryas. Subsequently, the occupation is recorded in the central area of the plain, near ponds and marshes where there were wetland vegetation with grasses such as *Panicum, Echinochloa/Setaria*, as well as Cyperaceae, *Amaranthus* type, and Chenopodiaceae during the early and mid-Holocene (Mercuri in Giraudi *et al.* 2012).The earliest layers at Site SJ-03-75, near Auenat Dagher (El Jawsh), are dated to 8210 ± 110 / 9350-9020 cal. BP (GdA-1501). This aspect, related to the wetter phase of the early Holocene, represents the direct continuation of the Upper LSA frequentation of Jebel Gharbi by groups of hunter-gatherers similar to those of the so-called 'Libyco-Capsian' (McBurney 1967). Evidence for an unfavorable tread during this period is visible in the landscape, culminating in the arid event at 8.2 ka BP, making the plains and locations near ponds and marshes increasingly sought after. However, the possibility that the interior of the Jebel continued to be frequented for hunting purposes cannot be excluded.

Neolithic

Early Neolithic sites (Jifarāh A, between 8.1 ka BP and 7.7 ka BP) develop during the mid-Holocene wet phase (Wet Phase 1: <8.2 ka BP). They are located in spring areas, in a wide stretch between the El Jawsh town and Wadi Alohim-Bazina, where hydromorphic soils are documented (Chighini 2012), and the environment shows a geomorphological stability thanks to the presence of continuous vegetation cover (Barich 2009, 2014; Barich *et al.* 2015). The subsistence model was probably still that of a broad-spectrum economy with collection of wild plants and molluscs (e.g. *Melanoides tuberculata* Müller, at Site SJ-06-87, Esu in Giraudi *et al.* 2012), although it is possible that it also included sheep / goat herding: evidenced at Jebel Akhdar with overlapping dating (e.g. Haua Fteah Layer VIII, Abu Tamsa Layer IV) (Higgs 1967; Klein, Scott 1986; de Faucamberge 2014). Geometrics, backed bladelets, and ostrich eggshells characterize the industry of this phase. Sites SJ-00-58H, SJ-00-67 (lower layer) near Shakshuk and the bottom horizon of SJ-06-87 at El Jawsh belong to this phase.

The beginning of the Holocene in the Libyan Jifarāh shows a continuation with the previous tradition. This emphasizes the extent to which people were rooted in the territory, and their ability to integrate and re-elaborate external stimuli. A greater transformation is observed in the next phase (Jifarāh B, Middle Neolithic, 6.7 ka BP to 5.7-5.4 ka BP) during which hearths and hearth-mounds appear across the region. The sites of western Jifarāh (Wadis Bazina, Alohim, Serwis) seem to be linked to movements of shepherds between the plain and the plateau pastures. Next to these are larger camps (base-camps), such as site SJ-00-59 on one side of Wadi Bazina, or the structures of site SJ-10-96 at Wadi Allohim, where plant foraging activities may also have been carried out. Following the spell of 5.4 ka BP (Jifarāh C: Late Neolithic, 5.4-4.4 ka BP), evidence for nomadic pastoralism increases and probably became the pre-eminent economic model.

Haua Fteah - Cyrenaica

Haua Fteah - with its stratigraphic section of about 14 m - is one of the most important North African sites. The importance of this huge cave, located a short distance (about 1.5 km) from the Mediterranean littoral at 67 m asl (Figure 10), is all the greater given the presence of two mandibles of *Homo sapiens* inside the section (McBurney 1967; Hublin 2001; Douka *et al.* 2014), which makes it a crucial site for studying the 'Out of Africa' phenomenon of modern humans.

The results of new research from the University of Cambridge (Cambridge Prehistoric Project, CPP), regarding the sequence highlighted by McBurney in the 1950s, produced an important chronostratigraphic revision and it's more precise correlation with known global climate change. These results were based on sedimentological and micromorphological analyzes (Hunt *et al.* 2010; Jones *et al.* 2011; Inglis 2012), on an extensive dating program (Barker *et al.* 2012; Russel, Armitage 2012; Farr *et al.* 2014; Jacobs *et al.* 2017) and on the study of the cultural aspects of the new artifacts highlighted during these works (Reynolds 2013; Scerri 2013; Lucarini, Mutri 2014). The revision took into consideration the first 7.5 m of the stratigraphy highlighted by McBurney (1967), reaching the base of McBurney's 'Middle Trench'. The lower part of the stratigraphy ('Deep Sounding'), containing the Levallois-based industry which McBurney termed 'Pre-Aurignacian', was not involved in the review. However, samples collected from a 1.25 m-deep trench (Trench S), excavated below the basal level of the Deep Sounding, analyzed with the OSL method on potassium-rich feldspar grains, indicate that modern humans had begun visiting the cave in about 150 ka BP (i.e. MIS 6). However, the greatest human presence at the cave began during MIS 5 (ca. 120 ka BP), as documented by the Deep Sounding (Jacobs *et al.* 2017).

The re-evaluation of the deposit led to the establishment of a series of five sedimentary facies, which are mainly palaeoclimatic indicators and only partially correspond to the global Marine Isotope Stages (Farr *et al.* 2014: 164; Douka *et al.* 2014: 55-57). Facies 5, identified at the base of the 'Middle Trench' (between 7.5 and 6.5 m below the current surface), is dated between 75 and

Figure 10.
Jebel Akhdar,
Libya.a-b: The
entrance of the
Haua Fteah Cave,
Cyrenaica (from
Barich 2010).

a

b

65 ka BP. Therefore, it was almost completely formed during MIS 4, whilst only a small part of its thinner and more compact sediments, such as those of the Deep Sounding which represent its continuation, can be attributed to MIS 5 (Douka *et al.* 2014: 58). Subsequently, both Facies 4 (between 68 and 47 ka BP), which is correlated with the Levalloiso-Mousterian occupation, and Facies 3 (48-34 ka BP), in which the Dabban blade-based industry is contained, show signs of climatic instability associated with MIS 4 and MIS 3. Facies 2 (35-12 ka BP) is made up of the sediments produced by frost-fracturing of the cave vault, together with greyish clays deposited in a period of cold and arid climate typical of MIS 2. Finally the sediments of

Facies 1 (13.6 ka BP - present) show reduced climatic alterations due to the re-establishment of the regular precipitations characteristic of Holocene (Farr *et al.* 2014; Douka *et al.* 2014).

Revision of the chronology (using ^{14}C, ERS, OSL and tephrochronology), allows for a more precise identification of the cultural succession. The sequence of the 35 layers is still attributed to the six phases already established by McBurney (from the bottom): Pre-Aurignacian, Levalloiso-Mousterian, Dabban, Eastern Oranian, Capsian, Neolithic. However, there are some chronological adjustments, with generally modest deviations from McBurney's chronology (see Farr *et al.* 2014: Table 1). The major corrections concern the Dabban whose development does not go beyond 18 ka BP (compared to the previous date of 15 ka BP); while the Eastern Oranian appears to start from ca. 17 (or 16) ka BP, compared to the previous dating of 15 ka BP.

Homo sapiens were certainly responsible for the Levalloiso-Mousterian industries, confirmed by the discovery of the two mandibles in the Levalloiso-Mousterian basal levels (Layer XXXIII), dated between 73 and 65 ka BP (Hublin 2001; Douka *et al.* 2014: 59). As for the pre-Aurignacian, recognized by McBurney in the Deep Sounding, the thesis supported by the CPP researchers is that those artifacts show many convergences with the Levallois industry or, at least, that from the very early stages of the human occupation at Haua Fteah there were characters similar to the Levalloiso-Mousterian sequence recognized in the Middle Trench (Reynolds 2013; Douka *et al.* 2014; Jones *et al.* 2016). This is an undoubtedly important observation: if the pre-Aurignacian can be assimilated to the Levalloiso-Mousterian industry, and as such be attributed to modern humans, considering the latest dates obtained from Deep Sounding which date back to about 150 ka BP, it can be said that *Homo sapiens* has been present in Cyrenaica as early as the Aterian in the Magheb (Jacobs *et al.* 2017).

Middle Stone Age- Levalloiso-Mousterian

Modern research has confirmed the difficulty of recognizing the presence of the Aterian in the Haua Fteah sequence, as already indicated by McBurney (1967). Haua Fteah lacks elements that would enable us to identify a true Aterian horizon, despite its alleged recognition not only at Haua but also in the pre-desert (Reynolds 2013; Douka *et al.* 2014). However, the Aterian is known at Hagfet et-Tera just 175 km away from Haua Fteah. On this point, the interpretative model adopted by CPP appears more inclined to look at the Nile Valley and the Levant, rather than the Sahara, as territories of interaction with the Jebel Akhdar. In this perspective, both the MSA complex (or Levalloiso-Mousterian) as well as the subsequent LSA (or Dabban) would be attributed to a *sapiens* whose path out of Africa would have passed from the Levant. "...... *there is at least as strong a case for Egypt* and/or the Levant (where early modern humans at Qafzeh and Skhul possibly date to ca. 130-70 ka.) as the source for the Haua Fteah's modern humans"… (Douka *et al.* 2014: 59). Another opinion seems to be that of Jones *et al.* (2016) who detail the environmental conditions, specifically the network of wet sources (lakes and rivers) through which it would have been quite possible for Saharan groups to reach Cyrenaica, and reject the idea of an isolation of the Jebel Akhdar.

Later Stone Age – Dabban

The Dabban, the oldest Upper Palaeolithic or Late Stone Age industry in North Africa, is an important presence in Jebel Akhdar, although limited to the Haua Fteah cave and the Hagfet et-Dabba site (and maybe the Haji Crem site), from which the industry takes its name. The Levalloiso-Dabban transitional level (XXV) is placed between 46 and 41 ka BP: a date that correlates the appearance of this industry with that of the first *Homo sapiens* industries in Europe. However, it should be noted that the transition between Levalloiso-Mousterian and Dabban is not clear (Douka *et al.* 2014: 5), since there is no clear separation between the two horizons.

At Haua Fteah, Dabban occupation is attested between the XXV and the XVI layers, from ca. 46 (or 41) to 18 ka BP. During this period the climate was cold, particularly in the early phase, in MIS 4, during which the occupation of the cave appears rather rarefied. The late Dabban, on the other hand, includes a greater wealth of artifacts and the appearance of innovative typologies close to MIS 2 (Jones *et al.* 2016). All this may be the consequence of movements of groups and their concentration in the localities that continued to be habitable in difficult climatic conditions. Concentration phenomena during critical and, conversely, expansion in favorable phases, appear to be a characteristic of these periods of climatic instability (Jones *et al.* 2016). In the long Dabban series there are various sub-phases separated from each other by moments of abandonment.

The Dabban appears with a revolutionary innovation in the panorama of Mediterranean North Africa. With it, the flaking method of direct percussion with soft hammer, or with indirect percussion (punch technique), from unipolar or bipolar cores is generalized. Long blades, burins, and endscrapers become widespread.

A characteristic type of the Dabban industry, especially in the early Dabban phase, are the chamfered blades, blades that show a kind of transversal burin. The chamfrein is a typical element of this industry, which in North Africa is apparently limited to the Jebel Akhdar (it is however present in the Nile Valley, where it was recognized by Vignard (1920) near Nag Hamadi). Due to its formal similarities with industries of the Levant (Ksar 'Akil, Lebanon and Ahmarian, Israel), many authors have considered the Dabban industry and culture to be a result of migrations from the

Levant along the Mediterranean coast (McBurney 1967; Clark 1992). Other authors have noted that the typical chamfrein can be used in the same way as a transversal burin, so much so that in industries where it appears the presence of burins, and also endscrapers, decreases. Chamfrein could therefore fulfill specific purposes by replacing both burins and endscrapers in some industries (Iovita 2002). Even at Haua Fteah the final phase of the Dabban industry, in contact with Eastern Oranian levels, shows a decrease in chamfered blades and an increase in endscrapers and burins. The presence of backed bladelets with characters that anticipate Upper LSA remains high (Lucarini, Mutri 2010b).

Upper Later Stone Age – Eastern Oranian

With the appearance of the Upper Later Stone Age industry there is an alignment of the two sequences of Jebel Gharbi and Jebel Akhdar. Both reflect the rapid increase in presence of microliths and backed bladelet industries not only within the Maghreb but also in the Nile Valley.

Facies 2-1 of the revised Haua Fteah sequence include layers XV-IX, characterized by industries well-differentiated from the Dabban. Due to the similarity with the Iberomaurusian tradition (initially called Oranian) McBurney called the products of the XV-XI layers 'Eastern Oranian'. In fact, one can recognize typological and technological parallels between these two complexes, particularly due to the emphasis placed on the microlithic backed pieces which reach a high percentage among the tools. A few polished bone tools are also present. The sediments associated with these layers show the influence of an arid climate with accentuated physical weathering, probably in relation to a prolonged cold period (Douka *et al.* 2014: 43-44). However, despite the harsh climatic conditions then prevailing, no interruptions occurred in the sequence at the transition between Facies 3 and Facies 2 (i.e. between Dabban and Eastern Oranian). In the transitional units, Dabban technological indicators are mixed with the innovative Eastern Oranian ones (Lucarini, Mutri 2010b: 73, fig. 9).

During the 'Oranian' phase (ca.18/17-13.5 ka BP) the human groups primarily practiced hunting of Barbary sheep (*Ammotragus lervia*), while in other sites closer to the desert (such as Hagfet et-Tera) they hunted the gazelle. Other resources were provided by Equidae, hartebeest (*Alcelaphus* sp.), rabbits and the collection of marine molluscs (*Patella* and *Osilinus*) along with land snails (*Helix melanostoma*) (Klein, Scott 1986; Barker *et al.* 2010: 86). The few remains of avifauna (e.g. *Alectoris barbara, Tringa* cf. *nebularia, Columba livea / oenas, Turdus* sp. (MacDonald 1997: 88) indicate a lack of interest in hunting birds, certainly not favored by the rarity of wooded areas due to the cold-arid climate. The presence of avifauna is also very limited in the Dabban collection, probably for the same environmental reasons (MacDonald 1997: cit.).

The latest 'Eastern Oranian' layer (XI), is followed by layers X and IX, showing different characteristics to the former ones. McBurney (1967) attributed these two layers to the new 'Libyco-Capsian' occupation (12.7-7.9 ka BP), considering them to be a local aspect of the Maghrebian Capsian, and evidence of an overlap of new people with the former inhabitants. The difficulties related to this hypothesis was already highlighted by A.E. Close, because of the great affinity in the lithics coming from the two transitional XI/X units (Close 1986), and has been further emphasized by the new CPP works. The analytical study of the transition between Eastern Oranian and Libyco-Capsian has confirmed that, for both environment utilization and technological profile (Figure 11),the two horizons have to be considered as a *continuum*: a single episode of occupation, which gradually developed on-site between the final Pleistocene and Early Holocene (Lucarini, Mutri 2014: 117). These results therefore bring new evidence in favor of the continuity of occupation in this territory, although the possibility of contacts and contributions from outside cannot be excluded (Lucarini, Mutri 2014, cit.).

Figure 11. Jebel Akhdar, Libya. Libyco-Capsian tools from the Haua Fteah Cave, Cyrenaica (from Lucarini, Mutri 2014).

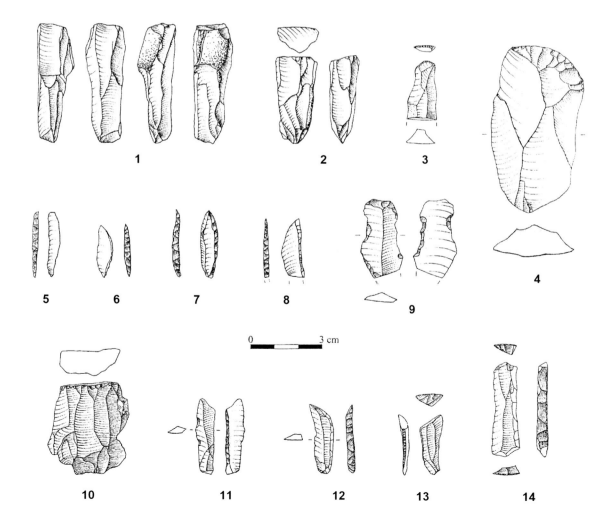

At the beginning of the Holocene, the Libyco-Capsian layers contain evidence for exploitation of the same resources found in the Oranian layers. Barbary sheep (*Ammotragus lervia*) and *Alcelaphus* continue to be present together with marine molluscs (*Patella caerulea, Patella ferruginea, Osilinus turbinatus*), whose abundance testifies to the greater proximity of the coast (Klein, Scott 1986). The presence of the land snail *Caracollina lenticula* (ca. 8 ka BP) indicates greater aridity and a reduction in forest cover (Hunt *et al.* 2011). The study of the botanical remains (Morales, Barker 2009) indicates an intensive collection of pine cones (*Pinus halepensis*) for the extraction of seeds, along with wild pulses and legumes, juniper and myrtle fruits, illustrating the role of wild plants and berries associated with hunting.

Neolithic

Data so far published by the CPP do not add much to what McBurney (1967) had already reported about the Holocene occupation of the cave and the appearance of the Neolithic features. The initial dating of the two layers that McBurney attributed to the Neolithic have now been corrected, so that Layer VIII, Early Neolithic, falls between 6917±31/ 8300-7670 cal. BP (OxA-18673) and 6115±31/ 7160-6890 cal. BP (OxA- 18667). Layer VI, Late Neolithic, is instead dated between 5759±28/ 6650-6480 cal. BP (OxA-18675) and 5462±30 / 6310-6200 cal. BP (OxA-18676) (Barker *et al.* 2009: Table 6; Douka *et al.* 2014: Table 1). Beginning with Early Neolithic, pottery is present alongside lithics (at ca. 7.4-6.9 ka BP, Douka *et al.* 2014: 59). Pottery from Layer VIII includes fragments of rims and body sherds. Analysis has shown the use of limestone and shell fragments, probably as temper for clay plasticity. The surfaces are smoothed and burnished, and some fragments have a decoration on dotted bands. Study of the lithic industry shows a continuation in the presence of the types most commonly occurring in the preceding Capsian layers, but with a marked decrease of the backed component, and increase in endscrapers, burins and sidescrapers. The raw material is made up of chert and flint, to which for the first time limestone is added, in the form of polishing and grinding implements. Two rounded pebbles have also traces of ocher, and bone tools are also attested. During the Late Neolithic (Layer VI), tanged bifacial arrowheads appear in the lithic industry, while the presence of microliths decreases. Limestone was still use to produce hoes and, importantly, grinding stones for milling grains. Pottery is similar to that of layer VIII, and there are also numerous ostrich eggshell beads.

The archaeozoological remains are quite numerous, particularly in the Early Neolithic. As regard wild species the Higgs (1967) list, later revised by Klein and Scott (1986), indicates Barbary sheep, hartebeest, *Hystrix* sp., *Microtus* sp., Cats (*Felix* sp., *Felix ocreata*) and some remains of jackal. The presence of marine molluscs (*Patella caerulea, O. turbinatus*) and land snails even after the 8.2 ka BP arid interval, does not testify for dryness, or decisive economic transformations, despite the presence of the arid episode. The main evidence of an economic change is represented by the presence of goats that Higgs (1967) reported at 6800 unc. bp (which would now be corrected to 7.8-7.6 ka BP). To these are added some remains of bovids, none of which are referred to as domestic.

The archaeobotanical study ruled out the presence of domestic cereals, in contrast to what had formerly been reported (Morales, Barker 2009: 29-30). All the remains are of wild taxa: Poaceae, Cyperaceae, *Chenopodium murale*, *Galium* sp., *Phalaris* sp., *Trifolium* sp., small seed legumes (Morales, Barker, cit: Table 5). The identification of remains belonging to the Cenchrinae family through residue analysis on stones used for grinding, allows us to add *Cenchrus* sp. to the list of taxa at the site. *Cenchrus* is widespread in the Sahara and very often found in Holocene deposits (Barich 1992; Wasylikowa 1992; Di Lernia 1999; Mercuri 2001), but has not been recorded as charred remains in the sites of the North African coast so far (Lucarini *et al.* 2016).

Beginnings of the Holocene in the Libyan Sahara

In southern Libya the sandstone Tadrart Akakus mountain (Figure 1), one of the central Saharan massifs, offers clear evidence of abandonment of the area during MIS4 and re-occupation at the beginning of the Holocene. The long investigation by the Sapienza University of Rome is well known thanks to the extensive bibliography (see Mori 1965, 1998; Barich 1974, 1978, 1987,1992,2013; Cremaschi, Di Lernia 1998; Di Lernia 1999, 2002, 2013, 2019; Di Lernia, Zampetti 2008; Biagetti, Di Lernia 2013; Garcea 2001).

Investigations at the two sites of Uan Tabu (Garcea 2001) and Uan Afuda (Di Lernia 1999) revealed the presence of thick aeolian sand formations below the Holocene horizon. Uan Tabu, in particular, also made it possible to date *in situ* the presence of the Aterian at around 61 ka BP (Garcea 2001), the last occupation limit of the region. Therefore the Holocene resumption of occupation appears here as a real colonization of a territory abandoned for millennia.

Inside the numerous shelters, the wet phase sediments appear to have been deposited starting from the very beginnings of the Holocene (i.e.11.5 ka BP). However, it is possible that a very early reoccupation, superficial but extensive, may have taken place in the central Sahara during the wet interstadial preceding the Younger Dryas (approx.16-13 ka BP). This would justify the emergence of LSA macro blade industries that have long been highlighted in the Libyan Sahara (in the Messak, Cancellieri, Di Lernia 2014: 45). However, evidence for more humid conditions has also been suggested for the end of MIS 3, as observed at Sodmein Cave (Red Sea) with a Upper Palaeolithic/LSA layer at ca. 25 ka BP (Moeyerson *et al.* 2002), and for Late

Palaeolithic/LSA industries recorded in the Eastern Sahara and Red Sea Mountains (Van Peer 2014: 30-31).

The Early Holocene occupation (or 'Early Akakus' as it was named by Di Lernia, Garcea 1997) occurred between ca. 11.2 ka BP and 9.2 ka BP. The phase is known at a large number of sites (N=88), which were mainly open air sites, specialized in hunting activities or the procurement of stone raw materials. Some sites have been identified in the flat territory of Erg Uan Kasa, east of the Tadrart Akakus (Cancellieri, Di Lernia 2014; Di Lernia 2019). Within the Tadrart only four sites, Ti-n-Torha Two Caves, Ti-n-Torha East, Uan Afuda, Uan Tabu (investigated through systematic excavations), can be considered as residential sites, due to the presence of stone dwelling structures. Their study made it possible to adequately reconstruct the characteristics of the region's reoccupation phase (Barich 1987; Di Lernia 1999; Garcea 2001). While the two sites of Wadi Ti-n-Torha, a tributary of Wadi Auis, are located in the north-eastern region of the Tadrart Akakus, the other two are in the central Tadrart's area. Uan Tabu overlooks Wadi Teshuinat, while Uan Afuda is at the bottom of Wadi Kessan, not far from Teshuinat. In all these locations the sediments hint at the presence of a moderately humid climate at the very beginning of the Holocene.

In both Ti-n-Torha Two Caves and Ti-n-Torha East, the Early Holocene layers appeared directly in contact with the wadi bedrock, at the base of the two sequences highlighted, both relating only to the Tadrart's pre-pastoral occupation. A third deposit, identified at the Ti-n-Torha northern end (named North Shelter), yielded a much more extensive sequence developing up to the Mid-late Holocene. At this latter site, layers belonging to the pastoral phase of the Tadrart overlie those relating to the earlier pre-pastoral occupation by hunter-gatherers-fishermen (Barich 1974, 1987).

At Two Caves the occupation took place in the area in front of the two openings of the shelter, consisting of two communicating cavities (hence the name of the site). On the external front, the anthropic section rests on an accumulation of large boulders that played a useful function for its protection. The stratigraphic section showed a considerable thickness, especially at its western end where it measured 1.40 m. Radiocarbon dating put the occupation interval between 9350±110 (R-1402) and 8230 ± 50 (R-1403), i.e. between 10.5 ka BP and 9.2 ka BP (Barich 1978, 1987a); therefore it is placed in the Early Holocene. The East Shelter had a base layer (RInf-base) made up of reddish sand mixed with the wadi stones and a grayish sediment of ash and coals, mixed with plant remains, lay upon it. In the section's upper part, two carbonate crusts were detected, the most recent of which was associated with stone structures. The stones lay at a depth of - 0.60 / 0.80 m from the modern surface and formed separate domestic spaces ('huts'). At least ten circular hut bases were identifiable, aligned against the sandstone wall of the shelter (Barich 1974, 2013: 452, Fig.31.5). The overall sequence is longer here than at Two Caves and extends up to ca. 8.0 ka BP. The date of 9080±70/ 10,298-10,181 cal. BP (R-1036α) records the initial occupation of the site, while later dates refer to the central part of the sequence during the 10th millennium cal. BP. From the RInf layer, dated 8640 ± 60 / 9673-9536 cal. BP (R-1035α) ceramics become more abundant.

Unlike the Ti-n-Torha sites, at Uan Tabu and Uan Afuda, inside the Tadrart, the Early Holocene (or Early Akakus) horizon stands above an older Upper Pleistocene occupation dating to between 90 and 60 ka BP (MIS 4), related to an Aterian occupation. The two, Pleistocene and Holocene horizons are separated by a thick deposit of aeolian sands which was formed before the start of deglaciation and the re-establishment of the monsoon rains (Di Lernia 1999). In the Uan Tabu stratigraphic section the Early Akakus reoccupation is recorded by Unit III (dated 11.2 - 9.9 ka BP), between -130 and -180 cm from modern surface. The soil is described as a unit composed of sand and silts reworked by colluvial process (Garcea 2001a:10). At Uan Afuda shelter, the occupation took place both in the external area and inside the cave. The Early Akakus levels (Unit 2: 11.1-10.4 ka BP) appeared only in the excavations outside the cave (Excavation I), above the Aterian layer, at a depth of ca. -150 cm from the surface. No hearth is present, only a structure formed of stones and interpreted as a wind break was detected (Di Lernia 1999: 21). Stone raw materials most commonly used in the Early Akakus are chert and silicified sandstones that may have been found locally. There is also a high presence of quartzites which could have been mined in Wadi Auis, or in the Messak Settafet, not far eastward. There are some rare examples of flint that is not present locally but can be found about a hundred kilometers away to the west, toward the Tassili. At Ti-n-Torha we found also quartz nodules (formations incorporated in the sandstone) particularly frequent in the basal part of the deposits. In the various assemblages, the lithic industry is formed of over 50% of working products, while non-retouched and microlithic splinters together make up the largest part of the debitage. This factor, together with the strong presence of cores, attests that the production and repair was carried out on site. The typological diversity is represented by seven lithic classes (endscrapers, perforators, backed pieces, notches/denticulates, truncations, segments, microburin technique), which in Uan Afuda are reduced to five: burins, backed pieces, notches/denticulates, sidescrapers, scaled pieces. Among these, backed blades/bladelets are the main classes. The conventional spit (ca. 5 cm) distinguished at the base of Torha East yielded 1432 items, 96% of which was represented by debitage, a similar situation was observed at Two Caves (Figure 12)

Uan Afuda's Early Akakus layers did not yield pottery, and only three ceramic sherds were collected from layers pertaining to the same phase at Uan Tabu (dated ca.10.1-9.8 ka BP). Ceramics become more numerous at Torha East- RInf starting from 8640±60/9673-9536 cal. BP (R-1035α). The oldest pottery is reddish in color, homogeneous and thin, and refers to small vessels. The decorations were created using the impression technique, and using stamps such as combs, punches, spatulas and vegetable cords (Barich

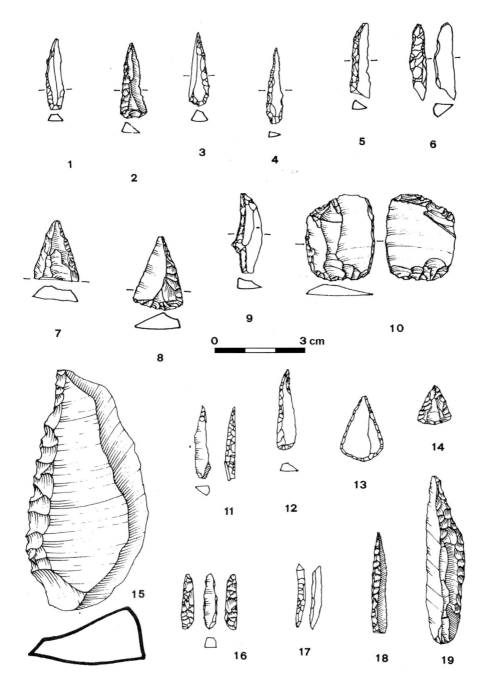

Figure 12. Tadrart Akakus, Libya. Lithic tools from Ti-n-Torha Two Caves (nn. 1-10) and Ti-n-Torha East (nn.11-19) (from Barich 1987a).

1974, 1987:107, Fig.5.6). Among the motifs adopted, the packed rocker and the 'Dotted Wavy Line' (DWL) motif characterizes this pre-pastoral phase of the central Sahara (Figure 13). From these sites, very elaborate and high quality bone implements (Figure 14) were collected (Petrullo, Barich 2020) as well as grinding equipment.

Figure 13.
Tadrart Akakus,
Libya. Pottery
from Ti-n-Torha
East RInf (from
Barich 1974).

Figure 13. Tadrart Akakus, Libya. Pottery from Ti-n-Torha East RInf (from Barich 1974).

In the layers post-dating 9.5 ka BP the number of remains increases significantly in the deposits, and in some of them with a particularly high density thus indicating the different use of the sites. However, there is no difference in the composition of the assemblages, where the flakes outnumbered the blades and the percentage of microlithic debris is very

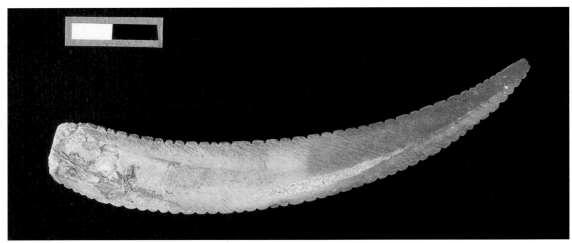

Figure 14.
Tadrart Akakus,
Libya. Decorated
spatula on a
warthog tusk
from Ti-n-Torha
Two Caves
(photo Barich).

high. Blades/bladelets and flakes with backed retouch represent the largest component of the assemblage, attesting to the importance of hunting activities.

The relatively good state of preservation of faunal and plant remains allowed the economic model of these ancient hunter-gatherer-fishermen of the Akakus to be reconstructed. Faunal remains from both Two Caves/Layer III and Torha East/Rinf-Base are very fragmented and their identification was difficult. Out of the large mammals only *Ammotragus lervia* is attested in both sites. In the Two Caves some small rodents, fish remains (*Clarias* sp., *Tilapia* sp.) and small birds are represented. Torha East Base has no fish remains, but an example of *Equus africanus* and a higher number of *Ammotragus* remains are present, forming 70.6 % of the faunal assemblage (Cassoli, Durante 1974; Gautier, Van Neer 1977-1982; Gautier 1987). Even in the contemporary layers of Uan Afuda (Unit 2) most of the faunal remains belong to *Ammotragus* (Di Lernia 1999).

The archaeobotanical study reveals the presence of a wide range of grasses (*Brachiaria*, *Urochloa* type, *Panicum* sp., *Setaria*, *Echinochloa*, *Pennisetum* (*elatum/setaceum*), and *Cenchrus* cf. *ciliaris*) (Wasylikowa 1992; Mercuri 2001) attesting to an intense use of wild plants. The evidence for wild grasses in the sites corresponds to the presence of grinding equipment. An initial hypothesis regarding ceramic utilization was that they were used for the preparation of vegetables, and cooking them on fire. This hypothesis was recently confirmed by the analysis of residues still present on the container walls found in the Tadrart Akakus. Biomolecular and stable carbon isotope analysis on Takarkori and Uan Afuda pottery confirmed that the vessels were used for the preparation of vegetable-based soups (Dunne *et al.* 2016).

A high level of territorial control has been suggested for the Early Akakus groups. It is possible to reconstruct movements between the inland Tadrart and the surrounding valleys, where stone raw materials could have been found in abundance (e.g. Titersine, Erg Uan Kasa, Tanezzuft) (Cancellieri,

Di Lernia 2014). These movements appear to have been undertaken within a well-established network. Comparisons between the lithic production of Ti-n-Torha Two Caves and the El Adam horizon at Bir Kiseiba and Nabta Playa in Nubia (Close 1987), suggests a network of exchanges and contacts across the central and eastern Sahara.

Discussion and Conclusions

We have analyzed three examples located in different ecosystems, one on the littoral strip, one in the pre-desert, and a third located in a Saharan environment which has rarely been compared to the former two. The Saharan example, inside a territory that was depopulated for a long time, provides evidence for re-occupation and for cultural dynamics that differentiate it from the northern region.

The comparison between the two sequences of Jebel Gharbi and Jebel Akhdar, at the western and eastern ends of the Libyan coastal strip, separated from each other by the Gulf of Sirte, has highlighted significant concordances in climatic alternations and in the dating of the human presence that develops from MIS5 with generalized MSA Levallois technology. Differences between the two regions are found in the presence / absence of Aterian and Dabban industries, which can be generally associated with the MIS 4 and MIS 3 phases of climatic instability and variability. The evidence for Aterian presence in the Jebel Gharbi seems absent from Haua Fteah, while for the Dabban we have a reversed situation with an important presence at Haua and absence in the western territory.

The recognition of the Aterian presence in the Maghreb is important. This MSA industry widespread in the Central and Egyptian Sahara is especially noteworthy in the region of Atlantic Morocco (Témara), where recent OSL dating has considerably aged its chronology, pushing it back to the initial MIS 5 stages (Barton *et al.* 2009). The trans-Saharan diffusion of the Aterian towards Northern Africa may have been favored by the presence of numerous water basins, such as the Kufra and Megafezzan systems (Drake *et al.* 2011) in the Central Sahara during MIS 5, and by a dense hydrographic network which may have served as corridors for human societies during their expansion to the north. It has been ascertained that the bearers of the Aterian industries were modern humans, displaying symbolic behavior that derives from an era as ancient as the first South African *sapiens* (Bouzouggar *et al.* 2007; Henshilwood *et al.* 2004). All this taken together suggests that the emergence of anatomically and behaviorally modern humans is truly a 'pan-African' phenomenon (Dibble *et al.* 2013).

There is also an important presence of Aterian industries in the Jebel Gharbi. Expansion of the Aterian from the Sahara can be explained by the proximity of the accessible Hammada el-Hamra tabular structure, which separates the southern edge of the Tripolitanian plateau from the Libyan desert. Inside the Jebel Gharbi region the Aterian is present from the Jifarāh (LU2) up to the interior of the mountain - Ras el Wadi-Ain Zargha - where Aterian hunting sites appear strategically located not far from each other and near water sources where the fauna may have gathered. Here, evidence of Aterian industries was collected on open spaces and on the hill slopes between 650 and 700 meters asl, upon the erosion of the soil that originally contained them.

A large chronological span is estimated for the Aterian of Jebel Gharbi. The date of ca. 60 ka BP has to be understood as a *terminus ante quem* for its appearance which, based on stratigraphic and radiometric data, can be pushed back to approx. 80 ka BP (Table 1). It may also persisted later than 40 ka BP, with an on-site evolution and related technological developments that may have triggered the rise of the local long-blade Lower LSA with an emphasis on the unipolar and bipolar reduction method.

Based on evidence presented in publications, the Aterian presence in the Haua Fteah stratigraphic section is uncertain. Although some artifacts can be traced back to this industry, this has not allowed the identification of a true Aterian assemblage in the cave (Reynolds 2013; Jones *et al.* 2016).

This problem is related to the difficulties concerning a precise definition of the Aterian industry and its separation from the Mousterian / Middle Palaeolithic industry with which it shares numerous features. In the absence of the specific traits related to the Aterian, i.e. tanged pieces and foliate retouch, the distinction between these two complexes requires quantitative comparative studies on a sufficient sample available (on this subject see in particular Dibble *et al.* 2013).

Alternative explanations can be put forward regarding the absence, or lack of recognition, of the Aterian industry at Haua Fteah. This could be due to: sparse regional occupation at the time, the specific destination of the cave, the small size of the available sample and the absence of industrial markers compared to other less characteristic elements. Only a very limited area has been excavated, therefore future work may overturn current knowledge (Reynolds 2013: 180). It is clear, however, that this issue does not only concern the situation at Haua Fteah, but that on a more general ground it is necessary to take into account the very definition of the Aterian, its meaning and nature as an industry. It is especially important not to overlook the fact that the North African MSA includes a multiplicity of variants that are more or less contemporary with the Aterian industry: "*On typological grounds it is at least clear that diagnostic assemblages such as the various 'Mousterian' groups, Aterian and the Nubian Complex share specific types with each other. Without dedicated study on the attribute level, it is uncertain whether similarities and differences reflect culture diversity, task specialization, ecological adaptation and/or isolation by distance or merely inappropriate and incompatible analytical procedures*" (Scerri 2013:6).

In the Haua Fteah sequence the transition between the Levalloiso-Mousterian and the Dabban is unclear (Douka *et al.* 2014: 59). The boundary layer between these two horizons (Layer XXV) was dug in two spits (spits 32 and 33) (Douka 2014 *et al.*: idem). McBurney (1967: 125) noted that pieces with Levalloiso-Mousterian characters were present along with Dabban artifacts in both spits, although the Dabban presence was greater in the upper one. There were also occasional 'more or less levalloisoid' flakes in the overlying XXIV layer. However, there is no interruption in the section, and therefore today we do not know if the situation described so far can be interpreted as a transitional mixed technology, with Levallois-based core techniques changing into upper Palaeolithic technology (with an assemblage comprising endscrapers, burins, and points), as was observed in the transitional industries of the Levant (Iovita 2009: 139). This point should be clarified by the results of analyses of materials coming from recent investigations (Douka *et al.* 2014: 59).

From its earliest appearance the Early Dabban shows very distinct characteristics: intentional preparation of blades to be used as blanks, together with a remarkable variety of tools characterized by specific retouches (Moyer 2003). Among these are the chamfered blades whose 'foreign' origin was immediately pointed out by McBurney (1967: 147). This is one of the aspects that Jones *et al.* (2016) ascribe to the relatively high diversity of tools found in northern Cyrenaica in comparison to other North African regions. According to the authors, the diversity at Jebel Akhdar was the product of several elements, including demographic factors that may have caused expansion, and the diffusion of cultural-technological features from adjacent territories in climatically favorable periods. Considering the current dating of the Levalloiso-Dabban transition (46-41 ka BP), the cold-humid interval recorded at 44 ka BP (Drake, Breeze 2016) may be significant. Large stretches of the Mediterranean coastal plain were exposed during similar cold-humid episodes, making encounters easier, facilitating the consequent technological diffusion, without necessarily involving the transfer of peoples.

McBurney (1967) first pointed out comparisons with Abu Halka (Layer IVE), Lebanon, regarding the typical 'chanfrein'. Other sites of the Lebanese Emiran context (Ksar 'Akil and Abri Antelias) were later cited by Bar Yosef (Bar Yosef 1994: 242). For his part, Iovita (Iovita 2009: 139) considered the function of chanfrein, noting that the presence of this technological trait at both Ksar 'Akil and Abu Halka sites seemed related to a reduced presence, or disappearance, of burins and endscrapers from the assemblages. Conversely, when chanfreins disappear, the other types reappear. It is therefore possible that chanfreins were utilized in similar ways to both transversal burins and endscrapers. On the other hand, the recognition of the chamfered blade as a type in itself depends on the precise position of the facet, which is not always easy to establish. On a more general ground, the same author observes that imports / diffusions from the outside could not be hypothesized based on a single technological feature such as, in this case, the chanfrein. A careful analysis of the Haua Fteah and Ksar 'Akil stone production, led Iovita (2009:

idem) to recognize considerable differences between them, beginning with the core reduction modalities [unidirectional (Ksar' Akil) vs. opposed bidirectional (Haua)], and extending to re-sharpening and core morphology (prismatic at Haua, pyramidal or triangular flat at Ksar 'Akil). In conclusion, the technological differences between the two repertoires, together with the large chronological space between them, make it difficult to argue that chanfreins were the product of a diffusion from one site to the other. It seems much more plausible to explain them in terms of a technological and functional convergence.

According to current research the Early Dabban is not attested in the Jebel Gharbi. During the late MIS 3 (<30 ka - 20 ka BP), contemporary with Jebel Akhdar's LSA (or Upper Palaeolithic), the industry in long-blades named Lower LSA (Garcea 2010 a and b) is recorded at Jebel Gharbi. However the post-Aterian phase at that site marks a gap before the appearance of the Lower LSA industry between 40 and 30 ka BP. Recognition of the continuation of the Aterian at the site, along with a better understanding of the transition to the blade industry may hopefully come with a resumption of fieldwork in Libya.

Conversely, relations between Jebel Akhdar and Gharbi can be established with respect to the final phase of Dabban and the subsequent emergence of the Upper LSA / Epipalaeolithic (or Eastern Oranian) bladelet industries. This evidence allows us to suggest that the minor humid spell recognized between 25 to 20 ka BP (Van Peer 2014: 31) may have favored a shift of the Dabban towards the western territory. In fact from that moment on, numerous morpho-typological and technological similarities in the assemblages from the two regions were recognized in some specific types (backed bladelets, endscrapers). The products of the base layer at Shakshuk Site SJ-00-56 (ca. 20 ka BP), show remarkable similarities with the transitional Late Dabban-Eastern Oranian 10,011 unit of Haua Fteah (Lucarini, Mutri 2010b: 74). Furthermore, later Upper LSA Gharbi assemblages - as SJ-98-26A - show close resemblances to Eastern Oranian backed tools (Barich, Conati 2003).

The Upper LSA of Jebel Gharbi is part of the so-called North African 'microlithic revolution' (Close 2002). The oldest aspects of the Upper LSA appear in the Last Glacial Maximum and allow comparisons between the complexes of the present-day regions of Morocco, Algeria, Libya and the Egyptian Nile Valley. This industrial explosion was diffused and continuous through the phase of climatic amelioration and the gradual 'settlement' of the groups inside the territories. Schild and Wendordf (2010) emphasize technological similarities between Arkinian and Iberomaurusian, suggesting the existence of possible contacts along the Mediterranean coast. The approach used in the analysis of the stone tool assemblages allows identification of orientation, preferences, and stylistic differences between groups on the basis of which multiple industries have been described, and hypothetically associated with social groups (Iberomaurusian, Capsian, Fakhurian, Kubbaniyan etc.) (Close 1977). However, we are not certain about the possible identification of specific industries with specific social groups. For this reason we prefer to emphasize the technological revolution that underlies these products, which are functional to a different resource procurement, and also to the new settlement model.

In light of this, the broader technological category of Upper LSA, or 'Epipalaeolithic', was preferred for the late Pleistocene-Early Holocene Jebel Gharbi industry, without ignoring the presence of a similarity with both the Iberomaurusian and the Capsian for specific morpho-typological characters. Instead we have set aside the generalized use of such definitions, which would be appropriate if referred only to the original contexts.

Greater presence of organic finds, especially in the Haua Fteah layers, allows us to trace a more complete economic model of the LSA hunter-gatherers. We can reconstruct a broad spectrum use of the environment with hunting and collection of terrestrial and marine molluscs as well wild fruits (and perhaps plants). In the Jebel Gharbi open air camps were always located at points of easy water supply. It is possible to recognize a systematic reoccupation of the same areas and a functional differentiation of the sites according to their distribution and assemblage composition (Barich, Conati 2003). Stone raw materials were procured locally or a short distance from the main habitation clusters.

The transition to the Holocene had different modalities in the North and South of the country. We have seen how in both the two northern areas the Pleistocene-Early Holocene transition occurred gradually and with the continuation of the main characters of the final LSA. Indeed, the first Holocene nuclei reproduce the lifeway model of LSA hunters and gatherers, for which the change has the aspect of integrating new cultural traits, perhaps also through the absorption of small human entities, into the local socio-economic-demographic base. The groups continued hunting local species (Barbary sheep, gazelle), and collecting molluscs, plants and wild fruits. Herding was the earliest form of food production. At Haua Fteah Higgs (1967) indicated the presence of goat dated between 7.8 and 7.6 ka BP. This dating is in agreement with what has come to light from Abu Tamsa (Layer IV, 7275 ± 40 /7881-7512 cal. BP (Pa2467), de Faucamberge 2014: Fig.5), a site not far from Haua Fteah, and with the hypothesized path for the spread of the goat in Mediterranean North Africa (Barich 2014). At the moment there is evidence for a wide use and management of wild grasses, but no evidence for any domesticated plant species.

Compared to this situation, in the southern Libyan territory, in the Saharan environment, evidence shows that the advent of the moist climate of the Holocene was a time of repopulation by new peoples, thus the beginning of a totally new cultural cycle after the long phase of abandonment. Where did these ancient colonizers come from? It is evident that the phenomenon must have occurred through successive arrivals of people, from different origins. However, two main paths can be identified.

On the one hand, it could not be excluded that there was a shift of populations from the north towards the areas that were previously deserted (Cancellieri, Di Lernia 2014:59; Di Lernia 2019: 187). A link to the northern region may be recognizable in the Early Akakus industry which continues Libyco-Capsian types. There is a chronological correspondence with the Libyco-Capsian (about 11 ka BP), in addition to the similarity in the ephemeral character of the settlements associated with a hunting-gathering model, dominated by hunting and, or, *Ammotragus* management. On the other hand, the ceramic production known to the north shows a different fabric and typology.

Another hypothesis, instead, looks to the east (or south-east), focusing on the movements of groups from the Saharo-Sudanese belt. Both lithic and ceramic technology could have originated in this context. The latter, the oldest African pottery, well known in the Akakus, is widespread between Niger, Chad, Nubia up to Sudan (Jesse 2010). Pottery decorated with impressions obtained by means of pointed tools, spatulae and cords, can be compared to the Tagalagal examples in Niger, dating back to the same period of repopulation of the central-Saharan area (i.e. 11-10th millennium cal. BP). The comparisons that Close (1987) established for the Early Holocene lithic industry of Ti-n-Torha Two Caves with the El Adam horizon in Nubia, may also include comparisons with ceramic production. We can cite as an example pottery sherds coming from El Adam Site E-79-8 at Bir Kiseiba dated ca. 9180 ± 140 /10,733-10,114 cal. BP (SMU-914, Connor 1984:239-240) in full agreement with the oldest Akakus pottery, also as regards the decorative techniques. The Nubian region has been suggested as the probable area of origin of domestic animal species in the mid-Holocene Akakus (Barich 2013: 450-451; Gautier 2001). To this should also be added the anthropological characteristics recognized in HI-H3 Uan Afuda individuals, pointing to a sub-Saharan population (Manzi 2001).

Obviously, by imagining the repopulation phase as a complex and long-term process, as already mentioned, each of the two hypotheses seems plausible and does not exclude the other.

Conclusion

The long record of human presence in Libya allows us to reconstruct a scenario of constant development and change. The possibility of associating the most ancient aspects of this presence with the first modern humans, assigns it a leading position in the debate about the dynamics of diffusion of the first *H.sapiens* from sub-Saharan to Mediterranean regions. Equally important in the subsequent phases is the appearance of technological advances, both those of the Later Stone Age in stone manufacturing, and later in ceramics, rock art, and the new ways of exploiting habitat and resources. Several Neolithic aspects of the littoral and pre-desert still need to be explored in order to reach a satisfactory picture for numerous themes and, in particular relating to economic species and the possible acquisition of domesticated grain and livestock. However, current information places the development of the Northern region, beginning the Neolithic, in the context of other contemporary Maghrebian societies, open to contacts and exchanges with the Near East.

In Southern Libya, on the other hand, early Holocene groups established in the new territory appear to have had a great mastery of the landscape. They used a broad spectrum of resources, accessing diversified ecological niches. They were also very active in the use and manipulation of wild plants, similar to the northern groups, and had the same interest in hunting *Ammotragus* (here also with advanced forms of management). They built dwellings, and are associated with an artistic production of the highest level, from which we can infer aspects of their social sphere and recurring symbolism. Their aesthetic sense is also apparent in the ceramics, which are very different from the northern types, and more similar to Saharo-Sudanese models. It is a ceramic repertoire that from the beginning has very evolved characteristics, with large globular containers used to contain and prepare plant foods, providing further evidence of the early manipulation of native plants. A scenario of social groups with highly articulated practices, provides a premise for subsequent, even more articulated, food production practices focused on cattle pastoralism from ca. 8.3 ka cal. BP, thus apparently in advance of the northern region. Therefore, if the latter was oriented towards the Maghreb and the Mediterranean, since the Neolithic southern Libya has turned to the Sahara and sub-Sahara, thus prefiguring modern characteristics.

Acknowledgments

The research of the *Italian-Libyan Joint Project in the Jebel Gharbi* was started by B.E.Barich in the early 1990s at the invitation of the Department of Antiquities of Libya (DoA). Since 2003 up to the suspension of the works in 2011, E.A.A. Garcea, University of Cassino, was associated as co-director. Carlo Giraudi, Enea-Rome, was responsible for the geomorphological and environmental study; Giuseppina Mutri and Giulio Lucarini, at that time Sapienza University of Rome, were incharge for Later Stone Age and Neolithic study, respectively. Barich's activity in the Auis/Ti-n-Torha (1970-1983), was carried out as Deputy Director of the *Italian-Libyan Joint Mission for Saharan Research,* headed by S.M. Puglisi. Both research projects were organized in the framework of the Sapienza University of Rome which granted most part of the funds along with CNR, the Ministry for Foreign Affairs and the Ministry of Research and University (MIUR and PRIN Programs). I thank the Tripoli DoA Direction for having granted the official research license and acknowledge the support and collaboration of all the Libyan colleagues who took part in the fieldwork.

I would like to thank the President of the UISPP, Prof. François Djindjian, for the invitation to take part in the publication of this volume as a contribution of the Commission: 'Art and Civilizations in the Sahara during Prehistoric Times'. Finally, I am grateful to Giuseppina Mutri, who kindly provided me with the two figs. 7a,b (Site SJ-00-56) and to Claire J. Malleson for the linguistic revision of the English text.

References

Alhaique F., Marshall F. 2009. Preliminary report on the Jebel Gharbi fauna from site SJ-00-56 (2000 and 2002 excavations). *Africa* LXIV 3-4, p.498-507.

Armitage S.J., Drake N.A., Stokes S., El-Hawat A., Salem M.J. *et al.* 2007. Multiple phases of North African humidity recorded in lacustrine sediments from the Fazzan Basin, Libyan Sahara. *Quaternary Geochronology* 2, p.181–86.

Bar- Yosef O. 1994. Western Asia from the end of the Middle Palaeolithic to the beginnings of food production. In: S.J. de Laet, A.H. Dani, J.L. Lorenzo and R.B. Nunoo (eds.), *Prehistory and the beginnings of civilisation*, Vol. 1, London, History of Humanity 273, UNESCO, p.241-255.

Barca D., Mutri G. 2009. Caratterizzazione geochimica e determinazione della provenienza delle selci del Jebel Gharbi, Libia. *Africa* LXIV, p.488-497.

Barich B.E. 1974. La serie stratigrafica dell'Uadi Ti-n-Torha (Tadrart Acacus, Libia) – Per una interpretazione delle facies a ceramica saharo-sudanesi. *Origini* VIII: 7-184. https://www.antichita.uniroma1.it/sites/default/files/riv/all/2020/origini_008_1974_007-184

Barich B.E. 1978. Nuove evidenze nell'area del Tadrart Acacus (Fezzan). Missione congiunta libico-italiana per ricerche sahariane (anno 1978). *Libya Antiqua* 15-16, p.279–302.

Barich B.E. (ed.) 1987. *Archaeology and Environment in the Libyan Sahara: The Excavations in the Tadrart Acacus 1978-1983.* Oxford, British Archaeological Reports, International Series 368.

Barich B.E. 1987a. The Two Caves Shelter: an Early Holocene Site in the North-Eastern Acacus. In: B.E. Barich (ed.), *Archaeology and Environment in the Libyan Sahara: The Excavations in the Tadrart Acacus 1978-1983.* Oxford, British Archaeological Reports, International Series 368, p.13-61.

Barich B.E. 1992. The botanical collections from Ti-n-Torha/Two Caves and Uan Muhuggiag (Tadrart Acacus, Libya). An archaeological commentary. *Origini* XVI, p.109-123 https://www.researchgate.net/publication/293132864

Barich B.E. 1995. Industrie à lamelles de la region de Jado, Jebel Gharbi: modèle typologique et occupation humaine en Libye au Pleistocene Final. In: R. Chenorkian (ed.) *L'Homme Méditerranéen: mélanges offerts à Gabriel Camps, professeur émérite de l'Université de Provence.* Aix-en-Provence, p.67-74.

Barich B.E. 2009. Uso dell'acqua e continuità di occupazione nel Jebel Gharbi tra fine Pleistocene e Olocene. *Africa* LXIV 3-4, p. 422-432.

Barich B.E. 2010. *Antica Africa*. Roma, L'Erma di Bretschneider.

Barich B.E. 2013. Hunter-gatherer-fishers of the Sahara and the Sahel 12,000-4,000 years ago. In: P. Mitchell and P. Lane (eds.), *The Handbook of African Archaeology*. Oxford, Oxford University Press, p.445-459.

Barich B.E. 2014. Northwest Libya from the early to late Holocene: New data on environment and subsistence from the Jebel Gharbi. *Quaternary International* 320 (1-2), p.15-27.

Barich B.E., Capezza C., Conati Barbaro C., Giraudi C. 1995. Geoarchaeology of Jebel Gharbi, Outline of the research. *Libya Antiqua* N.S. I, p.11-35.

Barich B.E., Conati-Barbaro C. 2003. Ras el Wadi (Jebel Gharbi): New Data for the Study of the Epipalaeolithic Tradition in Northern Libya. *Origini* XXV, p.75-146. https://www.researchgate.net/publication/293755929_Ras_el_Wadi_Jebel_Gharbi_

Barich B.E., Bodrato G., Garcea E.A.A., Conati Barbaro C., Giraudi C. 2003a. Northern Libya in the final Pleistocene. The late hunting societies of Jebel Gharbi. *Studi in Memoria di Lidiano Bacchielli. Quaderni di Archeologia della Libia* 18, p.259-265.

Barich B.E., Garcea E.A.A., Conati Barbaro C., Giraudi C. 2003b. The Ras El Wadi sequence in the Jebel Gharbi and the Late Pleistocene cultures of Northern Libya. In: L. Krzyżaniak, K. Kroeper and M. Kobusiewicz (eds.), *Cultural Markers in the Later Prehistory of Northeastern Africa and Recent Research*. Poznan, Poznan Archaeological Museum, p.11-20.

Barich B.E., Garcea E.A.A. 2008. Ecological Patterns in the Upper Pleistocene and Holocene in the Jebel Gharbi, Northern Libya: Chronology, Climate and Human Occupation. *African Archaeological Review* XXV 1-2, p.86-96. www.researchgate.net/publication/225501153_Ecological_Patterns_in_the_Upper_Pleistocene_and_Holocene_in_the_Jebel_Gharbi_

Barich B.E., Garcea E.A.A., Giraudi C. 2006. Between the Mediterranean and the Sahara: Geoarchaeological Reconnaissance in the Jebel Gharbi, Libya. *Antiquity* 8, p.567–582.

Barich B.E., Garcea E.A.A., Giraudi C., Lucarini G., Mutri G. 2010. The Latest Research in the Jebel Gharbi (Northern Libya): Environment and Cultures from MSA to LSA and the First Neolithic Findings. *Libya Antiqua* N.S. V, p.237-252. https://www.researchgate.net/publication/241685383_The_latest_research_in_the_Jebel_Gharbi_

Barich B.E., Lucarini G., Mutri G. 2015. Libyan-Italian Joint Mission in the Jebel Gharbi (Tripolitania). The Holocene sequence of the Jifarāh plain. *Libya Antiqua* N.S. VIII, p. 145-159.

Barker G., Antoniadou A., Barton H., Brooks I. *et al.* 2009. The Cyrenaican Prehistory Project 2009: The third season of investigations of the Haua Fteah cave and its landscape, and further results from the 2007-2008 fieldwork. *Libyan Studies* 40, p.55-94.

Barker G., Antoniadou A., Armitage S., Brooks I., Candy I., Connell K. *et al.* 2010. The Cyrenaican Prehistory Project 2010: The fourth season of investigations of the Haua Fteah cave and its landscape, and further results from the 2007-2009 fieldwork. *Libyan Studies* 41, p.63-88.

Barker G., Basell L., Brooks I., Burn L., Cartwright C., Cole F. *et al.* 2008. The Cyrenaican Prehistory Project 2008: The second season of investigations of the Haua Fteah cave and its landscape, and further results from the initial (2007) fieldwork. *Libyan Studies* 39, p.175-221.

Barker G., Bennett P., Farr L., Hill E., Hunt C., Lucarini G., Morales J., Mutri G. *et al.* 2012. The Cyrenaican Prehistory Project 2012: The fifth season of investigations of the Haua Fteah cave. *Libyan Studies* 43, p.115–36.

Barton R.N.E., Bouzouggar A., Collcutt S., Gale R., Higham T.F.G. *et al.* 2005. The Late Upper Palaeolithic occupation of the Moroccan northwest Maghreb during the Last Glacial Maximum. *African Archaeological Review* 22(2), p.77-100.

Barton R.N.E., Bouzouggar A., Bronk Ramsey C., Collcutt S., Higham T.F.G. *et al.* 2007. Abrupt Climatic Change and Chronology of the Upper Palaeolithic in Northern and Eastern Morocco. In: P.A. Mellars, K.V. Boyle, C. Stringer, O. Bar-Yosef (eds.), *Rethinking the Human Revolution: New Behavioural and Biological Perspectives on the Origin and Dispersal of Modern Humans.* Cambridge, McDonald Institute for Archeological Research, p.177-186.

Barton R.N.E., Bouzouggar A., Collcutt S., Schwenninger J.L., Clark-Balzan L. 2009. OSL dating of the Aterian levels at Dar es-Soltan I (Rabat, Morocco) and implications for the dispersal of modern *Homo sapiens*. *Quaternary Science Reviews* 28, p.1914-1931.

Biagetti S., Di Lernia S. 2013. Holocene Deposits of Saharan Rock Shelters: The Case of Takarkori and other Sites from the Tadrart Acacus Mountains (Southwest Libya). *African Archaeological Review* 30 (3), p.305-338.

Bouzouggar A., Barton N., Vanhaeren M., d'Errico F., Collcutt S., Higham T. *et al.* 2007. 82,000 year-old shell beads from North Africa and implications for the origins of modern human behavior. *PNAS* 104 (24), p.9964–9969.

Cancellieri E., Di Lernia S. 2014. Re-entering the central Sahara at the onset of the Holocene: A territorial approach to Early Acacus hunter-gatherers (SW Libya). *Quaternary International* 320, p.43-62.

Carto S.L., Weaver A.J., Hetherington R., Lamand Y., Wiebe E.C. 2009. Out of Africa and into an ice age: on the role of global climate change in the late Pleistocene migration of early modern humans out of Africa. *Journal of Human Evolution* 56, p.139-151.

Cassoli P.F., Durante S. 1974. La fauna del Ti-n-Torha (Acacus, Libia). *Origini* VIII, p.159–161.

Chighini G. 2012. *Studio pedologico di alcuni siti archeologici del Jebel Gharbi (NW Libia).*Unpubl. MA Dissertation, Sapienza University of Rome.

Clark J.D. 1992. African and Asian perspectives on the origins of modern humans. *Philosophical Transactions: Biological Sciences 337* (1280), p.201-215.

Close A. E. 1977. *The identification of style in lithic artefacts from North East Africa*. Le Caire, Mémoires de l'Institut d'Egypte.

Close A.E. 1986. The place of the Haua Fteah in the Late Palaeolithic of North Africa. In: G. N. Bailey and P.Callow (eds.), *Stone Age prehistory: Studies in memory of Charles McBurney*. Cambridge, Cambridge University Press, p. 169-180.

Close A.E. 1987. The lithic sequence from the Wadi Ti-n-Torha (Tadrart Acacus). In: B.E. Barich (ed.), *Archaeology and Environment in the Libyan Sahara*. Oxford, British Archaeological Reports, International Series 368, p.63-85.

Close A.E. 2002. Backed bladelets are a foreign country. *Archeological Papers of the American Anthropological Association* 12(1), p.31-44.

Connor D.R. 1984. Report on Site E-79-8. In: F. Wendorf, R. Schild (assemblers), A.E. Close (ed.), *Cattle-Keepers of the Eastern Sahara – The Neolithic of Bir Kiseiba*. Dallas, Dept.of Anthropology, Southern Methodist University, p.217-250.

Cremaschi M., Di Lernia S. 1998. *Wadi Teshuinat. Palaeoenvironment and Prehistory in South-Western Fezzan (Libyan Sahara)*. Firenze e Milano, Edizioni All'Insegna del Giglio e CNR, Quaderni di Geodinamica Alpina e Quaternaria.

Crucifix M. 2019. Pleistocene Glaciations. In: E. Chiothis (ed.), *Climate Changes in the Holocene, Impacts and Human Adaptations*. London, Taylor & Francis, p.77-103.

De Faucamberge E. 2014. *Le site néolithique d'Abou Tamsa (Cyrénaïque, Libye) - Apport à la préhistoire du nord-est de l'Afrique*. Paris, Riveneuve éditions.

De Menocal P.B., Ortiz J., Guilderson J.T., Adkins J., Sranthein M. *et al*. 2000. Abrupt onset and termination of the African humid period: rapid climate responses to gradual insolation forcing. *Quaternary Science Reviews* 19, p. 347-361.

De Menocal P. B. 2015. End of the African Humid Period. *Nature Geoscience*, Advance Online Publication. www.nature.com/naturegeoscience

Dibble H.L., Aldeias V., Jacobs Z., Olszewski D.I., Rezek Z. *et al*. 2013. On the industrial attributions of the Aterian and Mousterian of the Maghreb. *Journal of Human Evolution* 64, p.194-210. http://dx.doi.org/10.1016/j.jhevol.2012.10.010

Di Lernia S. 1999. *The Uan Afuda Cave. Hunter-Gatherer Societies of Central Sahara*. Firenze, All'Insegna del Giglio.

Di Lernia S. 2002. Dry climatic events and cultural trajectories: Adjusting Middle Holocene Pastoral Economy of the Libyan Sahara. In: F.A. Hassan (ed.), *Droughts, Food and Cultures*. New York, Kluwer Academic/Plenum Publishers, p.225-250.

Di Lernia S. 2013. Places, monuments, and landscape: evidence from the Holocene central Sahara. *Azania* 48 (2), p.173-192.

Di Lernia S. 2019. From 'Green' to 'Brown'. The Archaeology of the Holocene Central Sahara. In: E. Chiothis (ed.), *Climate Changes in the Holocene, Impacts and Human Adaptations*. London, Taylor & Francis, p.183-219.

Di Lernia S., Garcea E.A.A. 1997. Some remarks on Saharan terminology. Pre-pastoral archaeology from the Libyan Sahara and the Middle Nile Valley. *Libya Antiqua* N.S. 3, p.11-23.

Di Lernia S., Zampetti D. 2008. *La memoria dell'arte. Le pitture rupestri dell'Acacus tra passato e futuro*. Firenze, Edizioni All'Insegna del Giglio.

Douka K., Grün R., Jacobs Z., Lane C., Farr L., Hunt C., Inglis R. *et al.* 2014. The chronostratigraphy of the Haua Fteah cave (Cyrenaica, northeast Libya). *Journal of Human Evolution* 66, p. 39–63.

Drake N.A., Blench R.M., Armitage S.J., Bristow C.S., White K.H. 2011. Ancient watercourses and biogeography of the Sahara explain the peopling of the desert. *Proceedings of the National Academy of Sciences* 108, p.458-462.

Drake N., Breeze P. 2016. Climate change and modern human occupation of the Sahara from MIS 6-2. In: S.C. Jones and B.A. Stewart (eds.), *Africa from MIS 6-2: Population dynamics and paleoenvironments.* Dordrecht, Springer, p.103-122.

Dunne J., Mercuri A.M., Evershed R.P., Bruni S., Di Lernia S. 2016. Earliest direct evidence of plant processing in prehistoric Saharan pottery. *Nature Plants* 3, p.1-6. 16194 DOI: 10.1038/nplants.2016.194.

Farr L., Lane R., Abdulazeez F., Bennett P., Holman J., Marasi A. *et al.* 2014. The Cyrenaican prehistory project 2013: The seventh season of excavations in the Haua Fteah cave. *Libyan Studies* 45, p.163–173.

Garcea E. A. A. (ed.) 2001. *Uan Tabu in the settlement history of the Libyan Sahara.* Firenze, All'Insegna del Giglio.

Garcea E. A. A. 2001a. The Pleistocene and Holocene archaeological sequence. In: E.A.A.Garcea (ed.), *Uan Tabu in the settlement history of the Libyan Sahara.* Firenze, All'Insegna del Giglio, p.1-14.

Garcea E.A.A. 2009. L'adaptation atérienne entre sources d'eau et sécheresse. *Africa* LXIV 3-4, p.412-421.

Garcea E.A.A. 2010a. The spread of Aterian peoples in North Africa. In: E.A.A. Garcea (ed.), *South-eastern Mediterranean peoples between 130,000 and 10,000 years ago.* Oxford, Oxbow Books, p.37-53.

Garcea E.A.A. 2010b. The Lower and Upper Late Stone Age of North Africa. In: E.A.A. Garcea (ed.), *South-eastern Mediterranean peoples between 130,000 and 10,000 years ago.* Oxford, Oxbow Books, p.54-65.

Garcea E.A.A., Giraudi C. 2006. Late Quaternary human settlement patterning in the Jebel Gharbi. *Journal of Human Evolution* 51, p.411-421.

Gautier A. 1987. The archaeozoological sequence of the Acacus. In: B.E. Barich (ed.), *Archaeology and Environment in the Libyan Sahara. The excavations in the Tadrart Acacus 1978-1983.* Oxford, British Archaeological Reports, International Series 368, p.283-308.

Gautier A. 2001. The Early to Late Neolithic Archeofaunas from Nabta and Bir Kiseiba. In: F. Wendorf, R. Schild and Associates (eds.), *Holocene Settlement of the Egyptian Sahara, Volume I: The Archaeology of Nabta Playa.* New York Kluwer Academic/ Plenum Publishers, p.609-635.

Gautier A., Van Neer T. 1977-1982. Prehistoric Fauna from Ti-n-Torha (Tadrart Acacus, Libya). *Origini* XI, p.87-127.

Giraudi C. 2005. Eolian sand in peridesert northwestern Libya and implications for Late Pleistocene and Holocene Sahara expansions. *Palaeogeography, Palaeoclimatology, Palaeoecology* 218, p.161–173.

Giraudi C. 2009. Le fasi umide del Pleistocene Superiore e dell'Olocene nel Jebel Gharbi (Libia nordoccidentale). *Africa* LXIV 3-4, p.405-411.

Giraudi C., Mercuri A.M., Esu D. 2012. Holocene palaeoclimate in the northern Sahara margin (Jefara Plain, northwestern Libya).*The Holocene*, p.1-14 (on line) 10.1177/0959683612460787

Henshilwood Ch., d'Errico F., Vanhaeren M., van Niekerk K., Jacobs Z. 2004. Middle Stone Age Shell Beads from South Africa. *Science* 304(5669), p.404. 10.1126/science.1095905

Higgs E. 1967. Environment and chronology: The evidence from mammalian fauna. In: McBurney C.B.M. (ed.), *The Haua Fteah (Cyrenaica) and the Stone Age of the south-east Mediterranean.* Cambridge, Cambridge University Press, p.16-44.

Hublin J-J. 2001. Northwestern African Middle Pleistocene hominids and their bearing on the emergence of Homo sapiens. In: L. Barham and K. Robson-Brown (eds.), *Human Roots: Africa and Asia in the Middle Pleistocene.* Bristol, England, Western Academic and Specialist Press, p.99-121.

Hunt C., Davison J., Inglis R., Farr L., Reynolds T., Simpson D. *et al.* 2010. Site formation processes in caves: The Holocene sediments of the Haua Fteah, Cyrenaica, Libya. *Journal of Archaeological Science* 37(7), p.1600-1611.

Hunt C.O., Reynolds T.G., El-Rishi H.A., Buzaian A., Hill E., Barker G.W. 2011. Resource pressure and environmental change on the North African littoral: Epipalaeolithic to Roman gastropods from Cyrenaica, Libya. *Quaternary International* 244(1), p.15-26.

Inglis R.H. 2012. *Human occupation and changing environments during the Middle to Later Stone Age: Soil micromorphology at the Haua Fteah, Libya.* Ph.D. Dissertation, University of Cambridge.

Iovita R. 2002. *A Comparison of the Earliest Upper Palaeolithic at Ksar Akil (Lebanon) and Haua Fteah (Libya).* MA.Phil. Dissertation, University of Cambridge.

Iovita R.P. 2009. Reevaluating Connections between the early Upper Paleolithic of northeast Africa and the Levant: Technological differences between the Dabban and the Emiran. In: J.J. Shea: D.E. Lieberman (eds.), *Transitions in prehistory: Essays in honor of Ofer Bar-Yosef.* Oxford, Oxbow Books, p.125–142.

Jacobs Z., Li B., Farr L., Hill E., Hunt C., Jones S., Rabett R., Reynolds T. *et al.* 2017. The chronostratigraphy of the Haua Fteah cave (Cyrenaica, northeast Libya) - Optical dating of early human occupation during Marine Isotope Stages 4, 5 and 6. *Journal of Human Evolution* 105, p.69-88. 10.1016/j.jhevol.2017.01.008. Epub 2017 Mar 18.

Jesse F. 2010. Early Pottery in Northern Africa – An overview. *Journal of African Archaeology* 8(2), p.219-238.

Jones S., Farr L., Barton H., Drake N., White K., Barker G. 2011. Geoarchaeological patterns in the pre-desert and desert ecozones of northern Cyrenaica. *Libyan Studies* 42, p.11–19.

Jones S., Antoniadou A., Barton H., Drake N., Farr L., Hunt C., Inglis R. *et al.* 2016. Patterns of Hominin Occupation and Cultural Diversity across the Gebel Akhdar of Northern Libya over the Last ~200 kyr. In: S.C. Jones and B.A. Stewart (eds.), *Africa from MIS 6-2: Population dynamics and Paleoenvironments.* Dordrecht, Springer, p.77-99. https://link.springer.com/chapter/10.1007/978-94-017-7520-5_5

Klein R.G. 1989. *The Human Career.* Chicago, Chicago University Press.

Klein R.G., Scott K. 1986. Re-analysis of faunal assemblages from the Haua Fteah and other Late Quaternary archaeological sites in Cyrenaican Libya. *Journal of Archaeological Science* 13, p.514-542.

Larrasoaña J.C, Roberts A.P., Rohling E.J. 2013. Dynamics of Green Sahara Periods and Their Role in Hominin Evolution. *PLoS ONE* 8(10): e76514. https://doi.org/10.1371/journal.pone.0076514

Lucarini G. 2009. Percorsi tra la montagna e il mare. Nuovi dati sui complessi olocenici della pianura della Gefara (Libia). *Africa* LXIV 3-4, p.433-447.

Lucarini G., Mutri G. 2010a. Site SJ-00-56: Débitage Analysis and Functional Interpretation of a Later Stone Age Campsite of Jebel Gharbi (Libya). *Journal of Human Evolution* xxv (1-2), p.155-165.

Lucarini G., Mutri G. 2010b. Preliminary observations on the Capsian, Oranian, and final Dabban occupation in the Haua Fteah (Trench M) - The lithic assemblage. In: G.Barker et al., The Cyrenaican Prehistory Project 2009: The third season of investigations of the Haua Fteah cave and its landscape, and further results from the 2007-2008 fieldwork. *Libyan Studies* 40, p.72-74

Lucarini G., Mutri G. 2014. Microlithism and Landscape Exploitation along the Cyrenaican Coast between the Late Pleistocene and the Holocene: A matter of continuity. In: K. Boyle, R.J. Rabett and C.O. Hunt (eds.).*Living in the Landscape. Essay in Honour of Graeme Barker.* Cambridge, Cambridge University Press, p.109-120.

Lucarini G., Radini A., Barton H., Barker G. 2016. The exploitation of wild plants in Neolithic North Africa. Use-Wear and residue analysis on non-knapped stone tools from the Haua Fteah cave, Cyrenaica, Libya. *Quaternary International* 410 Part A, p.77-92.

MacDonald K.C.1997. The avifauna of the Haua Fteah (Libya). *Archaeozoologia* 9, p.83-102.

Manzi G., Passarello P. 1999. Human remains –deciduous and permanent teeth, In: S. Di Lernia (ed.), *The Uan Afuda Cave. Hunter-Gatherer Societies of Central Sahara.* Firenze, All'Insegna del Giglio, p.203-207.

McBurney C.B.M. 1967. *The Haua Fteah (Cyrenaica) and the stone age of the south-east Mediterranean.* Cambridge, Cambridge University Press.

McBurney C.B.M., Hey R.W. 1955. *Prehistory and Pleistocene geology in Cyrenaican Libya.* Cambridge, Cambridge University Press.

Mercuri A.M. 2001. Preliminary analyses of fruits, seeds and few plant macrofossils from the Early Holocene sequence. In: E.A.A. Garcea (ed.) *Uan Tabu in the settlement history of the Libyan Sahara.* Firenze, All'Insegna del Giglio, p.189-210.

Moeyersons J., Vermeersch P.M., Van Peer P. 2002. Dry Cave deposits and their palaeoenvironmental significance during the last 115 ka, Sodmein Cave, Red Sea Mountains, Egypt. *Quaternary Science Reviews* 21, p.837-851.

Morales J., Barker G. 2009. The macrobotanical remains from the 2008 season. In: G. Barker et al., The Cyrenaican Prehistory Project 2009: The third season of investigations of the Haua Fteah cave and its landscape, and further results from the 2007-2008 fieldwork. *Libyan Studies* 40, p.30-33.

Mori F. 1965. *Tadrart Acacus. Arte rupestre e Culture del Sahara preistorico.* Torino, Einaudi.

Mori F. 1998. *The great civilisations of the ancient Sahara - Neolithisation and the earliest evidence of anthropomorphic religions.* Rome, L'Erma di Bretschneider.

Moyer C. 2003. *The organisation of lithic technology in the middle and early upper palaeolithic industries at the Haua Fteah, Libya.* Ph.D. dissertation, University of Cambridge.

Mutri G. 2008-2009. *Stili tecnologici nelle industrie litiche Tardo Pleistoceniche del Jebel Gharbi (Libia nord occidentale).Approvvigionamento delle materie prime e sequenze operative.* PH.D. Dissertation: Sapienza University of Rome.

Mutri G., Lucarini G. 2008. New data on the Late Pleistocene of the Shakshuk area, Jebel Gharbi, Libya. *African Archaeological Review* xxv (1-2), p.99-107.

Nicoll K. 2004. Recent environmental change and prehistoric human activity in Egypt and Northern Sudan. *Quaternary Science Reviews* 23, p.561-580.

Nicoll K. 2018. A revised chronology for Pleistocene paleolakes and Middle Stone Age and Middle Paleolithic cultural activity at Bîr Tirfawi and Bîr Sahara in the Egyptian Sahara. *Quaternary Inter*national 463, p.18-28. http://dx.doi.org/10.1016/j.quaint.2016.08.037

Petrullo G., Barich B.E. 2020. The bone artefact collection from Wadi Tin-Torha (northern Tadrart Akakus, Libya): a reappraisal using new methodologies. *African Archaeological Review*, doi: 10.1007/s10437-020-09374-x

Revel M., Ducassou E., Grousset E., Bernasconi M., Migeon S. *et al.* 2010. 100,000 Years of African monsoon variability recorded in sediments of the Nile margin. *Quaternary Science Reviews* 29 (11-12), p.1342-1362.

Reynolds T. 2013. The Middle Palaeolithic of Cyrenaica: Is there an Aterian at the Haua Fteah and does it matter? *Quaternary International* 300, p. 171-181. http://dx.doi.org/10.1016/j.quaint.2012.09.025

Russell N.J., Armitage S.J. 2012. A comparison of single-grain and small aliquot dating of fine sand from Cyrenaica, northern Libya. *Quaternary Geochronology* 10, p.62-67.

Scerri E.M.L. 2013. The Aterian and its place in the North African Middle Stone Age. *Quaternary International* 300. http://dx.doi.org/10.1016/j.quaint.2012.09.008

Schild R., Wendorf F. 2010. Late Palaeolithic hunter-gatherers in the Nile Valley of Nubia and Upper Egypt. In: E.A.A.Garcea (ed.), *South-eastern Mediterranean peoples between 130,000 and 10,000 years ago.* Oxford, Oxbow Books, p. 89-125.

Smith J.R. 2010. Palaeoenvironments of eastern North Africa and the Levant in the late Pleistocene. In: E.A.A. Garcea (ed.), *South-eastern Mediterranean peoples between 130,000 and 10,000 years ago.* Oxford, Oxbow Books, p.6-17.

Smith J.R., Gigengack R., Schwarcz H.P., McDonald M.M., Kleindienst M.R. *et al.* 2004. A reconstruction of Quaternary pluvial environments and human occupations using stratigraphy and geochronology of fossil-spring tufas, Kharga Oasis, Egypt. *Geoarchaeology* 19, p.407-439.

Smith J.R., Hawkins A.L., Asmerom Y., Polyak V., Giegengack R. 2007. New age constraints on the Middle Stone Age occupations of Kharga Oasis, Western Desert, Egypt. *Journal of Human Evolution* 52, p.690-701.

Spinapolice E.E., Garcea E.A.A. 2013. The Aterian from the Jebel Gharbi (Libya): New Technological perspectives from North Africa. *African Archaeological Review* 30(2), p.169-194.

Van Peer P. 2014. Hints at Middle Stone Age occupation in the Farafra Oasis. In: B.E. Barich, G. Lucarini, M.A. Hamdan and F.A. Hassan (eds.), *From Lake to Sand – The Archaeology of Farafra Oasis, Western Desert, Egypt.* Firenze, All'Insegna del Giglio, p.25-58.

Van Peer P., Vermeersch P. 2000. The Nubian complex and the dispersal of modern humans in North Africa. In: L. Krzyzaniak, K. Kroeper and M. Kobusiewicz (eds.), *Recent research into the Stone Age of Northeastern Africa.* Poznan, Poznan Archaeological Museum, p.47-60.

Van Peer P., Vermeersch P., Paulissen E. 2010. *Chert Quarrying, Lithic Technology and a Modern Human Burial at the Palaeolithic Site of Taramsa I, Upper Egypt.* Egyptian Prehistory Monograph 5. Leuven, Leuven University Press.

Vignard E. 1920. Une station aurignacienne à Nag Hamadi (Haute Egypte), station du Champ de Bagasse. *Bulletin de l'Institut Français d'Archéologie Orientale* 18, p.1-20.

Wasylikowa K. 1992. Holocene flora of the Tadrart Acacus area, SW Libya, based on plant macrofossils from Uan Muhuggiag and Ti-n-Torha/Two Caves archaeological sites. *Origini* XVI, p. 125-159.https://www.researchgate.net/publication/293132864.

Wendorf F., Schild R., Close A.E. *et al.* (eds.) 1993. *Egypt during the Last Interglacial: The Middle Paleolithic of Bir Tarfawi and Bir Sahara East.* New York, Plenum Press.

Le rôle du Sahara dans l'évolution humaine en périodes humides, lorsqu'il n'était pas un désert

Miguel Caparros[1]

Abstract

The rock art paintings from the Tassili, Algeria, first inventoried and made famous by Henri Lhote in 1958, are testimony of a most striking feature of the African climate history, namely the transformation of the increasingly arid Sahara during the Pleistocene into a humid and green landscape sprinkled of lakes during the Holocene (11 to 5.0 ka BP). The paintings are evidence of a Saharo-Sudanese neolithic culture contemporaneous with the onset of this humid period in the Sahara. Six trips in central Sahara motivated us to investigate the scientific proofs of a wet Sahara, whether humid periods were prevalent during the Pleistocene and to which extent a Sahara open to migrations might have played a role in human evolution.

After summarizing the topographic and climatic features of the Sahara in a broad geological context of the African continent, we review the five styles of rock art identified in central Sahara (Bubalin-Large african fauna, Têtes rondes, Bovidian, Caballin and Camelin), their approximate chronology and their significance. It is clear that these different styles follow the progressive changes of the ecosystems affected by the northward migration of the West African summer monsoon during the Holocene, and the later southward shift. Various climate change studies, based on analyses of run-offs of the Nile and Niger rivers, changes in lake shore levels in central Sahara, and Atlantic deep offshore sampling of pollens and terrigenous deposits, confirm the existence of a progressive humid period in the Sahara from 15 to 5 ka interspersed with small intervals of dessication and a peak of humidity at around 9 ka. Several archaeological studies support the advent of a pastoral culture in the Sahara as reflected in the rock art further to the Neolithic revolution that took place in the Near East. It is scientifically documented that the main cause of the humidification of the Sahara during the Holocene is the northward shift of the Intertropical Convergence Zone (ITCZ) linked to: the glacial-interglacial instability in northern latitudes, changes of insolation and variations of atmospheric conditions in the West Atlantic.

Further continental and marine paleoclimatic studies up to 1.2 Ma confirm the existence of numerous humid periods during the Pleistocene, in particular from the analysis of sapropels which are blackish sediments rich in organic matter deposited in the Ionian Sea further to large run-offs of the Nile river linked to increased pluvial intensity in East Africa and the Sahara. If one uses sapropels as proxies of humid periods in the Sahara, such wet events would have occurred 28 times over

---

1 UMR 7194 «Histoire naturelle de l'Homme préhistorique» CNRS-MNHN-UPVD, Alliance Sorbonne Université, Musée de l'Homme, Palais de Chaillot, Paris.

the last 1.2 Ma. Humid periods in the Sahara translate into a green ecosystem covered in its majority by savannas and steppes, littered of lakes, some of them gigantic like the Mega-Tchad and Mega-Fezzan, with aquatic corridors allowing the passage of human groups who were prevented from settling in the Sahara during hyperarid glacial periods. The human presence in the Sahara in prehistoric times is supported by numerous archaeological sites throughout its extension and in North Africa, with lithic industries ranging from Oldowayan, Acheulean up to Middle Palaeolithic techno-complexes like Mousterian and Aterian, within a time frame of approximately 2.4 Ma. This presence occurred without any doubt during humid periods, particularly considering that there are no Upper Palaeolithic sites in the Sahara connected with the last hyperarid MIS 2 cold period, proof that humans were not present there during glacial periods.

We present a simple graphical model of arid-humid-arid cycles we name A-H-A to illustrate the ecosystem changes during a full cycle of glacial to interglacial back to glacial periods. We show that during periods of extreme aridification, North Africa must have acted periodically as refugia to human groups that had migrated during humid periods. The numerous alternating closing and reopening phases of the Sahara during the Pleistocene might have led to multiple events of genetic exchanges between isolated human groups in the Sahara and Subsaharan Africa, which would favor a reticulate mode of evolution of the genus Homo. We show the comparative chronology of the various species of the genus Homo with an index of wet-dry phases, and offer as conclusion various suggestions of further research in palaeoanthropology by taking into consideration paleoclimatic factors in East Africa and the Sahara. This would help shed further light in unresolved issues such as the timing of the emergence of the genus Homo, its exit Out of Africa and clarification of the phylogenetic relationships of the diverse group of archaic human fossils from the Middle Pleistocene in Africa to unravel the evolutionary process that led to modern humans.

Résumé

Les peintures d'art rupestre du Tassili, en Algérie, inventoriées et rendues célèbres par Henri Lhote en 1958, témoignent d'une caractéristique frappante de l'histoire climatique africaine, à savoir la transformation du Sahara de plus en plus aride au cours du Pléistocène en un paysage humide et verdoyant parsemé de lacs pendant l'Holocène (11 à 5,0 ka BP). Ces peintures représentent l'évidence d'une culture néolithique saharo-soudanaise contemporaine du début de cette période humide au Sahara. Six voyages dans le Sahara central nous ont motivés à étudier les preuves scientifiques d'un Sahara humide, si les périodes humides étaient répandues pendant le Pléistocène et dans quelle mesure un Sahara ouvert aux migrations aurait pu jouer un rôle dans l'évolution humaine.

Après avoir résumé les caractéristiques topographiques et climatiques du Sahara dans le large contexte géologique du continent africain, nous passons en revue les cinq styles d'art rupestre identifiés dans le Sahara central (Bubalin-à grande *faune africaine, Têtes rondes, Bovidien, Caballin et Camelin), leur chronologie approximative et leur signification. Il est clair que ces différents styles suivent les changements progressifs des écosystèmes affectés par la migration vers le Nord de la mousson d'été Ouest-africaine pendant l'Holocène, et le déplacement ultérieur vers le Sud. Diverses études sur le changement climatique, basées sur des analyses des crues du Nil et du Niger, des changements dans les niveaux des rives des lacs dans le Sahara central et de l'échantillonnage des pollens et des dépôts terrigènes en mer de l'Atlantique, confirment l'existence d'une période humide progressive dans le Sahara de 15 à 5 ka entrecoupée de petits intervalles de dessication et d'un pic d'humidité à environ 9 ka. Plusieurs études archéologiques soutiennent l'avènement d'une culture pastorale au Sahara telle qu'elle se reflète dans l'art rupestre suite à la révolution néolithique qui a eu lieu au Proche-Orient. Il est scientifiquement documenté que la principale cause de l'humidification du Sahara au cours de l'Holocène est le déplacement vers le Nord de la zone de convergence intertropicale (ITCZ) liée à: l'instabilité glaciaire-interglaciaire dans les latitudes septentrionales, les changements d'insolation et les variations des conditions atmosphériques dans l'Atlantique Ouest.*

D'autres études paléoclimatiques continentales et marines jusqu'à 1,2 Ma confirment l'existence de nombreuses périodes humides au cours du Pléistocène, en particulier à partir de l'analyse des sapropels qui sont des sédiments noirâtres riches en matière organique déposés dans la mer Ionienne suite aux importantes crues du Nil liées à une intensité pluviale accrue en Afrique de l'Est et au Sahara. Si l'on utilise des sapropels comme indicateurs des périodes humides dans le Sahara, de tels événements humides se seraient produits 28 fois au cours des derniers 1,2 Ma. Les périodes humides du Sahara se traduisent par un écosystème vert recouvert en majorité de savanes et de steppes, jonchées de lacs, dont certains gigantesques comme le Mega-Tchad et le Mega-Fezzan, avec des corridors aquatiques permettant le passage de groupes humains empêchés de s'installer au Sahara pendant les périodes glaciaires hyperarides. La présence humaine dans le Sahara à l'époque préhistorique est soutenue par de nombreux sites archéologiques tout au long de son extension et en Afrique du Nord, avec des industries lithiques allant de l'Oldowayen à l'Acheuléen jusqu'à des techno-complexes du Paléolithique moyen comme le Moustérien et Atérien, dans un laps de temps d'environ 2,4 Ma. Cette présence s'est produite sans aucun doute pendant les périodes humides, d'autant plus qu'il n'y a pas de sites du paléolithique supérieur dans le Sahara liés à la dernière période froide hyperaride MIS 2, preuve que les humains n'y étaient pas présents pendant les périodes glaciaires.

Nous présentons un modèle graphique simple des cycles arides-humides-arides que nous nommons A-H-A pour illustrer les changements de l'écosystème au cours d'un cycle complet de périodes glaciaires à interglaciaires remontant à des périodes glaciaires. Nous montrons que pendant les périodes d'aridification extrême, l'Afrique du Nord a dû agir périodiquement comme refuge pour les groupes humains qui avaient migré pendant les périodes humides. Les nombreuses phases alternées de fermeture et de réouverture du Sahara au cours du Pléistocène auraient pu conduire à de multiples événements d'échanges génétiques entre groupes humains isolés au

Sahara et en Afrique subsaharienne, ce qui favoriserait un mode d'évolution réticulé du genre Homo.

Nous montrons la chronologie comparative des différentes espèces du genre Homo avec un indice de phases humides-sèches, et proposons en conclusion diverses suggestions de recherches supplémentaires en paléoanthropologie en tenant compte des facteurs paléoclimatiques en Afrique de l'Est et au Sahara. Cela aiderait à faire la lumière sur des questions non résolues telles que le moment de l'émergence du genre Homo, sa sortie d'Afrique et la clarification des relations phylogénétiques du groupe diversifié de fossiles humains archaïques du Pléistocène moyen en Afrique pour démêler le processus évolutif qui a conduit à l'homme moderne.

Introduction

L'émergence du genre *Homo* en Afrique est estimée chronologiquement s'être produite il y a 2,8 Ma, bien que les spécimens venant en support de cette date sont fragmentaires (Prat, 2017); pour des fossiles tels que *Homo rudolfensis* la date estimée serait d'environ 2,5 Ma. L'explication la plus communément proposée quant à sa diffusion géographique hors du continent africain est la suivante: depuis la vallée du rift, berceau de l'humanité, ses premiers représentants migrent à l'est le long de la vallée du Nil, poursuivent leur périple par le Moyen Orient, puis se répandent et colonisent l'Eurasie. Ce simple schéma de diffusion par vagues successives, s'appliquerait non seulement à la sortie d'*Homo ergaster* il y a environ 2 Ma, schéma connu sous l'appellation anglo-saxonne d'« Out of Africa 1 », mais aussi au début du Pléistocène Moyen à *Homo heidelbergensis*, ancêtre supposé des Néandertaliens et Hommes Modernes (Lahr, 2010), ainsi qu'à la sortie d'Afrique plus récente (« Out of Africa 2 ») de notre espèce *Homo sapiens* qui viendra substituer par assimilation les populations archaïques Néandertaliennes en Europe et celles d'*Homo erectus* en Asie (Smith *et al.* 2017). Dans ces modèles de dispersion des hommes hors d'Afrique il est presque toujours sous-entendu que le Sahara fut une barrière infranchissable de façon permanente jusqu'au Pléistocène Moyen. Il nous semble important de déterminer si le Sahara était impénétrable au cours de l'évolution humaine, et représentait vraiment une barrière à l'expansion des hommes depuis leur berceau d'origine africain. Pour ce faire nous remonterons dans le temps en examinant le Sahara verdoyant en périodes humides au cours de l'Holocène et du Pléistocène.

La dernière évidence paléoclimatique d'un Sahara vert se manifeste au début de l'Holocène il y a 11 ka par le magnifique art rupestre qui apparait lors de la transition de la dernière période glaciaire d'un désert hyperaride vers un environnement humide. Nous avons effectué six expéditions au Sahara central de 2006 à 2011, dont trois en Algérie (Tassili N'Ajjer et volcans, Immidir) et trois en Libye (Akakus, Messak et Wadi Aramat), pour étudier cet art admirable en tant que témoignage de la présence humaine lors de la dernière période pluviale, art rupestre par certains aspects spectaculaire et grandiose. Ceci nous a emmené à poser les questions: combien de périodes humides le Sahara a-t-il connu depuis l'émergence du genre *Homo*, et quelle incidence une éventuelle alternance cyclique de périodes hyperarides à périodes pluviales a-t-elle pu avoir sur le mode d'évolution du genre *Homo*; à savoir quelle hypothèse évolutive est la plus vraisemblable: anagénétique, cladogénétique ou réticulée. Nous examinerons donc les preuves scientifiques de l'humidification du Sahara durant l'Holocène, ainsi que d'un point de vue climatologique et préhistorique les preuves d'une alternance de périodes humides et hyperarides au cours du Pléistocène. Nous conclurons en choisissant une des deux hypothèses, à savoir le Sahara fut-il toujours un obstacle infranchissable aux diverses migrations de l'homme vers le Nord de l'Afrique et l'Eurasie, ou une zone géographique de passage de grande extension permettant à des groupes humains isolés dans son pourtour en périodes hyperarides de se rencontrer en périodes humides, et ainsi favoriser un mode d'évolution humaine réticulée par hybridation introgressive au cours du Pléistocène? Avant d'entrer dans le vif du sujet, nous résumerons brièvement dans un contexte continental africain les particularités géographiques, climatiques et topographiques du Sahara.

Particularités géographique, climatique et topographique du Sahara

Avec une superficie approximative de 9 million de km² représentant environ un tiers du continent africain, le Sahara est le plus grand désert chaud et hyperaride du monde. S'étendant de l'Atlantique à la mer Rouge, il se trouve limité au nord par le massif de l'Atlas au Maroc ainsi que par l'Atlas Saharien en Algérie et en Tunisie, puis par la plaine côtière en Tunisie, Libye et Egypte, et enfin au sud par le Sahel, région de transition biogéographique et semi-aride couvrant d'ouest en est la Mauritanie, le Mali, le Niger, le Tchad et le Soudan (Fig. 1). Son climat actuel se caractérise par de basses précipitations pluviométriques sujettes à une grande diversité régionale avec des apports annuels de 300 mm au sud à la limite sahélienne, 150 mm sur la bordure nord du littoral et jusqu'à 5 mm dans les zones hyperarides telles que le Tanezrouft, le Ténéré et le désert Libyque. L'anomalie de la permanence des hautes pressions atmosphériques renforce l'anticyclone, et résulte en un air surchauffé au sol ne pouvant pas s'échapper vers le haut, et écartant toute possibilité de pluie (Rognon, 1994). Cette anomalie peut changer à l'échelle de millénaires due aux variations latitudinales de la zone de convergence intertropicale ZCIT de la mousson d'été en Afrique de l'Ouest (intertropical convergence zone ITCZ).

D'un point de vue topographique, l'Afrique se caractérise par une dualité : une topographie élevée en moyenne entre environ 500 et 3000 m dans l'est (plateaux volcaniques d'Ethiopie et du rift Africain) et le sud (escarpement du Drakensberg en Afrique du Sud et plateau Bié en Angola), plus élevée dans le nord-ouest (Atlas atteignant 4167 m au Toubkal), contrastée avec une topographie plus basse au centre, à l'ouest et le nord-est, celle-ci entrecoupée d'élévations de nature volcanique (Hoggar, Tibesti, Cameroun,..) et de bassins hydrographiques (Congo, Tchad, Niger, Nil, ..). Deux hypothèses s'opposent quant à l'explication de cette dualité. La première suggère que la majorité de cette topographie s'est mise en place circa 30 Ma (Burke, 1996). L'autre (Doucouré et de Wit, 2003) propose une explication plus ancienne, à savoir l'Afrique suite à son détachement de Pangaea était un continent de basse altitude dominé par des cratons avec des élévations topographiques épisodiques dans le centre et le sud, et une inférence selon laquelle, en synchronie relative avec des changements des niveaux de la mer, la quasi bimodalité de l'Afrique est une ancienne caractéristique héritée au moins du Paléozoïque supérieur. En ce qui concerne le Sahara (Fishwick et Bastow,

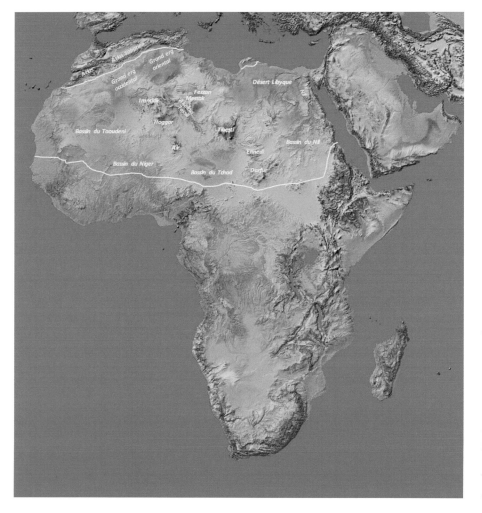

Figure 1. Carte de l'Afrique avec le Sahara délimité par les lignes jaunes. Les reliefs sont en couleur verte, la teinte plus obscure représentant une altitude plus élevée. Reproduite d'après NASA Super High Resolution Topographic Map of Africa.

2011), les hauts reliefs de l'Atlas dans la limite nord sont associés à la convergence des plaques tectoniques d'Afrique et d'Europe, alors que les zones d'élévations du métacraton Saharien en son centre (Hoggar, Tibesti, Air, Darfur) sont associées à du volcanisme actif de l'Oligocène jusqu'au Quaternaire avec des caractéristiques différentes des points chauds de l'Afrique de l'Est. Les bassins sahariens intracontinentaux ont été sujets à subsidence dont les causes sont sujettes à débat.

Malgré la variété géomorphologique apparente des diverses régions du Sahara, il existe des caractéristiques similaires stratigraphiques synchrones au cours du Cénozoïque sur toute son étendue (Sweeney, 2009). Selon cette étude, l'observation d'une transition lithologique générale coïncide avec une baisse eustatique à long terme du niveau de la mer depuis le Crétacé moyen et une transition climatique globale d'un Crétacé-Eocène moyen «chaud» à un Eocène supérieur-Quaternaire «froid» qui serait corrélé avec l'accumulation croissante de glace dans l'Antarctique et l'apparition de climats arides au Sahara. L'apparition d'un Sahara hyperaride est supposée coïncider avec les premières glaciations de l'hémisphère Nord il y a 2,5 Ma (Kröpelin, 2006); toutefois une étude de dépôts éoliens dans le bassin du Tchad témoigne du début de conditions désertiques récurrentes il y a 7 Ma (Schuster *et al.* 2006). Ces changements de désertification climatique du Sahara sont confirmés par une étude de simulation paléoclimatologique (Zhang *et al.* 2014) qui montre que la cause de l'aridification est due à un réarrangement de la masse terrestre autour de l'ancienne mer Téthys entre 7 et 11 Ma, laquelle diminua de taille avec une élévation de la péninsule arabique. Ces changements d'origine tectonique entraînèrent des changements climatiques par affaiblissement de la mousson Africaine d'été. En effet lorsque les vents de l'ouest déclinèrent, les flux d'humidité de l'Atlantique tropical qui couvraient le nord de l'Afrique jouissant d'un climat semi-aride se décalèrent vers le sud.

L'art rupestre, première évidence d'un Sahara humide à l'Holocène

Au congrès panafricain de 1952, l'abbé Breuil fit connaître au monde scientifique les merveilles de l'art rupestre du Tassili N'Ajjer, Algérie (Breuil, 1955). Pendant 16 mois de 1956 à 1957, Henri Lhote, explorateur naturaliste et ethnologue, dirigea une mission archéologique sous l'auspice du Musée de l'Homme de Paris pour inventorier et étudier cet art préhistorique (Lhote, 1958), puis en 1957-1958, il dévoila au grand public la beauté et la signification de ces peintures et gravures par une exposition au pavillon de Marsan. Mettant en garde que ses suggestions étaient provisoires et sujettes à critiques, Lhote proposa la classification stylistique classique et chronologique comprenant cinq phases, Bubalin, Têtes Rondes, Bovidien, Caballin et Camelin, l'antiquité des gravures du Bubalin et les peintures des périodes Têtes Rondes et Bovidien coïncidant avec le début d'une culture

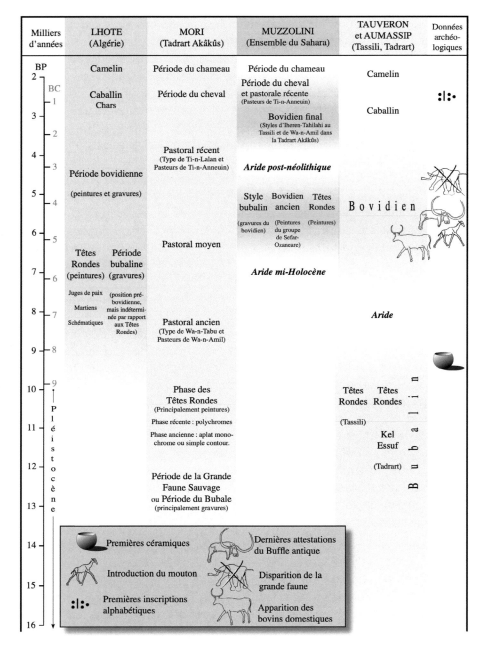

Figure 2. Classification stylistique de l'art rupestre du Sahara selon plusieurs auteurs (d'après Jean-Loïc Le Quellec).

pastorale Saharo-Soudanaise florissante datant du début de l'Holocène (Le Quellec, 1991; Lhote, 1976). Cette classification sujette à critiques (Le Quellec, 1993) est à contraster avec les classifications d'art rupestre saharien d'autres auteurs telles que résumées en Figure 2.

La classification stylistique s'appuie en général sur des animaux représentant des «fossiles directeurs» tels que le buffle antique pour le Bubalin, style principalement de gravures comprenant aussi un grand nombre de grands herbivores de faune africaine (éléphants, hippopotames, rhinocéros, girafes) et quelques félins (panthères, lions), et pour ce qui est essentiellement styles de peintures, les bovidés domestiqués du Néolithique pour le Bovidien, le cheval domestiqué suite à son introduction au Sahara pour le Caballin, et le chameau pour le Camelin. On peut distinguer une chronologie longue (Fig. 2, Mori et Tauveron- Aumassip) avec les styles Bubalin et Têtes Rondes représentant supposément des chasseurs cueilleurs prénéolithiques, et une chronologie courte qui correspondrait à l'introduction du pastoralisme néolithique au Sahara en période humide suite à la domestication de bovidés, avec le style Bovidien ancien étant contemporain des styles Bubalin et Têtes Rondes (Fig. 2, Muzzolini), ou ces derniers le précédant (Fig. 2, Lhote).

Le Quellec (Le Quellec, 2013), dans une synthèse de toutes les techniques utilisées pour dater aussi bien les gravures du Messak que les peintures de l'Akakus en Libye et celles en particulier du Tassili N'Ajjer, recommande d'abandonner l'appellation Bubalin qui regroupe des figures très disparates (faune africaine, faune pastorale et représentations anthropiques) au profit d'une classification naturaliste. On peut tirer de cette étude une périodisation très approximative pour les divers ensembles stylistiques qui semble valider la chronologie longue comme suit:

- Grande faune africaine (peintures et pétroglyphes) : 9.000 ⇒ 4.000 BP.
- Gravures styles Messak et Tazina : 7.000 ⇒ 4.000 BP.
- Peintures des Têtes Rondes (Tassili et Akakus) : 10.000-9.000 ⇒ 6.000-5.000 BP (Mercier, 2012; Hachid, 2016).
- Bovidien : 7.500 ⇒ 4.000 BP.
- Caballin : 4.000 ⇒ 2.000 BP.
- Camelin : 5ième siècle ⇒ à nos jours.

La diffusion du bétail domestiqué en Afrique depuis le Moyen-Orient lors du processus de néolithisation est mis en évidence à partir de 8.000 ans par une série d'isochrones émanant de données archéologiques de plusieurs sites de la dernière période humide au Sahara (Wright, 2017). Ces dates progressivement plus jeunes d'est en ouest viennent en support de la chronologie des peintures du Bovidien représentant diverses scènes pastorales. D'autre part, des datations par radiocarbone de 150 sites archéologiques dans la partie hyperaride de l'est du Sahara montrent un lien étroit entre les variations climatiques et l'occupation préhistorique humaine à partir de 12.000 ans, variations dont la cause est l'apparition de pluies de mousson avec une dessiccation graduelle à partir de 7.000 ans (Kuper et Kröpelin, 2016). Ces données montre clairement l'extension progressive vers l'ouest et le sud à partir de la vallée du Nil du pastoralisme suite à la domestication du bétail comme le témoigne les peintures du Bovidien. La période stylistique du Caballin coïncide chronologiquement avec la domestication du cheval dans les steppes eurasiatiques et l'Oural il y a

5 à 5,5 ka, et sa rapide diffusion géographique de par le monde (Olsen, 2006). Enfin la chronologie du Camelin n'est pas trop distante de la domestication tardive du chameau il y a environ 3.000 ans (Orlando, 2016).

Nous montrons en appendice quelques exemples de gravures et peintures correspondant aux différents styles d'art rupestre du Sahara de sites visités lors de nos voyages.

Preuves scientifiques de l'humidification du Sahara entre 15,000 et 5000 ans BP

Le grand saharien Théodore Monod fut le premier a présenté une synthèse multidisciplinaire (géologie, paléoclimatologie, préhistoire, anthropologie et primatologie) des recherches à l'époque couvrant les diverses régions «du plus grand et plus beau désert du monde», le Sahara. (Monod, 1961). Faisant référence dans cette étude à une publication de Mauny (Mauny, 1957, Figs. 3 et 4), il reproduit la répartition des représentations rupestres de l'éléphant et de la girafe sur toute l'étendue du Sahara central depuis l'Atlantique à la Méditerranée, qui témoignent d'un climat de savanes coïncidant avec la phase stylistique de Grande faune africaine. Il passe en revue diverses études d'évolution climatique du Sahara de plusieurs auteurs concernant l'alternance de périodes pluviales et inter-pluviales, et offre dans sa conclusion la remarque presciente «C'est évidemment aux pulsations humides, pendant lesquelles le Sahara se couvrait de steppes, de savanes et de lacs qu'il faut songer pour apprécier son rôle dans l'évolution humaine». Nous allons donc nous pencher sur ces pulsations humides depuis le dernier maximum glaciaire il y a 20 ka.

Une méta-étude des variations climatiques en basse latitude de la ceinture allant d'Afrique de l'ouest à la Chine (basée sur des analyses de grandes crues des fleuves Nil et Niger, sapropels émanant du delta du Nil, dépôts lacustres et changements de niveaux de lacs tels que le Shati et le Tchad, et diagrammes polliniques de plusieurs sites) identifie au Sahara jusqu'à alors hyperaride une augmentation des précipitations débutant il y a 15 ka (Fig. 3A; Yan et Petit-Maire, 1994) avec deux courtes périodes humides de 15 à 14 ka et 13 à 12 ka. Puis intervient une longue période très humide débutant à 10 ka avec un optimum pluvial entre 8,5 et 7 ka suivi par un changement marqué de moindre humidité à partir 6,7 ka progressant à partir de 4 ka vers un Sahara aride qui persiste jusqu'à nos jours. Une étude de fluctuations des niveaux du lac-basin Tchad (Fig. 3B; Gasse, 2006) montre des pics de haut niveau durant la période humide africaine de l'Holocène (African Humid Period - AHP) correspondant à des phases très humides en 12-9 et 8-5 ka entrecoupées d'intervalles de dessiccation cours approximativement en 9-8, 7-6, 6 et 6-5 ka.

Un carottage dans l'Atlantique au large de la Mauritanie a permis l'analyse de dépôts terrigènes marins provenant de déposition de poussière minérale apportée par les vents est-africains (Fig. 3C; de Menocal *et al.*

2000). Celle-ci montre une nette diminution d'apports éoliens, entre 14,8 et 5,5 ka, diminution causée par l'humidité croissante du Sahara durant la période pluviale africaine de l'Holocène. Cette série ne permet pas toutefois d'identifier avec précision les changements pluviométriques intervenus au sein de cet intervalle de temps; d'autre part, ce changement climatique abrupte en fin de période humide n'est pas partagé par d'autres auteurs qui démontrent l'existence d'un déclin graduel des précipitations dans cette région entre 5,5 et 4 ka suite à une étude sédimentologique, géochimique et palynologique du lac Yoa situé entre les montagnes du Tibesti et de l'Ennedi (Kröpelin, *et al.*, 2008). Ce changement graduel de pluviosité pourrait correspondre à la fin de la période pastorale d'art rupestre Bovidien dans le Sahara central, et au début de la période du style Caballin.

Une autre preuve de l'humidification du Sahara durant l'AHP est l'existence supposée de 5 Méga-lacs (Chotts et Ahnet en Algérie, Fezzan en Libye, Tchad, et Darfur au Soudan), les deux plus grands étant le Méga-Tchad avec une surface d'environ 350.000 km^2 (Schuster *et al.* 2005) et le lac Méga-Fezzan ou se trouve le Messak, la Chapelle Sixtine des gravures rupestres du Sahara, couvrant une surface à l'optimum pluvial de 76.250 km^2 (Armitage *et al.* 2007). Cependant autre que pour le Méga-Tchad, une étude (Quade *et al.* 2018) mets en doute l'existence de très grands lacs et propose plutôt pour le Méga-Fezzan une surface entre 10.000 - 20.000 km^2, étendue toutefois au moins 17 fois supérieure au lac de Genève. Ces auteurs suggèrent que l'environnement hydrologique du Sahara était composé de lacs de diverses tailles, de zones marécageuses et d'un réseau de cours d'eau endoréique, interprétation qui parait raisonnable vu la répartition géographique des sites d'art rupestre et archéologiques au Sahara central. Un Sahara humide vert-clair et discontinue aurait mieux favorisé l'existence de corridors aquatiques permettant le passage migratoire hors d'Afrique de groupes humains.

L'alternance observée de périodes sèches et humides depuis le dernier maximum glaciaire (Last Glacial Maximum -LGM) il y a 20 ka s'explique par les mouvements latitudinaux de la mousson ouest-africaine d'été, avec la limite Saharo-Sahélienne nord ITCZ située aujourd'hui à environ 17-19° N (Fig. 3D; Yan et Petit-Maire, 1994). Durant le LGM en période hyperaride du Sahara, l'ITCZ migre vers le sud à 13-14° N alors que durant l'optimum pluvial de l'AHP (8 ka) celle-ci remonte vers le nord à 22-23° N, c'est-à-dire une migration sud à nord d'environ 1000 km entre période glaciaire et maximum de période humide.

Quelles sont les causes de ces variations paléoclimatiques qui expliqueraient cette alternance des mouvements de l'ITCZ, et ces variations ce sont-t-elles produites aussi avant le LGM? L'intensité et la direction de la mousson sont influencées en grande partie par les différences de température et de pression atmosphérique du continent et de l'océan. Pour étudier ces variations passées il est utile de corréler la température des surfaces océaniques (Sea Surface Temperature - SST) et les archives de précipitations terrestres. Une étude de la mousson ouest-africaine (Weldeab

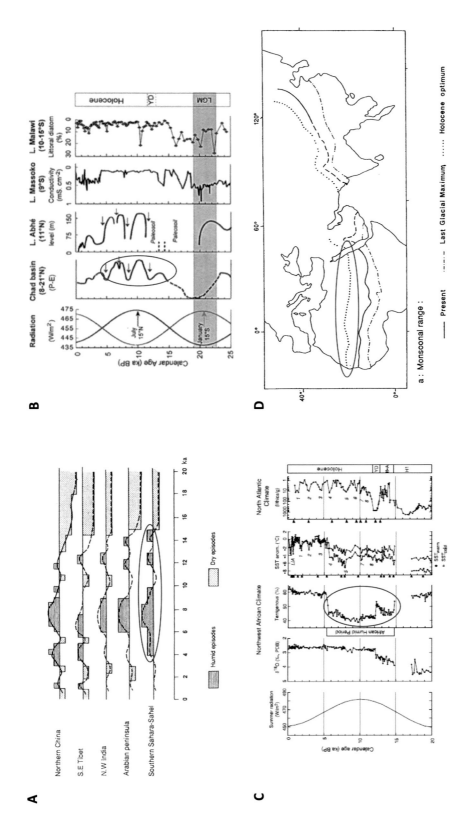

Figure 3. Changements climatiques depuis le dernier maximum glaciaire il y a 20 ka avec ellipses rouges soulignant les variations de la période humide durant l'Holocène au Sahara. A) reproduit d'après Yan et Petit-Maire, 1994. B) reproduit d'après Gasse, 2006. C) reproduit d'après de Menocal et al. 2000. D) reproduit d'après Yan et Petit-Maire, 1994, mouvements latitudinaux de la ITCZ.

et al. 2007) basée sur des données provenant d'une carotte océanique dans le Golfe de Guinée a permis d'analyser deux séries: le rapport Ba/Ca de plancton foraminifère (négativement corrélé à la salinité océanique) comme évidence de décharge d'eau douce du système fluvial du Congo en tant qu'indicateur de l'intensité de la mousson, et le rapport Mg/Ca qui permet d'évaluer l'évolution de la SST, sachant que ces deux séries varient en grande partie de façon indépendante. Les conclusions de l'étude sont les suivantes:

- Il existe un lien entre l'hydrologie de la mousson ouest-africaine et les oscillations climatiques glaciaires-interglaciaires dans les latitudes nord. En effet le rapport Ba/Ca ainsi que la série de l'isotope de l'oxygène $\delta^{18}O$ des foraminifères montrent que sur une échelle centenaire la décharge d'eau de rivière est synchrone avec la courbe isotopique glaciaire.
- Le rapport Mg/Ca (indicatif de la SST) est déphasé par rapport aux fluctuations millénaires en latitude nord, et répond essentiellement aux changements de CO_2 et à l'insolation solaire de basse latitude. En effet le début de l'augmentation des précipitations de la mousson intervient avec un retard de 7.000 ans durant les transitions glaciaire-interglaciaires.
- En conclusion, l'instabilité climatique glaciaire-interglaciaire des latitudes nord est le facteur dominant qui contrôle la dynamique de la mousson ouest-africaine par des liens atmosphériques complexes.

Au delà du LGM à 20 ka sur une échelle de temps allant jusqu'à 155 ka, cette étude met en évidence plusieurs oscillations climatiques humides entre 75 ka et 125 ka démontrant que l'AHP n'est pas un évènement unique. Ceci nous amène à examiner la problématique des périodes humides au Pléistocène?

Alternance de périodes humides et hyperarides au Pléistocène

En Fig. 4A nous faisons référence de nouveau à la méta-étude de Yan et Petit-Maire (1994) qui met en évidence une longue période humide au stade isotopique 5e de 125 à 140 ka, bien documentée au sud et centre du Sahara, avec plusieurs épisodes plus courts aux interstades 5a, 5b et 5d. Pas d'évidence pluviale n'existe entre 70 et 50 ka, alors qu'entre 45 et 40 ka puis entre 35 et 21 ka apparaissent quelques témoins de petits marécages et lacs peu profonds entrecoupant une phase semi-aride.

Le Messak, région possédant la plus grande densité de gravures rupestres de l'AHP (Le Quellec, 1998), est situé au sein du bassin versant du paléo Méga-lac Fezzan. Juste au nord en latitude 26-28° N dans le bassin endoréique Murzuq, quatre strates calcaires correspondant à des dépôts lacustres durant des périodes humides ont pu être identifiées et ont été datées par U/Th (Fig. 4B; Geyh et Thiedig, 2008). Les auteurs de cette étude arrivent aux conclusions suivantes:

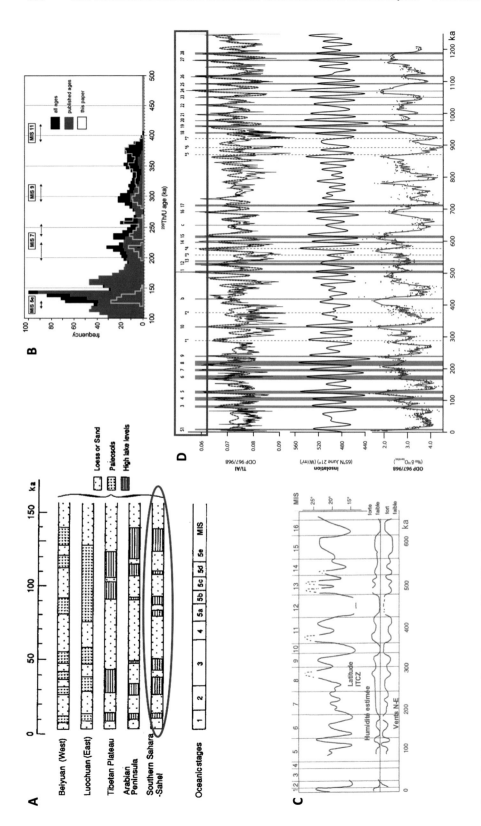

Figure 4. Changements climatiques et identifications de périodes humides du Sahara durant le Pléistocène jusqu'à 1.2 Ma. A) sur une échelle de temps de 150 ka, figure des variations de niveaux de lacs reproduite d'après la méta-étude de Yan et Petit-Maire, 1994. B) sur une échelle de temps de 500 ka, figure reproduite d'après Geyh et Thiedig, 2008 montrant les dates U/Th des fluctuations de dépôts lacustres dans le bassin de Murzuq, Libye. C) sur une échelle de temps de 670 ka, figure reproduite d'après Dupont et al. 1989 montrant, suite à une analyse de pollens de flore africaine à partir de carottage marin au large du Cap Blanc de Mauritanie, les fluctuations de l'ITCZ correspondant à des périodes humides du Sahara. D) sur une échelle de temps de 1,2 Ma, figure reproduite d'après Konijnendijk et al. 2015 mettant en évidence la présence de sapropels. (encadré rouge), marqueurs paléoclimatiques coïncidant avec les périodes pluviales d'un Sahara vert.

- Les datations des quatre formations calcaires mettent en évidence des épisodes de grande pluviosité correspondant à quatre stades isotopiques interglaciaires: > 420 ka (MIS 11), 380-290 ka (MIS 9), 260-205 ka (MIS 7) et 140-125 ka (MIS 5).
- Les épisodes arides ont pu être parfois interrompus par de brefs intervalles humides durant les périodes glaciaires MIS 10 et MIS 6.
- L'accumulation temporelle de ce calcaire est corrélée avec l'excentricité orbitale cyclique de 100 ka durant les dernières 500 ka.
- Comme ceci a été démontré dans d'autres études, les périodes humides précèdent de quelques milliers d'années la terminaison des périodes glaciaires MIS 10, 8 et 6.

Les phases pluviales documentées dans le bassin Murzuq sont confirmées plus à l'est au Soudan et en Egypte par les datations U/Th de plusieurs dépôts sédimentaires lacustres prouvant l'existence de paléolacs approximativement dans les intervalles 320-250, 240-190, 155-120, 90-65 et 10-5 ka, avec une plus grande extension des formations hydrologiques les plus anciennes (Szabo *et al.* 1995).

Durant les périodes humides les pollens de la flore africaine de basse latitude sont transportés par les vents d'est à ouest vers l'océan Atlantique. Le carottage marin du site 658 du Ocean Drilling Program à 160 km au large de Cap Blanc en Mauritanie à 21° N et profondeur de 2263 m a permis l'étude de pollens couvrant les stades isotopiques 1 à 17 de l'Holocène au Pléistocène Moyen (Fig. 4C; Dupont *et al.* 1989). La figure 4C résume les variations de l'ITCZ selon la chronologie de la stratigraphie basée sur les isotopes de l'oxygène extraits de foraminifères, ainsi que celles de l'humidité et alizés correspondants durant cet intervalle de temps. Les conclusions de l'étude sont les suivantes:

- Concordance entre la courbe isotopique de l'oxygène des cycles glaciaires-interglaciaires et le pourcentage de pollens Cyperaceae (espèces de marécages, localités humides et savanes) provenant de l'ITCZ ainsi que d'autres régions du nord et ouest de l'Afrique.
- Les pourcentages de pollens indicateurs d'alizés sont négativement corrélés à la courbe de l'oxygène.
- Huit changements temporels vers une période humide avec une durée moyenne de 15 ka sont mis en évidence avec une production élevée de pollens, en contraste aux périodes glaciaires avec une faible production de pollens.
- L'afflux de Poaceae indique un mouvement vers le nord de l'ITCZ, avec une amplitude de variation latitudinal entre 14° N en période glaciaire et 23° N en période interglaciaire durant l'intervalle de 670 ka couvert par l'étude.
- Les stades isotopiques 16, 12, 10 et le début du stade 6 furent des périodes d'extrême aridité en Afrique de l'ouest et du nord, ainsi qu'en Méditerranée.

Des sédiments noirâtres riches en matière organique, dénommés sapropels, se déposent de façon intermittente au fond de la Méditerranée dans la mer Ionienne en périodes de grandes crues du Nil. Ils résultent de conditions plus humides en Afrique du Nord-est suite à une croissante insolation d'été causant une augmentation d'humidité en provenance de l'Atlantique sud. Rossignol-Strick (Rossignol-Strick, 1983, 1985) démontra que ces sapropels s'accumulaient en période d'intense mousson ouest-africaine d'été en latitude nord avec des changements de précipitation en grande mesure déterminés par forçage orbital et en particulier par des minima de précession astronomique. Le lien entre les sapropels coïncidant avec les migrations vers le nord de l'ITCZ, et les périodes d'un Sahara vert et pluvieux est bien établi depuis le Miocène tardif (Larraosaña *et al.* 2013). Une étude de carottes marines des sites ODP 967 et 968 dans l'est de la Méditerranée a mis en évidence la présence de sapropels durant 1200 ka en relation avec trois variables paléoclimatiques pendant cette période: les changements d'humidité tels qu'exprimés par le rapport Ti/Al (Titane/ Aluminium), les variations d'insolation et la série des cycles glaciaires de l'isotope de l'oxygène extraite de foraminifères benthiques (Konijnendijk *et al.* 2015). La figure 4D illustre les liens existants entre les sapropels et les oscillations de ces trois variables, et permet de tirer les conclusions suivantes quant à l'évidence des sapropels en tant qu'indicateur de périodes humides du Sahara:

- Les changements du rapport Ti/Al liés à l'obliquité ne sont pas contrôlés par les cycles glaciaires, mais reflètent les oscillations climatiques de basse latitude. Cette indicateur de dépôts éoliens suit les changements d'humidité induit de la mousson du centre et nord de l'Afrique, et donc est inversement proportionnel à l'activité de la mousson nord-africaine (Konijnendijk *et al.* 2014). Les sapropels coïncident en grande partie avec des pics extrêmes de faible valeur de ce rapport, c'est-à-dire des périodes d'intense humidité de la mousson causée très probablement par une latitude plus au nord de l'ITCZ.
- Les sapropels sont toujours synchrones avec des pics d'insolation.
- Quelques sapropels apparaissent en périodes glaciaires telles que par exemple en stades isotopiques MIS 18, 14 et 6.
- Il existe plusieurs intervalles sans présence de sapropels, ce qui indiquerait des périodes d'extrême aridité soutenue telles qu'entre 720 et 940 ka, 490 et 420 ka, 420 et 330 ka, et 70 et 15 ka.
- Comme mis en évidence dans plusieurs études, les sapropels se produisent en majeure partie en fin de déglaciation précédant la terminaison des périodes glaciaires.

Sur une échelle de temps allant de l'Holocène à 1.2 Ma, les études continentales et marines de paléoenvironnement, en particulier les sapropels, montrent que les périodes humides du Sahara sont associées en grande mesure aux phases de déglaciation dans l'hémisphère nord, et d'autre part sont influencées par les variations climatiques océaniques dans l'Atlantique du sud-ouest et l'est de la Méditerranée. Les cycles d'un Sahara verdoyant qui se répètent irrégulièrement mais en grande fréquence se traduisent-t-il par une occupation humaine sur toute son étendue et quelles en sont les preuves?

Occupation humaine du Sahara à l'Holocène et au Pléistocène

L'occupation humaine du Sahara durant l'Holocène est bien documentée par les différentes phases stylistiques de représentation rupestre que nous avons exposé (Grande faune africaine-Bubalin, Têtes Rondes, Bovidien, Caballin et Camelin), les datations par radiocarbone de 150 sites archéologiques dans l'est du Sahara couvrant la période 9 à 2 ka (Kuper et Kröpelin, 2006), et la diffusion du bétail domestiqué depuis le Moyen-Orient jusqu'à l'Atlantique établie à partir de la chronologie de données archéologiques entre 8 et 3 ka (Wright, 2017). Les progrès récents dans le domaine de la paléogénétique apportent aussi quelques éléments illustrant les mouvements de populations transsahariennes tels que par exemple: affinités génétiques des fossiles de Taforalt au Maroc (15 ka) avec la population Natoufienne du Levant et liens ancestraux avec des populations du sud du Sahara (van de Loosdrecht *et al.* 2018), identification de partage de caractères génétiques de populations transsahariennes datant de la dernière période du Sahara Vert (12 à 5ka) à partir du chromosome Y de 104 males (D'Atanasio *et al.* 2018), et étude montrant les liens génétiques de populations préhistoriques de la péninsule Ibérique avec contribution limitée venant d'Afrique du Nord il y a 5 ka, à la fin de la période humide du Sahara (Olalde *et al.* 2019; Gonzáles-Fortes *et al.* 2019).

La conservation de restes osseux fossiles dans un environnement hyperaride est problématique et très rare, si ce n'est de façon exceptionnelle par déposition en bord de lac ou de rivière. Si nous excluons les fossiles d'homininés datés du Pléistocène des régions côtières atlantique et méditerranéenne du Maroc à l'Egypte, il n'existe pas aujourd'hui au Sahara central de sites archéologiques avec restes humains de cette époque à l'exception du crâne de Singa au Soudan daté de 133 ka (McDermott, 1996). Ceci nous amène à nous pencher sur les sites préhistoriques du Sahara ayant fourni des industries lithiques au Pléistocène. Les premiers essais de recensement de sites Sahariens furent celui de L. Balout (Balout, 1955), et dans un cadre africain plus vaste celui d'H. Alimen (Alimen, 1955), suivis plus tard par une synthèse plus axée sur l'est de McBurney (McBurney, 1960). Cependant la synthèse la plus complète et exhaustive de sites préhistoriques du Sahara est celle de G. Aumassip (Aumassip, 2004). Nous reproduisons en Figure 5 quatre panneaux de ce dernier ouvrage montrant la distribution géographique des sites au Pléistocène.

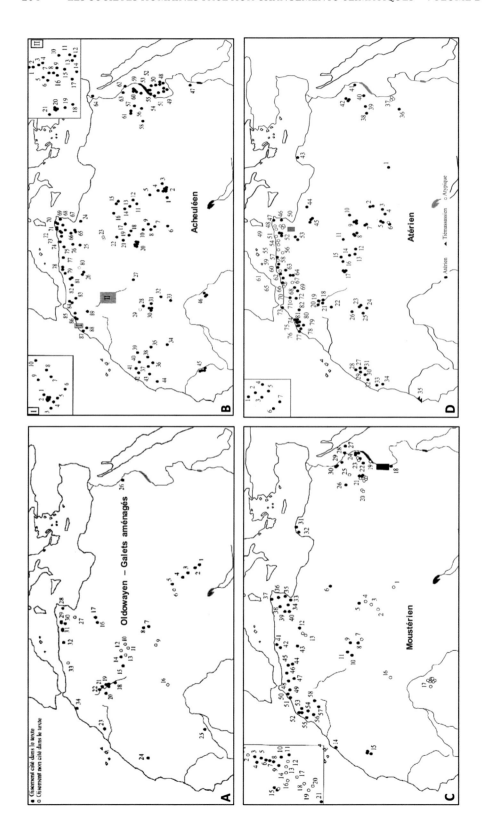

Figure 5. Distribution géographique au
Sahara et en Afrique du Nord des sites
d'industrie lithique datant du Pléistocène
reproduite d'après Aumassip (2004).
A) Oldowayen-Galets aménagés.
B) Acheuléen. C) Moustérien. D) Atérien.

Les industries de galets aménagés (Fig. 5A) connues aussi sous le nom d'Oldowayen, en référence au site d'Olduvai en Tanzanie, sont sans aucun doute les plus anciennes, et sont caractérisées par des pièces avec un tranchant, quelques éclats bruts avec retouche simple, des nucléus frustres dont certains en forme de polyèdre, et des pièces à encoches. La concentration des gisements suit une diagonale depuis le Tchad vers le nord en direction de l'ouest avec une répartition de plusieurs sites le long du littoral du Maroc et de l'Algérie. Chronologiquement, G. Aumassip (Aumassip, 2004) estime que les sites les plus anciens seraient contemporains du bed II d'Olduvai avec une antiquité couvrant un intervalle entre 1,0 et 1,7 Ma. Cependant le site à industrie Oldowayenne d'Ain Boucherit en Algérie daté à un maximum de 2,4 Ma (Sahnouni *et al.* 2018) est plus tardif, ce qui implique une présence très ancienne des homininés en Afrique du Nord qui coïnciderait avec l'apparition des premier hommes en Afrique de l'Est (Prat, 2017).La présence d'un seul gisement Oldowayen dans la vallée du Nil, comparée à une appréciable densité de sites de ce type au centre et nord du Sahara, laisse penser que les occupants d'Ain Boucherit en provenance d'Afrique de l'Est il y a 2,4 Ma seraient arrivés en traversant le Sahara en période humide, plutôt que par une migration le long de la vallée du Nil et du littoral méditerranéen.

L'Acheuléen (Fig. 5B), ensemble lithique caractérisé en Afrique par les bifaces et dans une moindre mesure par les hachereaux, est l'industrie la plus abondante au Sahara. Elle représente un techno-complexe de production de grands outils avec des formes standardisées et une première manifestation de symétrie, ce qui implique plus de prévoyance et planification dans sa manufacture que la technologie précédente de l'Oldowayen. Les sites peuvent se compter par centaines avec une présence impressionnante de milliers de pièces à même le sol sur toute son étendue. L'origine incontestée de l'Acheuléen se situe en Afrique de l'Est avec une première apparition datée à 1,7 Ma (Diez-Martín *et al.* 2015; Sanchez-Yustos *et al.* 2017). En Afrique du Nord les sites Acheuléens les plus complets et mieux étudiés sont ceux de Casablanca; ils couvrent une période allant d'environ1 Ma dans la Carrière Thomas à 0,3 Ma pour Sidi Abderrahmane (Mohib *et al.* 2019). La distribution des gisements montre une concentration le long du Nil, très intense au Sahara central et ouest ainsi qu'en Afrique du Nord. Les datations des sites au centre et à l'ouest sont incertaines, mais il est évident que l'Acheuléen perdure tardivement jusqu'à 0,3 Ma ce qui impliquerait théoriquement que durant 1,4 Ma des groupes humains acheuléens auraient pu traverser le Sahara en périodes humides.

Le Paléolithique Moyen au Sahara et en Afrique du Nord (Figs. 5C et 5D) manifeste une grande variabilité quant aux techno-complexes qu'il englobe, ce qui crée une grande incertitude en ce qui concerne les nomenclatures utilisées pour catégoriser ces ensembles lithiques, comme le témoigne l'appellation inappropriée anglo-saxonne de Middle Stone Age (Scerri, 2017). Le Moustérien est une appellation empruntée de la classification européenne; il consiste en un ensemble lithique avec technologie Levallois classique et discoïde, et comprend principalement des pièces retouchées latéralement, des pointes retouchées, des denticulés et encoches. Il couvre un intervalle chronologique approximatif entre 300 et 30 ka, et se trouve très présent quasiment sur toute l'étendue du Sahara (Fig. 5C). L'Atérien (Fig. 5D), individualisé par des outils avec pédoncules et par des retouches foliacées bifaciales, est caractérisé par certains auteurs comme un faciès spécifique avec substrat moustérien. Il s'étend sur toute l'étendue du Sahara d'est en ouest, avec une très haute concentration en Afrique du Nord mais n'est pas présent dans la vallée du Nil. Il se situe chronologiquement entre 145 ka et 30 ka (Scerri, 2017).

Le Paléolithique supérieur, industrie laminaire succédant en général au Moustérien et Atérien, est absent dans le Sahara central et apparait de façon sporadique dans la vallée du Nil, en Cyrénaïque et exceptionnellement dans la Saoura en Algérie. Dans sa phase finale, une industrie sur lamelles connue sous le nom d'Ibéromaurusien se généralise le long de la frange méditerranéenne et en vallée du Nil, mais est totalement absente en zone saharienne (Aumassip, 2004). Cette absence d'industries du Paléolithique supérieur au Sahara central s'explique par les conditions hyperarides persistant au cours de la dernière période glaciaire en stade isotopique 2.

L'art rupestre du Sahara central ainsi que les mouvements transsahariens de groupes humains mis en évidence par des études archéologiques et paléogénétiques durant l'Holocène montre sans équivoque que l'occupation humaine du Sahara durant la dernière période humide entre 12 et 5 ka est bien établie. Nous avons aussi observé que tout au long du Pléistocène depuis l'Oldowayen jusqu'au Moustérien-Atérien en passant par l'Acheuléen, les nombreux sites préhistoriques d'industrie lithique sont les témoins d'une présence humaine très répandue sur l'ensemble du territoire correspondant très probablement à des périodes humides d'un Sahara verdoyant. Ceci démontre en toute évidence que le Sahara ne fut pas toujours un obstacle infranchissable aux diverses migrations de l'homme vers le Nord de l'Afrique et l'Eurasie à partir de l'Afrique de l'Est, mais une zone de passage de grande extension géographique.

Au vu des évidences présentées, il nous semble utile de schématiser par un modèle simple les cycles aride-humide-aride que le Sahara a connu au Pléistocène afin de répondre à la question: quel mode d'évolution du genre *Homo* les alternances climatiques du Sahara ont-elles pu favoriser?

Hypothèse des cycles aride-humide-aride (A-H-A)

Les oscillations de périodes arides et humides au Sahara sont bien documentées par les études paléoclimatiques que nous avons présenté allant jusqu'à 1,2 Ma. On peut donc inférer qu'il y eut des cycles récurrents dont la fréquence, l'amplitude et la durée sont liées à plusieurs facteurs tels que les mouvements latitudinaux de l'ITCZ, les pics d'insolation, les changements climatiques glaciaires-interglaciaires dans les latitudes nord, et autres variables paléoenvironnementales continentales et marines. En Figure 6, nous résumons de façon illustrative l'alternance de périodes arides et humides au Sahara par un modèle graphique de cycles aride-humide-aride (A-H-A) en 6 phases temporelles t1 à t6. Cette illustration permet de conceptualiser les mouvements migratoires humains probables tels qu'ils auraient pu se produire au Pléistocène. Les cartes de cette figure sont extraites d'un projet de reconstruction d'écosystèmes des derniers 20 ka publié par Adams et Faure (Adams et Faure, 1997). Elles sont approximatives et ne tiennent pas compte de recherches publiées depuis leur édition; elles donnent toutefois une bonne estimation globale des écosystèmes à différentes époques, que nous estimons encore être valable dans ses grandes lignes.

A titre d'exemple de cycle aride-humide-aride nous utiliserons la chronologie de Adams et Faure (1997) des derniers 20 ka assignée aux 6 phases du modèle A-H-A, avec périodisation correspondante des différents styles d'art rupestre du Sahara (en italique) comme suit:

- t1 (20 à 16 ka): phase hyperaride du maximum de froid glaciaire puis début de la déglaciation.
- t2 (11 ka): fin de déglaciation et début de période humide avec *Grande faune africaine* témoignant de l'arrivée au Sahara de groupes humains en provenance de savanes africaines plus au sud.
- t3 (9 ka): période humide avec *Grande faune africaine* et *Têtes Rondes* confirmant la présence de populations de savanes.
- t4 (8-7 ka): maximum de période humide avec *Têtes Rondes* et *Bovidien* comme première manifestation de culture pastorale néolithique en provenance de l'est.
- t5 (5ka): fin de période humide avec *Bovidien et Caballin* correspondant à une probable réduction géographique du pastoralisme et l'introduction du cheval suite à sa domestication.
- t6 (période actuelle): retour de l'aridité sur une grande partie du Sahara avec *Camelin* suite à la domestication du chameau et réduction conséquente des aires d'habitat.

En remontant dans le temps, au delà de l'Holocène, est-ce que ce modèle A-H-A peut s'appliquer au Pléistocène? Vu le nombre très élevé de sites d'industrie lithique recensés sur toute l'étendue du Sahara allant du Paléolithique Inférieur au Paléolithique Moyen (Fig. 5; Aumassip, 2004), la réponse à la question est sans aucun doute positive. La présence de techno-complexes Oldowayens, Acheuléens, Moustériens et Atériens au Sahara ne peut s'expliquer que par l'arrivée de groupes humains en périodes humides depuis l'émergence du genre *Homo*, particulièrement si nous prenons en compte l'antiquité du gisement Oldowayen de Ain Boucherit (Sahnouni *et al.* 2018) dont la date la plus ancienne est de 2.4 MA. En effet, si l'industrie Oldowayenne a son origine en Afrique de l'Est, l'hypothèse la plus parcimonieuse en ce qui concerne la distance est que les homininés d'Ain

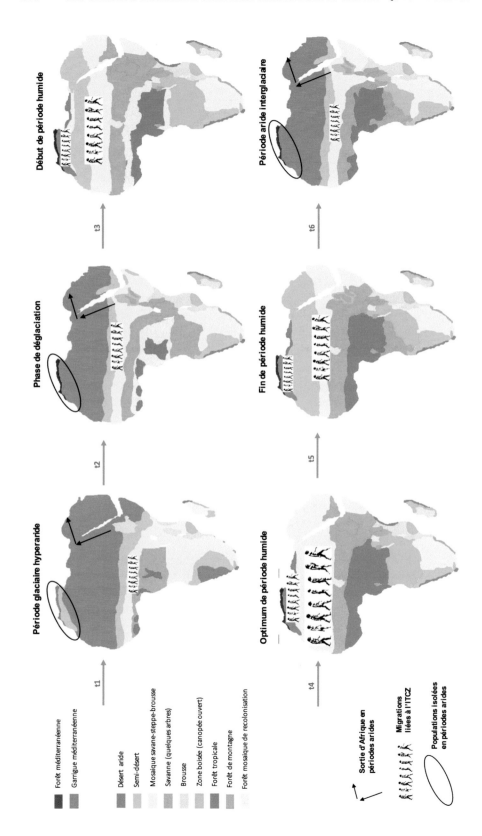

Figure 6. Modèle illustratif des cycles aride-humide-aride (A-H-A) du Sahara avec simulation de mouvements migratoires humains suite à l'alternance de périodes arides et humides. Cartes reproduites d'après Adams et Faure (1997).

Boucherit soient arrivés en traversant le Sahara en période humide, plutôt qu'en migrant par le corridor de la vallée du Nil puis longeant le littoral méditerranéen en période aride.

Quelles seraient les conséquences démographiques de l'application du modèle paléoclimatique A-H-A au Pléistocène? Pour répondre à cette question, considérons ce modèle comme un jeu hypothétique qui commence en période interglaciaire t4. Le jeu débute il y a 1,7 Ma (en t4 optimum de période humide), lorsque le Sahara et l'Afrique du Nord sont habités (Aumassip, 2004) par les premiers représentants du genre *Homo* émergeant d'Afrique de l'Est. Suite à des variations climatiques dues à plusieurs facteurs (forçage astronomique des cycles glaciaires, changements conséquents d'insolation et SST, migration vers le sud de l'ITCZ, etc.), avec le passage du temps les phases t5 (fin de période humide) et t6 (période aride interglaciaire) évoluent progressivement (ou subitement) vers la phase t1 (période glaciaire hyperaride). En t1 des isolats géographiques de populations se créent, en quelque sorte des refuges, et le Sahara devient une barrière infranchissable. Puis intervient un nouveau cycle avec les phases t2 (déglaciation), t3 (début de période humide) et un nouvel optimum de période humide t4 qui déclenche la réouverture du Sahara avec passage de groupes vers des isolats situés en Afrique du Nord et dans sa périphérie. Cette nouvelle phase t4 permet des échanges génétiques entre les groupes humains isolés en t1 et les nouveaux venus traversant le Sahara en t4. Ce phénomène d'ouverture et fermeture du Sahara aux mouvements de population aurait favorisé un processus complexe d'échanges génétiques entre les groupes isolés et les migrants. Si nous considérons qu'un sapropel correspond à un optimum de période humide, il s'avère qu'il y eut 28 cycles A-H-A de périodes humides ces derniers 1,2 Ma (Fig. 4D; Konijnendijk *et al.* 2015), et autant d'échanges démographiques potentiels.

Conclusions

Quelles conclusions pouvons-nous tirer du modèle A-H-A sur une échelle de temps allant jusqu'à l'apparition du genre *Homo* en Afrique de l'Est. Pour ce faire, nous avons superposé en Figure 7 les trois séries chronologiques suivantes sur 3 Ma:

1. Stratigraphie des séries de l'isotope de l'oxygène et sapropels de carottes marines des sites ODP 967 et 849 en Méditerranée reproduits d'après Fig. 3 de Kroon *et al.* 1998; la stratigraphie des sapropels est obtenue de Emeis *et al.* 1996. Les sapropels ne sont pas dessinés à l'échelle et leur positions sont déterminées par leur profondeur moyenne.

2. Série chronologique d'un index humide-sec (wet-dry) d'ODP Site 967 en Méditerranée (reproduit d'après Fig. 8 de Grant *et al*, 2017) représentant les périodes d'un Sahara vert (GSP-Green Sahara Periods). Les périodes humides sont en bleu et les sèches en orange. La

ligne rouge est une série de flux terrigènes d'ODP Site 659. Les barres vertes verticales indiquent les intervalles synchrones d'augmentation de l'humidité au Sahara (index GSP) et en l'Afrique de l'Est (estimés à partir de changements de niveaux de plusieurs lacs), et sont caractérisées comme étant des périodes d'humidité panafricaine (pan-African Humid Periods). L'index humide-sec combine une composante de sapropel indicative de migration vers le nord, et une composante de poussière éolienne indicative de mouvement vers le sud de la mousson Nord africaine. Cet index reflète donc avec une plus grande précision l'alternance de périodes sèches et humides du Sahara que celle basée seulement sur des sapropels.

3. Cadre chronologique conçu par Sandrine Prat (comprenant les espèces du genre *Homo*, celles du genre *Paranthropus*, *Australopithecus sediba*, *Australopithecus garhi* et *Australopithecus africanus*) sur la base de dates publiées par différents auteurs.

L'examen des trois séries permet de mettre en lumière les observations suivantes:

- Trois grandes périodes sèches au Sahara sont identifiées de façon continue: A1 = 2,59 à 2,35 Ma (240 ka), A2 = 2,09 à 1,95 Ma (140 ka) et la plus longue A3 = 0,96 à 0,70 Ma (260 ka). Ceci implique que le Sahara représentait une grande barrière à toute migration, et la vallée du Nil était donc la seule possible route de sortie d'Afrique durant ces intervalles. Les autres parties de la chronologie montrent une alternance de périodes sèches et humides au Sahara.

- On observe quatre phases durant lesquelles des périodes d'humidité intermittentes existèrent aussi bien au Sahara qu'en Afrique de l'Est de façon synchrone: H1 = 2,69 à 2,59 Ma (100 ka), H2 = 1,95 à 1,81 Ma (140 ka), H3 = 1,17 à 0,98 Ma (190 ka) et la plus longue avec des périodes de non concordance plus longues H4 = 0,64 à nos jours (640 ka). Dans le modèle A-H-A les périodes synchrones d'humidité au Sahara et en Afrique de l'Est sont représentées à titre d'exemple par la phase d'optimum de période humide t4. Durant ces périodes humides synchrones toute l'Afrique était potentiellement ouverte a des mouvements de population interrégionaux dans toutes directions y compris à travers le Sahara. La phase H4 semble être corrélée à une péjoration glaciaire-interglaciaire en latitudes nord comme le reflète la courbe des stades isotopiques montrant une pente négative avec une grande volatilité des cycles paléoclimatiques.

- En ce qui concerne l'émergence du genre *Homo* la question se pose de savoir, vues les incertitudes quant au diagnostique phénotypique et la chronologie des premiers représentants (Prat, 2017), si celle-ci eut lieu pendant la phase humide H1 de périodes d'humidité synchrones au Sahara et en Afrique de l'Est ou avant, ou durant A1 la longue période sèche du Sahara? De même il serait utile de déterminer si la première sortie d'Afrique supposée d'*Homo ergaster* s'est produite en période sèche du Sahara A2 ou durant la phase H2 de périodes d'humidité synchrones au Sahara et en Afrique de l'Est.

- La longue phase sèche A3 du Sahara est suivie par la phase H4 de périodes humides synchrones dans toute l'Afrique. Cette dernière correspond au Pléistocène Moyen, un long intervalle chronologique durant lequel apparaissent en Afrique un grand nombre de fossiles archaïques très divers dont la classification taxonomique reste encore à clarifier (Caparros, 1997).

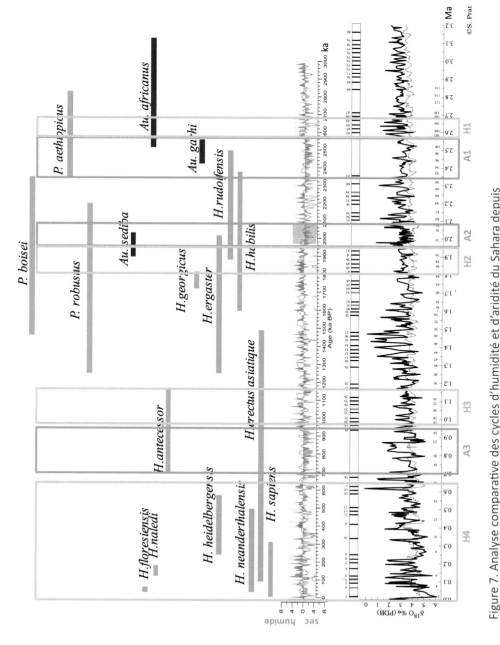

Figure 7. Analyse comparative des cycles d'humidité et d'aridité du Sahara depuis l'apparition du genre Homo. De haut en bas les séries sont: cadre chronologique du genre Homo conçu par Sandrine Prat, série chronologique de l'index humide-sec (bleu-orange) reproduit d'après la figure 8 de (Grant et al. 2017) avec périodes d'augmentation de l'humidité au Sahara et en l'Afrique de l'Est (barres vertes), et séries de l'isotope de l'oxygène et sapropels reproduites d'après la figure 3 de (Kroon et al. 1998).

Ces observations représentent autant de questions à résoudre en paléoanthropologie, et donc des axes de recherche qui intègreraient des données précises de changements climatiques et de paléoenvironnement en Afrique de l'Est avec celles du Sahara. Ces axes de recherche en particulier seraient: une plus précise chronologie de l'apparition du genre *Homo*, de sa sortie d'Afrique (« Out of Africa »), et clarification des relations phylogénétiques du groupe très divers de fossiles humains du Pléistocène Moyen en Afrique afin de déterminer plus précisément le processus évolutif menant aux hommes modernes.

Le paradigme de divergence évolutive dans le genre *Homo* par isolation reproductive et adaptation reconnait maintenant que les échanges génétiques sont de grande importance dans l'explication de la diversité, et joue un rôle fondamental dans le processus de spéciation (Ackerman *et al.* 2019). Le défi est de savoir comment reconnaître la diversité phénotypique émanant d'échanges génétiques pour savoir si d'un point phylogénétique toutes les espèces du genre *Homo* peuvent être reliées par des liens de parenté en forme de branches ou de réseau. Un premier essai basé sur des caractères phénotypiques semble montrer que le mode d'évolution du genre *Homo* suit un processus évolutif réticulé plutôt que dichotomique par embranchements (Caparros et Prat, 2021), ce qui indiquerait que sa diversité taxonomique peut s'expliquer par un modèle de spéciation par hybridation introgressive plutôt que par allopatrie (Arnold, 2006, 2009). Le modèle de spéciation par hybridation introgressive peut se justifier en Afrique, berceau de l'humanité, tout au long du Pléistocène par l'alternance de périodes humides et arides au Sahara, alternance qui aurait favorisé les échanges génétiques entre populations isolées lors de ses multiples réouvertures en périodes humides.

Références bibliographiques

Ackermann R.R, Arnold M.L., Baiz M.D., Cortés-Ortiz L., Evans B.J., Grant B. R., Hallgrimsson B., Humphreys R.A., Cahill J.A., *et al.* 2019. Hybridization in human evolution: Insights from other organisms, *Evol. Anthropol.,* 28, p. 189–209.

Adams J.M., Faure H. 1997. QEN members, eds, *Review and Atlas of Palaeovegetation: Preliminary land ecosystem maps of the world since the Last Glacial Maximum*, Oak Ridge National Laboratory, TN, USA, https://www.esd.ornl.gov/projects/qen/adams1.html

Alimen H. 1955. *Préhistoire de l'Afrique*, Paris, Editions N Boubée & Cie.

Armitage S.J., Drake N.A., Stokes S., El-Hawat A., Salem M.J., White K., Turner P., McLaren S.J. 2007. Multiple phases of North African humidity recorded in lacustrine sediments from the Fazzan Basin, Libyan Sahara, *Quaternary Geochronology,* 2, p. 181–186.

Arnold M.L. 2006. *Evolution through genetic exchange*, Oxford University Press, Oxford.

Arnold M.L. 2009. *Reticulate evolution and humans: origins and ecology*, Oxford University Press, Oxford.

Aumassip G. 2004. *Préhistoire du Sahara et de ses abords*, Paris, Maisonneuve & Larose.

Balout L. 1995. *Préhistoire de l'Afrique du Nord, Essai de Chronologie*, Paris, Arts et métiers graphiques.

Breuil H. 1955. Les roches peintes du Tassili-n-Ajjer. *In:* L. Balout (Dir.) *Actes du Congrès Panafricain de Préhistoire, 2ème session, Alger 1952.* Direction de l'Intérieur et des Beaux-arts - Service des Antiquités / Arts et Métiers graphiques, p. 5-161.

Burke K. 1996. The African plate, *South African Journal of Geology*, 99, p. 341-409.

Caparros M. 1997. *Homo sapiens archaïques: un ou plusieurs taxons (espèces)? Analyse cladistique et morphométrique*, thèse de doctorat, Muséum national d'Histoire naturelle, Paris.

Caparros M., Prat S. 2021. A Phylogenetic Networks perspective on reticulate human evolution, *iScience*.https://doi.org/10.1016/j.isci.2021.102359

D'Atanasio E., *et al.* 2018. The peopling of the last Green Sahara revealed by high-coverage resequencing of trans-Saharan patrilineages, *Genome Biology*, 19:20, p. 1-15.

deMenocal P., Ortiz J., Guilderson T., Sartnthein M. 2000. Coherent High- and Low-Latitude Climate Variability During the Holocene Warm Period, *Science*, 288, p. 2188-2202.

Diez-Martín F., Sánchez Yustos P., Uribelarrea D., Baquedano E., Mark D. F., Mabulla A., Fraile C., Duque J., Díaz I., Pérez-González A., Yravedra J., Egeland C. P., Organista E., Domínguez-Rodrigo M. 2015. The Origin of the Acheulean: The 1.7 Million-Year-Old Site of FLK West, Olduvai Gorge (Tanzania), *Sci. Rep.* 5, p. 1-9.

Doucouré C.M., de Wit M.J. 2003. Old inherited origin for the present near bimodal topography of Africa, *Journal of African Earth Sciences*, 36, p. 371-388.

Dupont L.M.,Beug H.-J.,Stalling H., Tiedemann R. 1989. First palynological results from site 658 at 21°N of northwest Africa: pollen as climate indicators, *Proceedings of the Ocean Drilling Program, Scientific Results,* Vol. 108, p.93-111.

Emeis K.C., Robertson A., Richter C., *et al.* 1996. Paleoceanography and sapropel introduction. *In: Proceedings of the Ocean Drilling Program, Initial Reports,* vol. 160. ODP, College Station, TX, USA, p. 21-28.

Fishwick S., Bastow I.D. 2011. Towards a better understanding of African topography: a review of passive-source seismic studies of the African crust and upper mantle, *Geological Society, London, Special Publications*, 357, p. 343-371.

Gasse F. 2006. Climate and hydrological changes in tropical Africa during the past million years, *C. R. Palevol*, p. 35-43.

Geyh M.A., Thiedig F. 2008. The Middle Pleistocene Al Mahrúqah Formation in the Murzuq Basin, northern Sahara, Libya evidence for orbitally-forced humid episodes during the last 500,000 years, *Palaeogeography, Palaeoclimatology, Palaeoecology*, 257, p.1–21.

Gonzalez-Fortes G. *et al.* 2019. A western route of prehistoric human migration from Africa into theIberian Peninsula, *Proc. R. Soc. B*, 286, p. 1-10.

Grant K.M., Rohling E.J., Westerhold T., Zabel M., Heslop D., Konijnendijk T., Lourens L. 2017. A 3 million year index for North African humidity/aridity and the implication of potential pan-African Humid period, *Quaternary Science Reviews*, 171, p. 100-118.

Hachid M. 2016. Chronostratigraphie, bandes pariétales de couleur sombre et claire des parois au Tassili-n-Ajjer et un possible «calage» chronologique des peintures rupestres. *In:* N. Honoré, M. Guterriez (eds.) *l'Art rupestre d'Afrique,* Actes du Colloque International, Paris 15-17 janvier 2014, Paris 1, Centre Panthéon et Musée du quai Branly, p. 65-110.

Konijnendijk T.Y.M., Ziegler M., Lourens L.J. 2014. *Newsletter on Stratigraphy*, 47/3, p. 263-282.

Konijnendijk T.Y.M., Ziegler M., Lourens L.J. 2015. On the timing and forcing mechanisms of late Pleistocene glacial terminations: Insights from a new high-resolution benthic stable oxygen isotope record of the eastern Mediterranean, *Quaternary Science Reviews*, 129, p. 308-320.

Kroon D., Alexander I., Little M., Lourens L.J., Matthewson A., Robertson A.H., Sakamoto T. 1998. Oxygen isotope and sapropel stratigraphy in the eastern Mediterranean during the last 3.2 million years, *Proceedings of the Ocean Drilling Program, Scientific Results*, vol. 160, p. 181-188.

Kröpelin S. 2006. Revisiting the Age of the Sahara Desert, *Science*, 312, p. 1138

Kröpelin S., *et al.* 2008. Climate-Driven Ecosystem Succession in the Sahara: The Past 6000 Years, *Science*, 320, p.765-768.

Kuper R., Kröpelin S. 2006. Climate-Controlled Holocene Occupation in the Sahara: Motor of Africa's Evolution, *Science*, 313, p. 803-807.

Lahr M.M. 2010. Saharan Corridors and Their Role in the Evolutionary Geography of 'Out of Africa'. *In:* J. G. Fleagle *et al.* (eds), *Out of Africa I: The First Hominin Colonization of Eurasia*, Springer Netherlands, p. 27-46.

Larrasoaña J.C., Roberts A.P., Rohling E.J. 2013. Dynamics of Green Sahara Periods and Their Role in Hominin Evolution, *PLoS ONE*, 8, 10, p. 1-12.

Le Quellec J.-L. 1991. L'apport d'Henri Lhote au problème de la chronologie des figurations rupestres sahariennes. *Sahara*, 4, p. 107-113.

Le Quellec J.-L. 1993. *Symbolisme et Art Rupestre au Sahara*, Paris, L'Harmattan, 638 p.

Le Quellec J.-L. 1998. *Art rupestre et préhistoire du Sahara*, Paris, Editions Payot & Rivages, 616 p.

Le Quellec J.-L. 2013. Périodisation et chronologie des images rupestres du Sahara central, *Préhistoires Méditerranéennes*, 4, p. 1-47.

Lhote H. 1958. *A la découverte des fresques du Tassili*, Paris, Arthaud.

Lhote H. 1976. *Vers d'autres Tassilis*, Paris, Arthaud.

Mauny R. 1957. Répartition de la grande faune éthiopienne du Nord-Ouest africain du Paléolithique à nos jours, *Bull. Institut Français d'Afrique noire*, 18, 1, p.246-79.

McBurney C.B.M. 1960. *The stone age of Northern Africa*, Pelican, Baltimore.

McDermott F., Stringer C., Grün R., Williams C.T., Din V.K. 1996. New Late-Pleistocene uranium–thorium and ESR dates for the Singa hominid (Sudan), *Journal of Human Evolution*, 31, p. 507–516.

Mercier N., Le Quellec J.-L., Hachid M., Agsous S., Grenet M. 2012. OSL dating of quaternary deposits associated with the parietal art of the Tassili-n-Ajjer plateau (Central Sahara), *Quaternary Geochronology*, p. 1-7.

Mohib A., Raynal J-P.,Gallotti R., Daujeard C., El Graoui M., Fernandes P., Geraads D., Magoga L., Rué M. , Sbihi- Alaoui F.-Z., Lefèvre D. 2019. Forty Years of Research at Casablanca (Morocco): New Insights in the Early/Middle Pleistocene Archaeology and Geology, *Hespéris-Tamuda* LIV (3). p. 25-56.

Monod T. 1964. The Late Tertiary and Pleistocene in the Sahara. *In:* F. Clark Howell and F. Bourlière (Eds.) *African Ecology and Human Evolution*, Viking Publications in Anthropology No 36, New York, p.666.

Olalde I., *et al.* 2019. The genomic history of the Iberian Peninsula over the past 8000 years, *Science*, 363, p. 1230–1234.

Olsen S.L. 2006. Early horse domestication on the Eurasian steppe. *In:* M. A. Zeder, D. G. Bradley, E. Emshwiller, B. D. Smith (Eds.) *Documenting Domestication: New Genetic and Archaeological Paradigms*. Univ. of California Press, p.245–269.

Orlando L. 2016. Back to the roots and routes of dromedary domestication, *PNAS*, 113, p. 6588-6590.

Prat S. 2017. First hominin settlements out of Africa. Tempo and dispersal mode: Review and perspectives, *C. R. Palevol*, p. 1-11.

Quade J., Dente E., Armon M., Ben DorY., Morin E., Adam O., Enzel Y. 2018. Megalakes in the Sahara? A Review, *Quaternary Research,* 90, p.253–275.

Rognon P. 1994. *Biographie d'un désert*, Paris, L'Harmattan.

Rossignol-Strick M. 1983. African monsoons, an immediate climate response to orbital insolation, *Nature* 30, p.446–449.

Rossignol-Strick M. 1985. Mediterranean Quaternary sapropels, an immediate response of the African monsoon to variations of insolation, *Palaeogeography, Palaeoclimatology, Palaeoecology*, 49, p. 237-262.

Sahnouni M., *et al.* 2018. 1.9-million and 2.4-million-year-old artifacts and stone tool–cutmarked bones from Ain Boucherit, Algeria, *Science*, 362, p.1297-1301.

Sanchez-Yustos P., Diez-Martin F., Dominguez-Rodrigo M., Duque J., Fraile C., Diaz I., de Francisco S., Baquedano E., Mabulla A. 2017. The origin of the Acheulean. Techno- functional study of the FLK W lithic record (Olduvai, Tanzania), *PLoS ONE*, 12(8), p. 1-30.

Schuster M., Roquin C., Duringer P., Brunet M., Caugy M., Fontugne M., Mackaye H.T., Vignaud P., Ghienne J.-F. 2005. Holocene Lake Mega-Chad palaeoshorelines from space, *Quaternary Science Reviews*, 24, p. 1821-1827

Schuster M., Duringer P., Ghienne J.-F., Vignaud P., Mackaye H.T., Likius A., Brunet M. 2006. The Age of the Sahara Desert, *Science*, 311, p. 821.

Scerri E.M.L. 2017. The North African Middle Stone Age and its place in recent human evolution, *Evol. Anthropol.*, 26, p. 119–135.

Smith F.H., Ahern J.C.M., Janković I., Karavanić I. 2017. The assimilation model of modern human origins in light of current genetic and genomic knowledge, *Quaternary. International* 450, p.126–136.

Swezey C.S. 2009. Cenozoic stratigraphy of the Sahara, Northern Africa, *Journal of African Earth Sciences*, 53, p. 89–121.

Szabo B.I., Haynes Jr C.V., Maxwell T.A. 1995. Ages of Quaternary pluvial episodes determined by uranium-series and radiocarbon dating of lacustrine deposits of Eastern Sahara. *Palaeogeography, Palaeoclimatology, Palaeoecology*, 113, p. 227-242.

van de Loosdrecht M., *et al.* 2018. Pleistocene North African genomes link Near Eastern and sub-Saharan African human populations, *Science*, 360, p. 548–552.

Weldeab S., Lea D.W., Schneider R.R., Andersen N. 2007. 155,000 Years of West African Monsoon and Ocean Thermal Evolution, *Science*, 316, p. 1303-1307.

Wright D. 2017. Humans as Agents in the Termination of the African Humid Period, *Front. Earth Sci.,* 5, p. 1-14.

Yan Z., Petit-Maire N. 1994. The last 140 ka in the Afro-Asian arid/semi-arid transitional zone, *Palaeogeography, Palaeoclimatology, Palaeoecology*, 110, p. 217-233.

Zhang Z., Ramstein G., Schuster M., Li C., Contoux C., Yan Q. 2014. Aridification of the Sahara desert caused by Tethys Sea shrinkage during the Late Miocene, *Nature*, 513, p. 401-407.

Appendice - Exemples des styles d'art rupestre du Sahara

Grande faune africaine-Bubalin

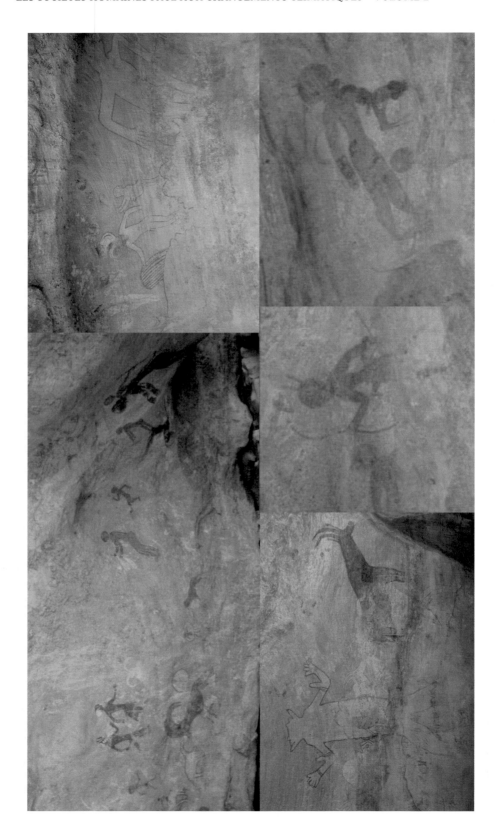

Têtes Rondes

Bovidien

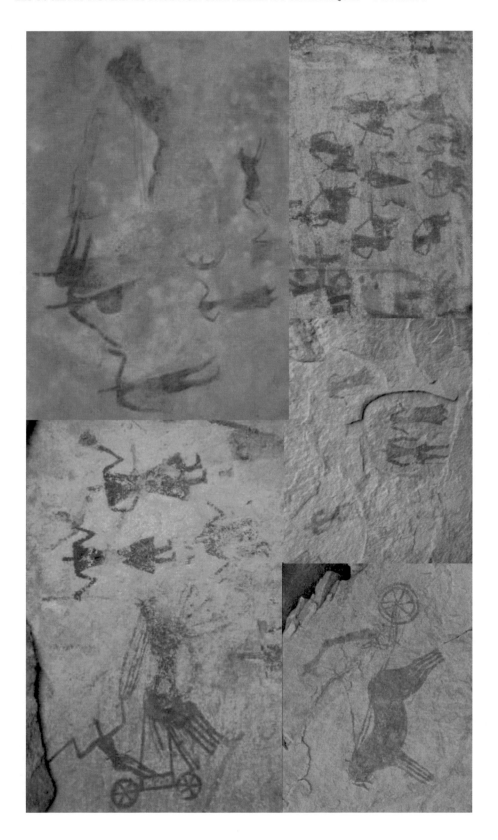

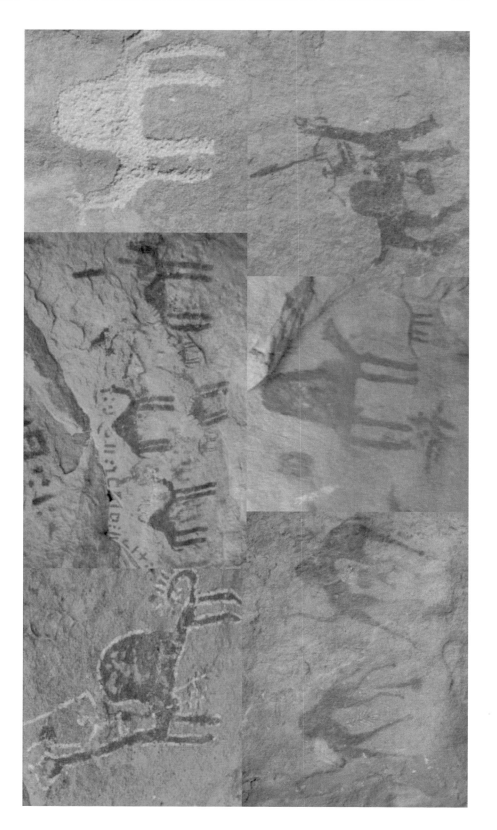

Camelin

Le Tilemsi et ses abords de la préhistoire à nos jours

Christian Dupuy[1]

Résumé

Le Tilemsi est une longue et large dépression, partant des confins algéro-maliens, rejoignant le Niger au niveau de Gao, bordée par les massifs de l'Adrar des Iforas et du Timétrine. Si la préhistoire ancienne de cette vallée demeure encore largement méconnue, en revanche sa préhistoire récente fournit quelques jalons clés qui permettent de suivre le développement de l'économie agro-pastorale en Afrique de l'Ouest subtropicale, et ses conséquences sur l'organisation de la société.

Les expressions rupestres anciennes situent le début de l'élevage dans la région à un moment ou à un autre entre le VIe millénaire et la fin du IVe millénaire av. J.-C. Par ailleurs, les empreintes d'épillets de mil observées sur des poteries archéologiques révèlent une cueillette de cette graminée à des fins alimentaires dans le nord du Sahara malien au Ve millénaire av. J.-C. La mise en culture de cette céréale au cours du millénaire suivant entraine un grossissement des grains, puis l'acquisition de caractères domestiques dans le bas Tilemsi à la fin du IIIe millénaire av. J.-C. Parallèlement, la construction d'enceintes à double parement de pierres et l'édification de tombes à superstructures lithiques imposantes attestent d'un niveau de complexification sociale déjà élevé à la charnière des IVe-IIIe millénaires av. J.-C.

Le développement de l'économie agropastorale au cours de la première moitié du IIe millénaire av. J.-C. est consubstantielle de l'apparition d'artisanats spécialisés (métallurgie, taille des pierres dures, charronnage), de l'acquisition d'animaux exotiques (chevaux, zébus), de la circulation d'idées et de croyances sur de grandes distances (motifs symboliques complexes et thèmes caractéristiques de l'âge du bronze), de la transmission de cultivars de mil de signature génétique ouest-africaine jusqu'en Inde. Des aristocraties guerrières s'affirment par l'ostentation durant le Ier millénaire, avant que l'aridité ne finisse par contraindre les populations du Tilemsi à abandonner leurs terroirs pour des contrées plus favorisées par la géographie et le climat. À partir des IVe-Ve siècles apr. J.-C., les espaces libérés sont investis par des éleveurs nomades porteurs de traditions nouvelles qu'illustre l'art rupestre : monte des chevaux et des dromadaires, port de plusieurs javelots et de vêtements amples et bien couvrants, chasses à courre, rédaction de courts messages au moyen d'une écriture composée de signes très semblables aux tifinagh des Touaregs. Ces éléments conjugués attestent du rattachement de la région au domaine amazighophone, avant que ne se développe le commerce transsaharien arabo-berbère. La vallée du Tilemsi devient alors un axe caravanier de première importance. Du XIe au XIVe

---

1 Institut des Mondes Africains (IMAf Paris, UMR 8171)

siècles, la ville marchande d'Essouk-Tadmakkat au sud-ouest de l'Adrar des Iforas est florissante. La cité est abandonnée au XVᵉ siècle pour des raisons qui restent à préciser.

Abstract

The Tilemsi is a long and wide depression, starting from the Algerian-Malian borders, joining the Niger at Gao, bordered by the Adrar des Iforas and Timetrine massifs. While the ancient prehistory of this valley remains largely unknown, its recent prehistory provides some key milestones that allow us to follow the development of the agro-pastoral economy in subtropical West Africa and its consequences on the organization of society.

Ancient rock art expressions make it possible to situate the beginning of livestock breeding in the region at some point between the sixth millennium and the end of the fourth millennium BC. Furthermore, the imprints of millet spikes observed on archaeological pottery reveal that this grass was gathered for food in the northern Malian Sahara in the fifth millennium BC. The cultivation of this cereal during the following millennium led to an increase in the size of the grains, and then to the acquisition of domestic characteristics in the Lower Tilemsi at the end of the third millennium BC. In parallel, the construction of enclosures with double stone facing and the building of tombs with imposing lithic superstructures attest to a level of social complexity that was already high at the turn of the fourth-third millennium BC.

The development of the agro-pastoral economy during the first half of the second millennium B.C. is consubstantial with the appearance of specialized crafts (metallurgy, hard stone cutting, manufacture of chariots), the acquisition of exotic animals (horses, zebus), the circulation of ideas and beliefs over great distances (complex symbolic motifs and themes characteristic of the Bronze Age), and the transmission of pearl millet cultivars with a West African genetic signature as far as India. Warrior aristocracies asserted themselves through ostentation during the first millennium before aridity forced, at the end, the populations of Tilemsi to abandon their lands for regions more favoured by geography and climate. From the fourth-fifth centuries AD, the liberated spaces were taken over by nomadic herders who brought with them new traditions that are illustrated by rock art: riding horses and dromedaries, carrying several javelins and wearing ample and well-covered clothing, hunting with hounds, writing short messages using a script composed of signs very similar to the Tuareg tifinagh. These combined elements attest to the attachment of the region to the Amazigh-speaking domain, before the development of Arab-Berber trans-Saharan trade. The Tilemsi valley then became a caravan route of primary importance. From the eleventh to the fourteenth centuries, the merchant city of Essouk-Tadmakkat in the southwest of the Adrar des Iforas flourished. The city was abandoned in the fifteenth century for reasons that remain to be clarified.

Introduction

Le Tilemsi est une large dépression de faible pente qui s'étend sur près de 400 km du sud du Tanezrouft jusqu'au fleuve Niger au niveau de Gao (Figure 1). Cette vallée constitue l'artère principale d'un réseau hydrographique ayant pour affluents majeurs les oueds de rive gauche issus de l'Adrar des Iforas, et pour affluents secondaires ceux de rive droite issus du Timétrine. Le biotope est saharien au nord, sahélien au sud. Il n'en a pas toujours été ainsi. Le climat n'a cessé d'évoluer au cours du quaternaire (Petit Maire, Riser 1983), attirant tantôt les hommes, puis les repoussant.

1. Le paléolithique et l'épipaléolithique

Il est probable que la plupart des gisements pléistocènes soient encore pris dans les alluvions. On doit à Jean et Michel Gaussen (1988) un inventaire des artefacts lithiques retrouvés épars en surface. L'Acheuléen occupe au moins 3 km² à Lagreich dans le bas Tilemsi. D'autres gisements de moindre étendue sont connus dans ce secteur et, plus au nord, jusqu'au sud du Tanezrouft. Les outils mis au jour par l'érosion consistent essentiellement en bifaces, hachereaux, trièdres et racloirs. Aucune trace de foyer, ni aucun reste organique n'ont été observés. Par conséquent, les activités domestiques et les modes de prédation s'avèrent insaisissables. Le post-acheuléen est mal connu. Lui sont attribués des éclats levallois retouchés ou non, des pièces foliacées à enlèvements bifaciaux plus ou moins envahissants, quelques pointes pédonculées que l'on peut qualifier d'« atériennes ». L'abondance des lamelles à bords abattus à Kreb in Karoua en rive droite du haut Tilemsi suggère une occupation épipaléolithique. En l'absence de chrono-stratigraphie, il est malaisé de dater ces industries. Les provenances des roches taillées ainsi que les lieux de débitage restent souvent à préciser.

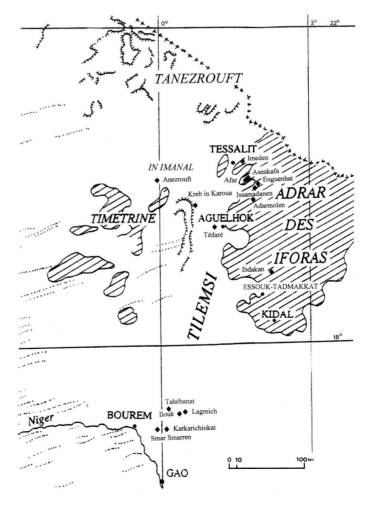

Figure 1. La vallée du Tilemsi, ses bassins versants – Adrar des Iforas et Timetrine – et les sites archéologiques mentionnés dans le texte (fond de carte d'Y. Assié, tiré de M. Raimbault 1994).

2. L'avènement de l'économie agro-pastorale

2.1. Le début de l'élevage

Les plus anciens témoins d'une pratique du pastoralisme dans la région résident dans la présence, au nord-ouest de l'Adrar des Iforas, de gravures rupestres de taurins (*Bos taurus taurus* ou bœuf domestique à dos droit) aux robes et aux cornes variées attestant d'un stade de domestication avancé. Apparaissent à leurs côtés des représentations d'éléphants, de rhinocéros, de girafes, d'autruches, de deux grands carnivores, probablement des

Figure 2. Gravures anciennes. *Sources :*
a. Éléphant d'Afar près du puits d'Assawa situé à une trentaine de kilomètres au sud-est de Tessalit. Les membres du pachyderme sont bien proportionnés et leurs positions correctement transcrites. Son corps et sa ligne dorsale sont recouverts par des inscriptions en caractères *tifinagh* de réalisation beaucoup plus récente (cliché V. Adam) ;
b. Taurin silhouetté aux traits polis à Issamadanen au nord-ouest de l'Adrar des Iforas. Son corps bien proportionné est sous-tendu de membres aux perspectives bien rendues. Ses longues cornes inclinées vers l'arrière et sa robe bigarrée attestent de son statut domestique (longueur du sujet : ~ 1,40 m ; cliché C. Dupuy).

lionnes, d'un archer face à une autruche, d'un personnage masqué en marche avec deux girafes à son contact, d'un humain en prise avec un guib harnaché (Dupuy 1989, 1999 ; Lhote, Tomasson 1972 ; Mauny 1954). Ce bestiaire témoigne de l'existence d'une savane arborée avec plans d'eau permanents. Les corps des sujets sont bien proportionnés et les épaisseurs, segments et perspectives de leurs membres bien rendus (Figure 2). Plusieurs de ces réalisations sont sous-jacentes à des expressions riches de taurins, d'autruches et de girafes, marquées par le schématisme. L'ordre inverse de recouvrement ne s'observe jamais ; ce qui suppose que les auteurs des figures les plus naturalistes furent les premiers à s'exprimer avec des préoccupations distinctes de celles des graveurs qui, à des époques plus récentes, se remirent à inciser les rochers.

Des milliers de gravures réalisées plus au nord dans les Tassilis algériennes, l'Ahaggar, les Messaks Mellet et Settafet, l'Aramat, les Tadrarts Akoukas et méridionale, le Djado et le Tibesti peuvent être rapprochées, selon des critères de styles et de thèmes, des expressions anciennes de l'Adrar des Iforas nord-occidental (Dupuy 1989, 1999). Dans ces diverses régions, les taurins furent les sujets de prédilection des graveurs, suivis par les animaux de la grande faune sauvage. Vient ensuite un éventail plus ou moins large de sujets selon les localités parmi lesquels figurent parfois quelques chèvres et moutons (Van Albada, Van Albada 2000).

Vingt-six monuments cultuels en pierres sèches incorporant des stèles décorées de pétroglyphes caractéristiques de cet art gravé, ont été fouillés dans le Messak Settafet (di Lernia *et al.* 2010, 2013). Ils renfermaient des ossements de taurins, d'ovins ou/et de caprins ainsi que des restes organiques datés au ^{14}C de la seconde moitié du Ve millénaire av. J.-C. On sait, d'autre part, que les plus anciens ossements de taurins intentionnellement enterrés au nord-est de l'Aïr, soit sensiblement aux mêmes latitudes que le haut Tilemsi, se rattachent à la fin du VIe millénaire av. J.-C. (Paris 1998). Les sciences de la terre indiquent, en outre, que l'aridité s'est intensifiée dans le Sahara libyen à la fin du VIIe millénaire av. J.-C. (Cremaschi, di Lernia 1998 ; Cremaschi *et al.* 2014). À l'aube du IIIe millénaire av. J.-C., elle était à tel point marquée au nord du tropique du Cancer que les hippopotames, les rhinocéros blancs et les autres animaux inféodés à l'eau que l'on retrouve gravés dans les Messaks libyens et la Tassili-n-Ajjer, ne pouvaient y survivre, excepté peut-être en quelque zones refuges favorisées par l'hydrographie. Les connaissances archéologiques et paléoécologiques invitent ainsi à dater l'art à gravures naturalistes du Sahara entre le VIe millénaire et la fin du IVe millénaire av. J.-C. duquel participent les expressions anciennes de l'Adrar des Iforas nord-occidental. Les figurations de taurins composants ces expressions imposent l'idée selon laquelle le début du pastoralisme dans le haut Tilemsi remonte à cette époque ; ce qui ne signifie pas pour autant que tous les groupes de la région se soient alors convertis à l'élevage. Des chasseurs, pêcheurs, cueilleurs ont très bien pu conserver leur mode de vie basé sur la prédation pour peu que les ressources naturelles aient été suffisantes pour couvrir leurs besoins en protéines.

2.2. Le début de l'agriculture du mil

L'observation sous fort grossissement de dizaines tessons issus de sites du Tilemsi et du Sahara malien voisin se rapportant aux Ve, IVe et IIIe millénaires av. J.-C., alliée aux micro-tomographies réalisées sur certains de ces artefacts, ont révélé, tant à leur surface que dans leur épaisseur, des empreintes de résidus de mil. Leur présence est due à l'utilisation de la balle de cette graminée comme dégraissant pour faciliter le modelage de l'argile, laquelle balle s'est consumée au moment de la cuisson des poteries et y a laissé ses empreintes (Fuller *et al.* 2021). Le choix de ce dégraissant résulte de l'exploitation alimentaire qui était faite de cette céréale qui pousse encore de nos jours spontanément sur les sols sableux du Sahara méridional. Il était en effet commode pour les potiers et/ou les potières de récupérer les résidus de décorticage des épillets qui s'accumulaient sur le sol des campements, là où les grains étaient vannés. La morphologie des empreintes révèle qu'au IVe millénaire av. J.-C., leur largeur moyenne a augmenté de 28% par rapport à celle du morphotype sauvage en raison vraisemblablement de la mise en culture du mil. Elle lui est de 38% supérieure un millénaire plus tard dans le secteur de Karkarichinkat à l'est de Bourem (bas Tilemsi). Le syndrome de la domestication s'affirme alors. Il consiste en une perte de caducité des épillets, en un appariement fréquent des grains par épillet, et en une réduction des enveloppes et des soies. Le rayonnement ^{14}C émis par les restes non consumés du dégraissant végétal conservé dans certaines poteries et, parallèlement, les mesures OSL effectuées sur les petits cristaux inclus au départ dans la terre modelée, permettent de dater cette céramique de la fin du IIIe millénaire av. J.-C. (Manning *et al.* 2011). La domestication du mil se trouve ainsi indirectement documentée, il y a plus de quatre mille ans, dans le bas Tilemsi. Ces données archéométriques et paléobotaniques s'accordent avec les résultats des généticiens qui ont dressé l'arbre phylogénétique de *Pennisetum* (cette graminée est rattachée depuis peu au genre *Cenchrus*) duquel il ressort que les premières opérations agricoles menées sur cette graminée se seraient déroulées aux alentours du IVe millénaire av. J.-C. quelque part entre la Mauritanie sud-orientale et les confins de l'Algérie, du Mali et du Niger (Burgarella *et al.* 2018 ; Clotault *et al.* 2012 ; Tostain 1998).

Les apports de ces recherches confèrent un regain d'intérêt aux onze enceintes circulaires et elliptiques inventoriées au cours des années 1980 dans le haut Tilemsi (Raimbault 1994, 1995). Les murets de pierres qui les délimitent, circonscrivent des surfaces de quatre à onze hectares (Figure 3). Ceux-ci étaient destinés à renforcer l'ancrage au sol de clôtures et/ou de haies derrière lesquelles des groupes s'adonnaient à leurs activités. Trois datations sur charbons issus de sondages situent leurs occupations aux IVe-IIIe millénaires av. J.-C., soit à l'époque présumée des premières opérations agricoles menées sur le mil dans leur zone nucléaire supposée. L'épaisseur décimétrique des dépôts anthropiques témoigne d'installations, sinon

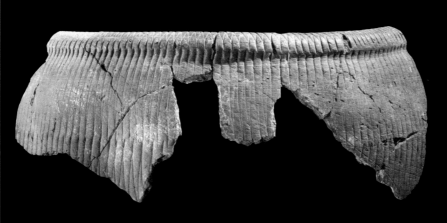

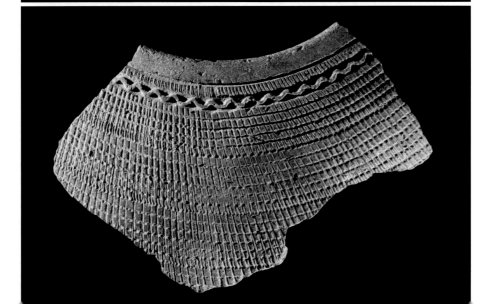

◄ Figure 3. En *a.*, vue partielle de la base d'un mur à double alignement de pierres circonscrivant un espace habité à Anezrouft à la charnière des IVe-IIIe millénaires av. J.-C. (d'après M. Raimbault 1995). Vestiges récoltés dans les espaces clos de haies du haut Tilemsi (d'après M. Raimbault (ed.) 2009, © Musée National de Préhistoire, Eyzies-de-Tayac, collections Gaussen, clichés P. Jugie). *Sources :*
b. Poterie à bord épaissi décorée par impressions pivotantes (diamètre de l'ouverture : 28 cm ; inv. MNP 2007 01 27 20) ;
c. Poterie décorée d'un quadrillage et d'un cordon ondulé obtenu par impressions alternées d'une spatule (diamètre de l'ouverture : 15 cm ; inv. MNP 2007 01 27 20) ;
▼*d.* Haches en roche verte de petit module (longueur de la plus grande : 5 cm ; inv. MNP 2007 01 27 20).

permanentes, du moins saisonnières ou répétées. Les artefacts consistent en des haches et herminettes, en des meules et molettes, en des disques et anneaux, en des percuteurs polyédriques et boules de pierre, en des lames et éclats retouchés, auxquels s'ajoutent quelques microlithes géométriques et armatures de pointes de flèche ainsi que des poteries sphériques aux bords souvent épaissis, décorés de quadrillages, chevrons, traits pointillés, sous tendus de lignes ondulées et aux corps fréquemment ornés de motifs flammés obtenus par impression pivotante (Figure 3). Des ossements et des dents de taurins, de chèvres et/ou de moutons et des restes d'animaux sauvages attestent de la pratique de l'élevage et de la chasse au gros gibier. Des vestiges comparables se retrouvent rassemblés dans des sites voisins dépourvus de murets, mais aux limites cependant bien visibles, suggérant par-là la présence de barrières végétales sans pierres de calage (Gaussen, Gaussen 1988). Seules les découvertes de grains carbonisés et/

ou d'empreintes d'épillets sur tessons lors de futures fouilles permettront de savoir si du mil était consommé en ces lieux et si les enclos servaient à protéger quelques champs de mil contre l'appétit des grands herbivores. En attendant, notons que la nature même des vestiges archéologiques inventoriés alliée aux datations radiocarbones obtenues et aux données de la génétique, plaident en faveur de l'existence dans ce secteur d'un terroir agricole ayant compté parmi les plus anciens de l'Ouest africain. Peut-être fut-il même le premier d'entre eux ?

2.3. Des inhumations sous constructions monumentales

Michel Raimbault (Raimbault, 1994: 932-940) a inventorié neuf tumulus en croissant à l'occasion d'une mission menée dans le haut Tilemsi : huit ouverts vers l'est, un vers l'ouest. Le plus grand s'étend sur 50 m avec des pierres accumulées dans sa partie centrale sur 7,5 m de largeur et 2 m de hauteur. Ces monuments de dimensions imposantes sont visibles sur les images satellitaires Google Earth. Plus de 600 de ces constructions ont été ainsi repérées dans le nord de l'Adrar des Iforas et ses alentours, dix fois plus dans une bande large de plusieurs centaines de kilomètres centrée sur le tropique du Cancer allant des confins libyco-nigériens au Sahara atlantique (Gauthier, Gauthier 2008 ; Jarry 2018). Quinze fouilles dans le nord du Niger ont révélé des inhumations individuelles d'homme, une de femme et une sépulture collective d'au moins huit individus (Bernus *et al.* 1999: 136 ; Paris 1996: 274,545). Seule une tombe a livré un matériel d'accompagnement modeste, à savoir un pilon de dolérite et une poterie brisée. La plus ancienne de ces sépultures date du milieu du IVe millénaire av. J.-C., la plus récente du début du IIe millénaire av. J.-C.

Cette architecture funéraire voit donc le jour au moment des premières manipulations menées sur le mil, dans une aire géographique qui intègre probablement le berceau de ces expérimentations. Le nombre limité de ces monuments (six mille inventoriés recouvrant plus de deux mille ans d'histoire, soit une moyenne de trois tumulus en croissant édifiés par an sur une superficie à peu près égale à trois fois le territoire de la France) et la main d'œuvre importante mobilisée pour leur construction, permettent d'y voir les sépultures de sujets au statut social hors du commun. Qui étaient-ils ? S'agissait-il de céréaliculteurs, d'éleveurs ou de chasseurs-pêcheurs-cueilleurs émérites, ou bien d'individus qui occupaient des fonctions politiques, religieuses et/ou économiques importantes ? Des groupes aux modes de subsistance spécialisés et complémentaires se côtoyaient-ils ? Ou bien faut-il se ranger à l'idée de populations relativement autarciques qui pratiquaient chasse, pêche, collecte, cueillette, élevage et agriculture, là où les biotopes le permettaient ? Il est d'autant plus délicat de se prononcer que ces deux types d'organisation ne s'excluaient pas forcément et qu'au début de l'agriculture, les situations ont été changeantes et évolutives. La classification en faciès des artefacts collectés dans le Tilemsi par Jean et Michel Gaussen (Gaussen *et al.* 1988) rattachables à cette époque, si utile soit elle, ne permet pas de se déterminer.

3. Une nouvelle dynamique sociétale

3.1. La diffusion l'agriculture du mil dans un contexte d'aridité croissante

Les déchets de repas mis au jour sur différents tells de la basse vallée du Tilemsi occupés à la charnière des IIIe-IIe millénaires av. J.-C. révèlent que l'agriculture du mil était pratiquée au côté de l'élevage des bovins et des ovi-caprins, de la chasse, de la pêche, de la cueillette des produits spontanés de la savane, de la collecte de mollusques d'eau douce (Finucane *et al.* 2008 ; Manning 2011). Au début du IIe millénaire av. J.-C., le mil n'est plus seulement cultivé dans ce secteur ; il l'est aussi désormais au sud-est de la Mauritanie (Amblard-Pison 2006 ; Amblard, Pernès 1989 ; Fuller *et al.* 2007), en pays dogon (Ozainne 2013: 42, 61) et dans le nord du Ghana (D'Andréa *et al.* 2001). À la fin de ce même millénaire, son exploitation est attestée au Burkina Faso (Neumann 1999 ; Vogelsang *et al.* 1999) et au sud-ouest du lac Tchad (Breunig, Neumann 2002 ; Klee, Zach 1999). Comment expliquer l'extension considérable de cette céréaliculture alors que sa pratique s'avère somme toute contraignante puisqu'elle implique, conservation des semences, semailles, surveillance des champs, stockage des grains, protection des récoltes ?

Les grains carbonisés et les tessons supportant des empreintes d'épillets retrouvés dans les sites archéologiques du IIe millénaire av. J.-C., montrent que le mil a dès lors acquis sa morphologie domestique. Celle-ci consiste en des grains fermement attachés aux épis et entourés d'enveloppes réduites que prolongent de rares soies végétales. Ces caractères sont très différents de ceux du mil sauvage pourvu, lui, de grains déhiscents, c'est-à-dire tombant au sol de façon très échelonnée dès leur arrivée à maturité, et que protègent d'épaisses enveloppes munies de longues soies végétales (Figure 4). La morphologie domestique s'avère, de fait, très utile à l'agriculteur qui, le temps des moissons venu, récolte le maximum de grains en un seul passage dans ses champs, contrairement au cueilleur qui ne récupère dans les prairies à graminées spontanées que les grains encore sur tiges, le rendement de son travail pouvant s'avérer décevant s'il intervient trop tard. De plus, les grains domestiques sont plutôt faciles à décortiquer comparativement à ceux des céréales sauvages dont l'élimination des pédicelles, des soies végétales et des enveloppes riches en cellulose indigestes pour l'homme, nécessite un long et fastidieux travail (Pernès 1983). Ainsi les hasards de la génétique alliés à l'observation et à l'ingéniosité des hommes, ont permis au mil d'acquérir des caractères extrêmement avantageux qui ont encouragé la diffusion de sa culture. La baisse généralisée de la pluviosité que connaît

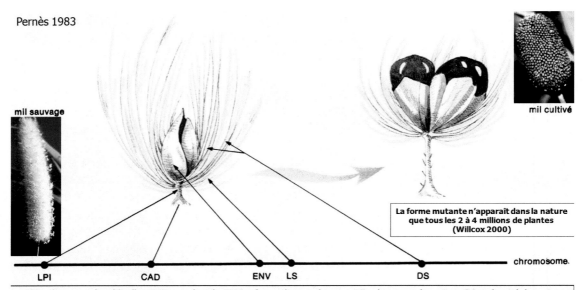

Pernès 1983

mil sauvage

mil cultivé

La forme mutante n'apparaît dans la nature
que tous les 2 à 4 millions de plantes
(Willcox 2000)

chromosome.

LPI   CAD   ENV LS   DS

LPI = longueur du pédicelle ; CAD = caducité ; ENV = forme des enveloppes ; LS = longueur des soies ; DS = densité des soies

Figure 4. Du mil sauvage au mil domestique : les cinq gènes responsables du syndrome de la domestication – LPI, CAD, ENV, LS, DS – sont portés par le même chromosome et transmis en bloc à chaque génération (adapté de J. Pernès 1983).

l'Afrique de l'Ouest à partir du début du IIe millénaire av. J.-C a sans doute aussi favorisé la propagation de la nouvelle économie sur de vastes espaces (Le Drezen 2008 ; Lézine *et al.* 2011 ; Petit-Maire, Riser 1983). Cette péjoration des conditions climatiques a entraîné une ouverture des paysages accompagnée, là où les pluies de mousson étaient fortement déficitaires, d'une raréfaction du gibier et du poisson et d'une faible régénération des pâturages et des plantes nourricières spontanées. Les craintes de disettes auraient-elles incité quelques communautés à privilégier l'agriculture du mil ? Il est tentant de répondre par l'affirmative tant l'adoption de cette stratégie pouvait s'avérer payante. En semant et en récoltant plus que nécessaire à leurs besoins, les céréaliculteurs, encore de nos jours, se placent au cœur du tissu social par les ventes, les dons ou les prêts de grains qu'ils peuvent consentir aux pêcheurs, chasseurs et éleveurs friands de mil ou contraints de s'en procurer pour éviter la famine, mais aussi aux personnes trop accaparées par leurs activités religieuses, politiques, marchandes ou artisanales pour produire leur propre nourriture végétale.

Des cultivars de mil de signature génétique ouest-africaine atteignent l'Inde au cours de la première moitié du second millénaire av. J.-C. (Fuller 2003 ; Manning *et al.* 2011). L'exportation de cette graminée a pu être encouragée par la connotation positive dont elle était chargée, une plante symbole d'abondance car source de prospérité pour ceux qui, en échange des excédents de production agricole qu'ils contrôlaient et redistribuaient, acquerraient des bijoux de pierres fines, des objets coudés à lame métallique, des chars, des bœufs à bosse, autant de produits qui étaient alors en circulation dans le Tilemsi et ses abords et qui contribuaient à l'affichage de la richesse. Cernons plus précisément les faits.

3.2. Des lapidaires, des métallurgistes et des charrons à l'ouvrage

Des grains d'enfilage en calcédoine rouge à orange appelée cornaline de forme standardisée et soigneusement polis, ont été retrouvés, à Karkarichinkat, au cou d'individus inhumés dans deux tells occupés à la charnière des IIIe-IIe millénaires av. J.-C. (Smith 1974). Ces perles proviennent probablement des ateliers voisins de Lagreich et d'Ilouk. Les milliers de mètres cubes de déchets de taille accumulés en ces lieux témoignent de productions qui dépassaient de loin la demande locale (Gaussen 1993 ; Gaussen, Gaussen 1988). Ni les termes, ni l'étendue des échanges desquels participaient ces atours, n'ont été saisis jusqu'ici. Peut-être les sept perles annulaires plates en cornaline retrouvées mille kilomètres à l'ouest, en Mauritanie sud-orientale dans les villages aux murs de pierres sèches construits sur les Dhars Tichitt et Oualata au cours des deux derniers millénaires avant l'ère chrétienne, examinées par Sylvie Amblard-Pison (Amblard-Pison, 2006: 219-220), proviennent-elles de ces centres de production ? Des analyses chimiques devraient permettre de se prononcer.

Des haches, pics et ciseaux en silex étaient aussi fabriqués en d'abondantes quantités dans le bas Tilemsi à proximité immédiate des gites de roches siliceuses. L'étaient par ailleurs des armatures perçantes de différents types et un large éventail de parures (Figure 5), à savoir en plus des perles en cornaline dont il vient d'être question, des anneaux, des pendeloques et des labrets, de tailles et de formes diverses, aux couleurs et aux éclats variés, tirés soit de roches dures telles que le quartz hyalin, le quartz opaque, la quartzite, le jaspe, l'amazonite, le gneiss, les calcédoines, soit de matériaux tendres tels que le schiste, la stéatite, le calcaire, l'œuf d'autruche, les oursins et coquillages fossiles, les vertèbres de poisson, les phalanges de mammifères (Chancerel 2009 ; Duhard 2007 ; Lhote 1942, 1943 ; Mauny 1952).

L'usage d'objets coudés à lame de métal avec crochet en partie proximale est documenté à cette même époque par des dizaines de gravures représentées dans l'art rupestre de l'Adrar des Iforas. Leurs profils variés, sans équivalent connu en Afrique septentrionale, plaident en faveur d'une fabrication locale (Dupuy 1994). Les associations de deux de ces objets, respectivement à un char et par ailleurs à un bœuf à bosse (un zébu – *Bos indicus* – ou un sanga, hybride obtenu par croisement d'un zébu et d'un taurin), permettent de rattacher cette pratique de la métallurgie au IIe millénaire av. J.-C. En effet, le char et le zébu (et, conjointement, le cheval) furent introduits dans la vallée du Nil depuis le Proche-Orient aux alentours du XVIe siècle av. J.-C. À partir de là, leur diffusion de proche en proche vers des régions toujours plus occidentales paraît avoir été rapide comme donnent à le penser plusieurs motifs complexes gravés et peints de façon marginale au Sahara, et qui témoignent d'une ouverture de l'Afrique septentrionale au monde méditerranéen de l'âge du bronze (Dupuy 2005, 2006, 2016-2017). Le métier de métallurgiste semble donc bien se développer

Figure 5. Outils, armatures et éléments d'apparat en pierre du bas Tilemsi (d'après M. Raimbault (ed.) 2009, © Musée National de Préhistoire, Eyzies-de-Tayac, collections Gaussen, clichés P. Jugie). *Sources :*
▲ *a.* Haches taillées en silex provenant d'un dépôt qui en comptait dix-sept (longueur de la plus grande : 18,5 cm ; inv. MNP 2007 01 02 02) ;
◄ *b.* Armature de flèche à ailerons et pédoncule taillée à la pression après chauffe probable du silex (longueur : 4,5 cm ; inv. MNP 2007 01 41 51) ;

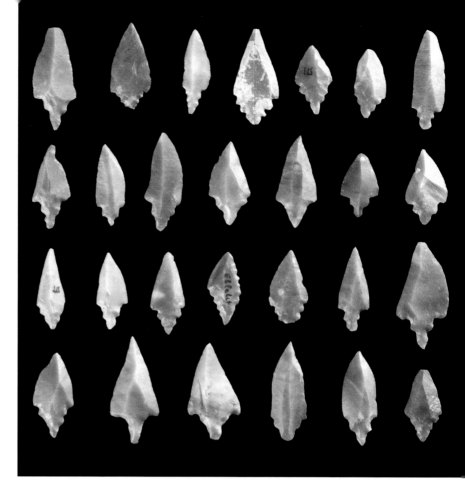

Figure 5 ▶ *c.* Armatures de flèche triangulaires pédonculées (longueur moyenne : 4 cm ; inv. MNP 2007 01 58 25) ; ▼ *d.* Perles et pendeloques tirées de diverses roches, d'os et de coquilles d'œuf d'autruche (longueur de la pendeloque en haut à gauche : 2,4 cm ; inv. MNP 2007 01 06 08) ;

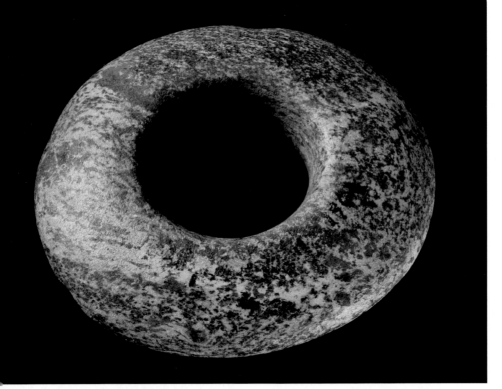

Figure 5 *e.* Anneau en gneiss (diamètre extérieur : 13,7 cm ; inv. MNP 2007 01 06 21) ;

Figure 5 *f.* Labrets en quartz (longueur du plus grand : 5,5 cm ; inv. MNP 2007 01 05 09).

dans le Tilemsi et ses abords sensiblement à la même époque que celui tout aussi spécialisé de lapidaire.

Un autre métier faisant appel à des connaissances éprouvées tant dans le travail du bois que dans celui des peaux animales, voit le jour dans ce contexte : celui de charron comme l'attestent au Sahara de nombreuses représentations rupestres de chars détélés et d'attelages de taurins et de chevaux montrant des dispositifs inconnus dans la vallée du Nil et autour de la Méditerranée (Dupuy 2006, 2019 ; Camps 1993 ; Spruytte 1977, 1996). À type d'exemple, certains des véhicules fabriqués sont équipés de plates-formes surmontées d'arceaux latéraux contre lesquels les cochers appuyaient leurs jambes pour assurer leur équilibre lors des déplacements à vive allure. De nombreux autres artifices techniques existent que nous ne pouvons présenter pas dans le cadre de cet article.

Dispensés de la contrainte de devoir produire leur nourriture grâce à la redistribution des surplus vivriers générés par l'économie agro-pastorale, des lapidaires, des métallurgistes, des charrons ont pu s'adonner à temps complet à leur activité et fabriquer des objets toujours plus nombreux et performants, pendant que quelques élites affichaient leur richesse, biens de luxe à l'appui, confortant ainsi leur pouvoir politique par le prestige à les posséder.

3.3. De nouveaux gestes, de nouvelles préoccupations

La décoration de la céramique par impressions roulées de cordelettes de fibres végétales torsadées fait son apparition dans le bas Tilemsi durant la seconde moitié du IIIe millénaire av. J.-C. (Figure 6). Ce type d'ornementation va connaître ensuite un large succès au Sahel (Haour *et al.* 2010). Sont aussi décorés de cette manière d'énigmatiques boudins de terre à base plane, de section ronde à rectangulaire, hauts de 4 à 20 cm et larges de 27 à 58 mm (Gaussen, Gaussen 1988 ; Smith 1978). Apparaissent concomitamment

Figure 6. Céramiques décorées au moyen de cordelettes de fibres végétales torsadées (d'après M. Raimbault (ed.) 2009, © Musée National de Préhistoire, Eyzies-de-Tayac, collections Gaussen, clichés P. Jugie). *Sources :* ▼ *a.* Vase sphérique (diamètre de l'ouverture : 21 cm ; inv. MNP 2007 01 38 01) ; ▶ *b.* Cylindre de fonction inconnue (hauteur : 13,3 cm ; inv. MNP 2007 01 56 04).

des figurines zoomorphes et anthropomorphes en terre cuite de taille centimétrique (Gaussen, Gaussen 1988 ; Gaussen *et al.* 2002 ; Smith 1978). L'une d'elle collectée à Smar Smarren dans le bas Tilemsi mérite une attention particulière en raison de son pouvoir à évoquer des dizaines de motifs gravés à quelques trois cents kilomètres de son lieu de découverte sur six éminences granitiques de l'Adrar des Iforas.

Les grains de quartz déci-millimétriques à millimétriques visibles à la surface de cette pièce, ainsi que les petits cratères marquant les emplacements de ceux s'étant déchaussés, révèlent que l'argile utilisée contenait du sable fin (Figure 7). Il est impossible de savoir si ce sable était présent dans la couche sédimentaire où l'argile a été prélevée ou s'il a été rajouté comme dégraissant minéral pour faciliter son modelage. Quelle qu'ait été la recette suivie pour sa préparation, la pâte argilo-sableuse obtenue a été façonnée, alors qu'elle était encore humide et malléable, en une galette sub-rectangulaire aux angles et aux bords arrondis, plane d'un côté, légèrement convexe de l'autre, haute de 50 mm, large de 30 mm, épaisse de 10 à 17 mm. Cette petite tablette a été ensuite minutieusement lissée, puis percée en deux points au tiers de sa hauteur. Les contours anguleux des trous et les bavures visibles à leurs débouchés révèlent que le percement s'est effectué par enfoncement d'une fine tige dans la pâte encore molle. Douze autres trous borgnes de diamètre millimétrique plutôt bien alignés ont été réalisés, peut-être au moyen de la même tige, sur le bord le plus proche de l'axe commun aux deux perforations, tandis qu'un sillon peu profond a été creusé par rainurage sur le bord opposé. La couleur grisâtre de la face bombée peut traduire une cuisson en atmosphère réductrice, à moins qu'elle ne soit due à l'application d'un enduit. Le soin apporté au lissage de cette pièce et la maîtrise subséquente de sa cuisson rendent peu crédible l'idée qu'un enfant l'ait réalisée pour en faire son jouet. En raison de l'habileté dont relève sa confection, il est plus logique de l'attribuer à un adulte qui maîtrisait l'art de la terre cuite. Quelle était sa destination ? Si les deux trous sont demeurés bruts de perçage après cuisson, c'est parce qu'ils n'ont jamais servi de passage à une ficelle ou à une lanière de suspension. Dans le cas contraire, le frottement du lien dans les orifices en aurait ovalisé et ébavuré les contours. Cette figurine n'a donc jamais été portée en pendeloque. De plus, elle ne montre aucun stigmate d'usure qui puisse la faire identifier à un outil. Sa taille réduite et sa solidité rendaient sa manipulation aisée. Elle pouvait être posée à même le sol sur son bord rainuré ou dressée sur un support, ou bien couchée, ou encore bridée en hauteur sur une armature, avec fichée dans sa partie supérieure une douzaine de tiges végétales (paille, brindilles, épines...) ou animales (plumes, piquants, os longs et menus...). Que représentait-elle ? Un visage humain avec ses yeux, surmonté de cheveux en brosse ou d'un couvre-chef, dépourvu de bouche, de nez et d'oreilles ? Un personnage vivant ou mort, au corps enveloppé dans un sac en peau ou en cuir duquel n'émergeait que le haut de la tête coiffée ? Un être invisible – divinité,

Figure 7. Figurines (échelle centimétrique ; © Musée National de Préhistoire, Eyzies-de-Tayac, collections Gaussen, clichés P. Jugie). *Sources : a.* Représentation schématique d'un être sensible d'essence anthropomorphe (Smar Smarren ; Gaussen, Gaussen 1988 : fig. 94/6) ;

Figure 7 *b.* Protomé de taurin aux cornes et aux oreilles brisées. L'enfoncement d'une tige dans l'argile molle rend compte des yeux comme sur la figurine précédente (Tahébanat ; Gaussen, Gaussen 1988 : fig. 94/9).

esprit, génie, ancêtre, âme d'un défunt, double d'une personne – doté d'un arrangement céphalique, au cou et aux membres non dégagés d'un tronc sub-rectangulaire ? Un masque facial à ouvertures oculaires, hérissé d'appendices dans sa partie supérieure ? A défaut de disposer d'éléments déterminants pour se prononcer, on l'assimilera simplement à une entité au regard bien rendu, humanisée par sa coiffe.

Cette figurine fournit un repère chronologique pour l'art rupestre, eu égard à ses affinités formelles avec plusieurs dizaines de gravures rupestres du nord-ouest de l'Adrar des Iforas. Ces pétroglyphes montrent des contours sub-ovalaires à sub-rectangulaires et une double-ponctuation en décoration interne (Figure 8). Cet aménagement suggère deux yeux surtout lorsque se trouve gravé en partie opposée un U évoquant soit une

Figure 8. Motifs plus ou moins ovalaires à double ponctuation gravés au nord-ouest de l'Adrar des Iforas (*a*, *b* et *c* : Issamadanen ; *d* : Asenkafa ; clichés C. Dupuy). *Sources :*
▼*a*. Paroi oblique supportant trois motifs sub-rectangulaires décorés de doubles ponctuations. Les exemplaires disposés tête-bêche sont ornés d'un U qui pose question : s'agit-il d'une bouche, d'un collier, du menton délimitant le bas d'un visage, ou d'autre chose ? Le diamètre du cache objectif vaut 53 mm ;
▶*b*. Motif sub-rectangulaire bi-ponctué sur coussinet cintré muni d'antennes latérales. Hauteur du motif : ~ 40 cm ;

bouche, soit un collier, voire un menton délimitant le bas d'un visage. Des appendices latéraux sur quelques exemplaires évoquent des bras : l'un d'eux aboutit sous la tête d'un quadrupède, un autre semble armé d'une hache à lame en croissant fixée à un manche coudé (Dupuy 2006: 42-44). À rester dans cette logique de lecture, le double trait cilié délimitant l'un de ces motifs bi-ponctués pourrait représenter une coiffe, une coiffure ou bien la pilosité développée d'un être sensible tandis que l'étranglement médian sur deux autres motifs pourrait marquer la taille d'entités à discrète essence anthropomorphe.

Ces gravures ô combien intrigantes participent d'un art animalier non narratif, emprunt de schématisme, dominé par les taurins, les girafes et les autruches, pour partie contemporain de l'époque des chars et des objets coudés à lame métallique. Les actions de gravures répétées qui

Figure 8 *c*. Dalle décorée d'une spirale liée à deux motifs sub-rectangulaires. Celui de droite montre un double contour rempli de courts traits rayonnants et intègre deux cupules. Une large embase trapézoïdale piquetée le sous-tend. Hauteur du motif : ~ 35 cm. Ces caractères le rapprochent de la figurine de Smar Smarren, affublée ici graphiquement des tiges telles qu'elles devaient être fichées dans les trous borgnes ponctuant son bord supérieur (voir Fig. 7a). Peut-être son bord inférieur rainuré était-il posé sur un support à l'image de l'embase gravée ?

Figure 8 *d*. Motif sub-rectangulaire bi-ponctué s'inscrivant dans un arciforme entouré de figures schématiques de quadrupèdes. La branche montante gauche de l'arciforme aboutit sous la tête d'un quadrupède. Faut-il voir dans cette association une entité aux caractères anthropomorphes discrets touchant un animal ? Hauteur du motif : ~ 20 cm.

commandaient à ces réalisations sur des éperons rocheux en bordure de vallée, souvent à l'aplomb de mares se formant encore de nos jours durant les pluies de la mousson, relevaient peut-être de rites propitiatoires, ce à quoi pourraient aussi avoir été dévolus les modelages anthropomorphes et zoomorphes du Tilemsi réalisés à partir de la seconde moitié du IIIe millénaire av. J.-C. en contexte agro-pastoral. La sépulture d'un bœuf en pleine terre à Karkarichinkat est contemporaine de cette époque (Manning 2011).

3.4. Des architectures funéraires diversifiées

Des dizaines de tumulus coniques, tantôt associés à des alignements d'amas pierriers, tantôt circonscrits par des cercles de pierres, ont été érigés à In Imanal, petit massif situé 90 km à l'ouest de Tessalit faisant partie intégrante du haut Tilemsi (Raimbault 1994: 932-934). S'observent à leurs côtés des bazinas souvent flanquées sur leur face orientale d'une petite niche quadrangulaire. Deux d'entre elles ont été édifiées sur des socles polygonaux de pierres appareillées. Par ailleurs un tumulus à cratère a été construit sur la partie centrale d'un tumulus en croissant et un petit édicule sur l'appendice septentrional de ce même tumulus en croissant.

Un tumulus à cratère fouillé dans ce secteur a permis la mise au jour d'un squelette dont la bioapatite des ossements a fourni l'âge radiocarbone de 1980 ± 60 BP (Pa 1217), soit après correction dendrochronologique un âge calendaire compris entre 166 av. J.-C. et 134 apr. J.-C.

Ces diverses familles de tombes se rencontrent dans de nombreuses régions sahariennes (Gauthier 2009 ; Gauthier, Gauthier 2007 ; Jarry 2020). Plusieurs dizaines de ces sépultures fouillées et datées au nord du Niger délimitent les fourchettes chronologiques suivantes (Paris 1996) : la construction des tumulus à cratère s'échelonne entre 3340 av. J.-C. et 945 apr. J.-C., celle des bazinas entre 3050 et 1330 av. J.-C., celle des monuments à alignements entre 2030 av. J.-C. et 1040 apr. J.-C. On dispose d'une seule date pour un tumulus à cratère entouré d'un cercle de pierres, lui même circonscrit à des pierres alignées en arc de cercle : le collagène mal conservé des os de l'individu exhumé de cette construction fouillée dans l'Ighazer wan Agadez a fourni l'âge ^{14}C très imprécis de 3000 ± 1000 BP (Paris 1984: 173-175).

Les aires de répartition étendues de ces diverses familles de tombes, leurs imbrications spatiales et les quelques datations radiocarbones disponibles, dénotent une dynamique de peuplement complexe étalée sur près de quatre millénaires à laquelle ont pris part les populations du Nord Tilemsi.

4. Des premières aristocraties guerrières à nos jours

4.1. La fin de la protohistoire

L'emploi ostentatoire de la lance à large pointe métallique se généralise sur la frange saharo-sahélienne, Adrar des Iforas inclus, durant le premier millénaire av. J.-C. (Dupuy 1998, 2018 ; Paris 1990, 1996 ; Roset 1988, 2007). Dès lors l'iconographie rupestre donne primauté aux figures de porteurs de lance (Figure 9). Les deux tiers de ces guerriers sont très visiblement de sexe masculin. Leurs coiffures, coiffes, parures et vêtements sont diversifiés. Quelques uns dirigent leur arme contre les corps de girafes, d'éléphants et de rhinocéros aux tailles miniaturisées. De rares chevaux, absents jusque-là du bestiaire, les accompagnent. Au nord-ouest de l'Adrar des Iforas, deux étalons sont gravés de part et d'autre d'un char que deux personnages semblent se préparer à atteler (Figure 10). Conduire ces équidés à vive allure debout sur la plateforme d'un tel véhicule était à n'en pas douter un exercice périlleux qui répondait du même dessein que celui de se déplacer à pied, lance à la main, avec le projet d'affronter quelques grands animaux sauvages : célébrer à la fois sa bravoure et sa réussite sociale et s'en glorifier sur les rochers. L'art rupestre valorise ces gestes sportifs dans des contextes figuratifs riches de taurins. En cela, il témoigne du rôle important que jouent désormais les guerriers au sein d'une société pastorale qui considère les chevaux de trait comme des biens éminemment prestigieux (Dupuy 2016-2017). Alors que s'affirme cette aristocratie, les caprices des

Figure 9. Gravures de porteurs de lance datables du Ier millénaire av. J.-C. (relevés et clichés C. Dupuy, sauf mention contraire). *Sources :* *a.* À droite, relevés à l'encre de personnages visiblement masculins, têtes et corps figurés de face, réalisés en des temps plus récents que les porteurs d'objets coudés sans vêtements ni parures détaillés reproduits à gauche ; les objets qu'ils brandissent sont surdimensionnés par rapport à leur taille ;

Figure 9 *b.* Personnage appliquant la pointe de sa lance sous la queue d'une girafe (Asenkafa ; hauteur du personnage : ~ 35 cm) ;

Figure 9 ▲ c. Porteur de lance tenant en longe un taurin paré d'un collier (Issamadanen ; hauteur du personnage : ~ 45 cm) ;
► d. Personnage à coiffe volumineuse revêtu d'une sorte de culotte bouffante (Adarmolen ; hauteur du personnage : ~ 40 cm) ;

Figure 9 ▼ *e.* Personnages réunis autour d'une même lance (Ibdakan ; hauteur du personnage principal : ~ 1 m) ;

▲ *f.* Personnages aux coiffes volumineuses traités dans des dimensions très différentes (Issamadanen ; hauteur du personnage principal : ~ 1 m) ;

◄ *g.* Porteur de lance richement paré, tenant dans sa main gauche un bouclier à décors d'ocelles tiré peut-être d'une peau de girafe (Tédaré ; hauteur du personnage : ~ 1 m ; cliché R. Di Popolo).

Figure 10. Étalons figurés de part et d'autre d'un char, chaque timonier semble tenu par un porteur de lance : cette composition évoque une scène d'attelage (Asenkafa, hauteurs au garrot des chevaux : ~ 40 cm ; cliché C. Dupuy).

Figure 11. Gravures réalisées à partir des ɪvᵉ-vᵉ siècles apr. J.-C. (clichés C. Dupuy). *Sources :*
▲ *a.* Cavalier et méhariste coursant des gazelles (Imeden, longueur de la composition : ~ 1,50 m) ;
◄ *b.* Porteur de javelots entouré de personnages revêtus d'habits amples et bien couvrants et d'inscriptions libyco-berbères disposées verticalement (Enguenhat, hauteur du personnage principal : ~1 m).

pluies de la mousson dans le sud du Sahara rendent les récoltes céréalières de plus en plus aléatoires. L'aridité qui culmine aux alentours du début de l'ère chrétienne, finit par contraindre les populations agropastorales du Tilemsi à se replier vers la moyenne vallée du Niger et/ou vers d'autres secteurs favorisés par la géographie et le climat. À partir des ɪvᵉ-vᵉ siècles apr. J.-C., les espaces ainsi libérés sont investis par des éleveurs dont les manifestations d'art rupestre dans l'Adrar des Iforas attestent de traditions nouvelles : monte des chevaux et des dromadaires, port de plusieurs javelots et de vêtements amples et bien couvrants, chasses à courre aux girafes, aux antilopes, aux gazelles et aux autruches, rédaction de courts messages à base de signes dont la plupart se retrouvent dans l'écriture *tifinagh* des Touaregs (Casajus 2015 ; Drouin 2014). Ces figurations et inscriptions attestent du rattachement de la région au domaine *amazighophone* (= berbérophone) et simultanément de son basculement dans l'histoire (Figure 11).

4.2. Une aire de nomadisation traversée par des pistes caravanières

À l'époque médiévale, des marchands arabo-berbères parcourent régulièrement le Tilemsi et ses abords pour se rendre dans la ville d'Essouk-Tadmakkat où s'opèrent d'importants échanges commerciaux (Figure 1). L'or du Sahel ouest-africain raffiné sur place compte parmi les produits précieux qui se négocient dans la cité (Nixon 2017). Une inscription rupestre en écriture arabe à l'entrée de l'agglomération fait état d'un marché comparable à celui de La Mecque. Quelques épitaphes sur pierres tombales renvoient à des noms d'étrangers témoignant du cosmopolitisme de la ville : là apparaît le patronyme d'un chiite sud-marocain, plus loin celui d'un perse (Moares Farias 2017). Cette cité située aux portes du désert et à la croisée des principales pistes caravanières issues du Sahara et du Sahel est florissante du Xᵉ jusqu'au XIVᵉ siècle. Elle est abandonnée au XVᵉ siècle, peut-être parce que touchée par l'épidémie de peste noire qui frappe le bassin méditerranéen, l'Égypte et l'Éthiopie (Derat 2018 ; Gallagher, Dueppen 2018). Les sources orales évoquent des destructions liées à des conquêtes et à des rivalités politiques (Claudot-Hawad 1985). L'aridité et/ou la désertification et, corrélativement, les problèmes d'approvisionnement en eau ont pu aussi contraindre les habitants à quitter les lieux.

De nos jours, les pâturages dont se couvrent la vallée du Tilemsi et ses affluents après chaque saison des pluies attirent des éleveurs touaregs et arabes qui parcourent cette terre de nomadisme multiséculaire, durement frappée par l'obscurantisme depuis une dizaine d'années.

Bibliographie

Amblard-Pison S. 2006. *Communautés villageoises néolithiques des Dhars Tichitt et Oualata (Mauritanie)*. Oxford, BAR International Series 1546.

Amblard S., Pernès J. 1989. The identification of cultivated pearl millet (*Pennisetum*) amongst plant impressions on pottery from Oued Chebbi (Dhar Oualata, Mauritania). *The African Archaeological Review* 7, p. 117-126.

Bernus E., Cressier P., Durand A., Paris F., Saliège J.-F. 1999. *Vallée de l'Azawagh (Sahara du Niger)*. Saint-Maur, Editions Sépia, Etudes Nigeriennes 57.

Breunig P., Neumann K. 2002. From hunters and gatherers to food producers: New archaeological and archaeobotanical evidence from the West African Sahel. *In:* F.A. Hassan (ed.) *Droughts, food and culture. Ecological change and food security in Africa's later prehistory.* New York, Kluwer/ Plenum, p.123-153.

Burgarella P., Cubry C., Kane N.A., Varshney R.K., Mariac C., Liu X., Shi C., Thudi M., Couderc M., Xu X., Chitikineni A., Scarcelli N., Barnaud A., Rhoné B., Dupuy C., François O., Berthouly-Salazar C., Vigouroux Y. 2018. A Western Sahara origin of African agriculture inferred from pearl millet genomes. *Nature Ecology and Evolution, Brief communication* 2(9), p.1377-1380.https://doi.org/10.1038/s41559-018-0643-y

Camps G. 1993. Char. *Encyclopédie Berbère* XII, p.1877-1892.

Casajus D. 2015. *L'alphabet touareg.* Paris, CNRS Editions.

Chancerel A. 2009. L'outillage en pierre. La parure. *In:* M. Raimbault (ed.) *Le Sahara il y a 7 000 ans : des lacs, des rivières et des hommes.* Les Eyzies-de-Tayac, Musée National de Préhistoire, RMN, p.18-20.

Claudot-Hawad H. 1985. Adrar des Iforas. Histoire du peuplement. *Encyclopédie Berbère* II, p.147-153.

Clotault J., Thuillet A.C., Buiron M., De Mita S., Couderc M., Haussmann B.I.G., Mariac C., Vigouroux Y. 2012. Evolutionary history of pearl millet (*Pennisetum glaucum* [L.] R. Br.) and selection on flowering genes since its domestication. *Mol. Biol. Evol.* 29(4), p.1199-1212.

Cremaschi M., di Lernia S. (eds) 1998. *Wadi Teshuinat. Palaeoenvironment and Prehistory in South-western Fezzan (Libyan Sahara).* Roma-Milano, Consiglio Nazionale delle Ricerche, Quaderni di Geodinamica Alpina e Quaternaria 7.

Cremaschi M., Zerboni A., Mercuri A.M., Olmi L., Biagetti S., di Lernia S. 2014. Takarkori rock shelter (SW Libya): an archive of Holocene climate and environmental changes in the central Sahara. *Quaternary Science Reviews* 101, p.36-60.

D'Andrea A.C., Klee M., Casey J. 2001. Archaeobotanical evidence for pearl millet (*Pennisetum glaucum*) in sub-Saharan West Africa. *Antiquity* 75, p.341-348.

Derat M.-L. 2018. Du lexique aux talismans : occurrences de la peste dans la Corne de l'Afrique du XIIIe au XVe siècle. *Afriques,* http://journals. openedition.org/afriques/2090

Di Lernia S., Gallinaro M. 2010. The date and context of Neolithic rock art in the Sahara: engravings and ceremonial monuments from Messak Settafet (south-west Libya). *Antiquity* 84, p.954-975.

Di Lernia S., Tafuri M.A., Gallinaro M., Alhaique Fr., Balasse M., Cavorsi L., Fullagar P.D., Mercuri A.M., Monaco A., Perego A. 2013. Inside the "African Cattle Complex": Animal Burials in the Holocene Central Sahara. *PLoS ONE* 8(2): e56879. doi:10.1371/journal.pone.0056879

Drouin J. 2014. Les incipit des inscriptions rupestres. Corpus de l'Adrar des Ifoghas (Mali). *Epigraphie libyco-berbère, Répertoire des Inscriptions Libyco-Berbères* 20, p.11-16.

Duhard J.-P. 2007. À propos de labrets néolithiques en pierre du Sahara malien. *Sahara* 18, p.85-94.

Dupuy C. 1989. Les gravures naturalistes de l'Adrar des Iforas (Mali) dans le contexte de l'art rupestre saharien. *Travaux du LAPMO*, p.151-174.

Dupuy C. 1994. Signes gravés au Sahara en contexte animalier et les débuts de la métallurgie ouest-africaine. *Préhistoire et Anthropologie Méditerranéennes* 3, p.103-124.

Dupuy C. 1998. Réflexions sur l'identité des guerriers représentés dans les gravures de l'Adrar des Iforas et de l'Aïr. *Sahara* 10, p.31-54.

Dupuy C. 1999. L'art rupestre à gravures naturalistes de l'Adrar des Iforas (Mali). *Sahara* 11, p.69-86.

Dupuy C. 2005. Les gravures de bœufs à bosse de l'Aïr (Niger) et de l'Adrar des Iforas (Mali). *Bulletin de la Société d'études et de recherches préhistoriques des Eyzies* 54, p.63-90.

Dupuy C. 2006. L'Adrar des Iforas à l'époque des chars : art, religion, rapports sociaux et relations à grande distance. *Sahara* 17, p. 29-50.

Dupuy C. 2016-2017. Chars sahariens préhistoriques et araires africains actuels. I. L'alimentation des animaux de trait. *Les Cahiers de l'Association des Amis de l'Art Rupestre Saharien* 19, p. 29-44.

Dupuy C. 2017. La domestication du mil et ses implications sociétales. *Le Saharien*, 220, p.16-39.

Dupuy C. 2018. Du port d'objets coudés au port de la lance dans l'Adrar des Iforas: la traduction figurative d'un important virage sociétal. *In:* D. Huyge, F. Van Noten (eds), *What Ever Happened to the People ? Humans and Anthropomorphs in the Rock Art of Northern Africa.* Brussels, Royal Academy for Overseas Sciences, p.101-117.

Dupuy C. 2019. À propos des chars rupestres sahariens: "rendre à César ce qui appartient à César". *Portails académiques HAL SHS, Research Gate & Academia.*

Finucane B., Manning K., Touré M. 2008. Late Stone Age subsistence in the Tilemsi Valley, Mali: Stable isotope analysis of human and animal remains from the site of Karkarichinkat Nord (KN05) and Karkarichinkat Sud (KS05). *Journal of Anthropological Archaeology* 27, p.82-92.

Fuller D.Q. 2003. African crops in prehistoric South Asia: a critical review. *In:* K. Neumann, S. Kahlheber, A. Butler (eds), *Food, Fuel and Fields: Progress in African Archaeobotany.* Cologne, Heinrich-Barth Institut, p.239-271.

Fuller D.Q., Barron A., Champion L., Dupuy C., Commelin D., Raimbault M., Denham T. 2021. Transition From Wild to Domesticated Pearl Millet (*Pennisetum glaucum*) Revealed in Ceramic Temper at Three Middle Holocene Sites in Northern Mali. *African Archaeological Review*, https:// doi.org/10.1007/s10437-021-09428-8.

Fuller D.Q., MacDonald K., Vernet R. 2007. Early domesticated pearl millet in Dhar Nema (Mauritania): evidence of crop processing waste as ceramic temper. *In:* R. Cappers (ed.), *Fields of change. Progress in African archaeobotany*. Groningen, Barkhuis & Groningen University Library, p.71-76.

Gallagher D.E., Dueppen S.A. 2018. Recognizing plague epidemics in the archaeological record of West Africa. *Afriques*: //journals.openedition.org/afriques/2198

Gaussen J. 1993. Perles néolithiques du Tilemsi et du pays Ioullemedene (ateliers et techniques). *Memorie della Società di Scienze Naturali e del Museo Civico di Storia Naturale di Milano* XXVI (II), p.254-256.

Gaussen J., Gaussen M. 1988. *Le Tilemsi préhistorique et ses abords : Sahara et Sahel malien*. Bordeaux, Cahiers du Quaternaire 11, Editions du CNRS.

Gaussen J., Duhard J.P., Ridouard P. 2002. Trois modelages anthropomorphes du Mali nord-oriental (pays Ioullimeden, cercle de Gao). *Le Saharien* 161, p.43-45.

Gauthier Y. 2009. Nouvelles réflexions sur les aires de distribution au Sahara central. *Cahiers de l'AARS* 13, p.121-134.

Gauthier Y., Gauthier C. 2007. Monuments funéraires sahariens et aires culturelles. *Cahiers de l'AARS* 11, p.65-78.

Gauthier Y., Gauthier C. 2008. Monuments en trou de serrure, monuments à alignement, monuments en "V" et croissants : contribution à l'étude des populations sahariennes. *Cahiers de l'AARS* 12, p.105-124.

Haour A., Manning K., Arazi N., Gosselain O., Guèye N.S., Keita D., Livingstone Smith A., MacDonald K., Mayor A., MacIntosh S., Vernet R. 2010. *African Pottery Roulettes. Past and Present. Techniques, Identification and Distribution*. Oxford, Oxbow Books.

Jarry L. 2018. *Les monuments en croissant*. http://ressources.ingall-niger.org/documents/cartes/archeologie/les_croissants.pdf

Jarry L. 2020. *Inventaire archéologique satellitaire de la plaine de l'Ighazer (Niger)*. https://www.researchgate.net/project/Inventaire-archeologique-satellitaire-de-la-plaine-de-lIghazer-Niger

Klee M., Zach B. 1999. The exploitation of wild and domesticated food plants at settlement mounds in north-east Nigeria (1800 cal BC to today), *In:* M. Van Der Veen (ed.), *The exploitation of plant resources in Ancient Africa*. New York, Kluwer Academic/Plenum Publishers, p.81-88.

Le Drezen Y. 2008. *Dynamiques du paysage de la vallée du Yamé depuis 4000 ans. Contribution à la compréhension d'un géosystème soudano-sahélien (Ounjougou, Pays Dogon, Mali)*. Caen, Thèse de doctorat non publiée, Université de Basse-Normandie.

Lézine A.M., Hely, C., Grenier, C., Braconno P., Krinne G. 2011. Sahara and Sahel vulnerability to climate changes, lessons from Holocene hydrological data? *Quaternary Science Reviews* 30 (21-22), p.3001-3012.

Lhote H. 1942. Découverte d'un atelier de perles néolithiques dans la région de Gao (Soudan français). *B. S. P. F.* 39 (10-12), p.277-292.

Lhote H. 1943. Découverte d'un atelier de perles néolithiques dans la région de Gao (Soudan français). *B. S. P. F.* 40 (1-2-3), p.24-36.

Lhote H., Tomasson R. 1967. Gravures rupestres de la haute vallée du Tilemsi (Adrar des Iforas, République du Mali). *VIe Congrès panafricain de Préhistoire* (Dakar 1967), p.235-241.

Manning K. 2011. The First Herders of the West African Sahel : Inter-site Comparative Analysis of Zooarchaeological Data from the Lower Tilemsi Valley, Mali. *In:* H. Jousse, J. Lesur (eds), *Recent advances in archaeozoology*. Frankfort, Africa MagnaVerlag, p.75-85.

Manning K., Pelling R., Higham T., Schwenniger J.-L., Fuller D.Q. 2011. 4500-Year old domesticated pearl millet (*Pennisetum glaucum*) from the Tilemsi Valley, Mali: new insights into an alternative cereal domestication pathway. *Journal of Archaeological Science* 38, p. 312-322.

Mauny R. 1952. Les gisements néolithiques de Karkarichinkat (Tilemsi, Soudan français). *In:* L. Balout (ed.), *Congrès panafricain de Préhistoire. Actes de la IIe session*. Alger: Arts et Métiers Graphiques, p.616-629.

Mauny R. 1954. *Gravures, peintures et inscriptions rupestres de l'Ouest africain*. Dakar: Mémoire IFAN XI.

McIntosh S.K. (ed.) 1994. *Excavations at Jenné-Jeno, Hambarketolo and Kaniana (Inland Niger Delta, Mali), the 1981 Season*. Berkeley and Los Angeles: University of California Press 20.

Moraes Farias P.F. 2017. Arabic and Tifinagh Inscriptions. *In:* S. Nixon (ed.), *Essouk-Tadmekka. An Early Islamic Trans-Saharan Market Town*. Leiden/Boston: Brill, Journal of African Archaeology, Monograph Series, Volume 12, p.41-51.

Murray S. 2005. Recherches archéobotaniques. *In:* R.M.A. Bedaux, J. Polet, K. Sanogo, A.M. Schmidt (eds), *Recherches archéologiques à Dia dans le Delta intérieur du Niger (Mali) : bilan des saisons de fouilles 1998-2003*. Leiden: CNWS publications Vol. 144, Mededelingen van het Rijksmuseum voor Volkenkunde (RMV) 33, p.386-400.

Neumann K. 1999. Early plant food production in the West African Sahel: new evidence. *In:* M. Van Der Veen (ed.), *The exploitation of plant resources in Ancient Africa*. New York, Kluwer Academic/Plenum Publishers, p.73-80.

Nixon S. (ed.) 2017. *Essouk-Tadmekka. An Early Islamic Trans-Saharan Market Town*. Leiden/Boston: Brill, Journal of African Archaeology, Monograph Series, Volume 12.

Ozainne S. 2013. *Un Néolithique Ouest-africain. Cadre chrono-culturel, économique et environnemental de l'Holocène récent en Pays dogon (Mali)*. Frankfort, Africa Magna Verlag, Journal of African Archaeology Monograph Series 8.

Paris F. 1984. *Les sépultures du Néolithique final à l'Islam Atlas (III). La région d'In Gall-Tegidda-n-Tessemt (Niger). Programme archéologique d'urgence, 1977-1981*. Niamey, Paris, Études nigériennes 50.

Paris F. 1990. Les sépultures monumentales d'Iwelen. *Journal des Africanistes* 60 (1), p.44-74.

Paris F. 1996. *Les sépultures du Sahara nigérien du Néolithique à l'Islamisation*. Paris, Orstom Editions, collection études et thèses, 2 tomes.

Paris F. 1998. Les inhumations de *Bos* au Sahara méridional au Néolithique. *Archaeozoologia* IX (1-2), p.113-122.

Petit-Maire N., Riser J. (eds) 1983. *Sahara ou Sahel ?* Marseille, Imprimerie Lamy.

Pernès J. 1983. La génétique de la domestication des céréales. *La recherche* 146, p.910-919.

Raimbault M. 1994. *Sahara malien: environnement, populations et industries préhistoriques.* Aix-en-Provence: Université de Provence, Thèse de l'Université, 3 tomes.

Raimbault M. 1995. La culture néolithique des "villages à enceintes dans la région de Tessalit, au nord-est du Sahara malien. *In:* R. Chénorkian (ed.), *L'homme méditerranéen. Mélanges offerts à Gabriel Camps.* Aix-en-Provence, Publications de l'Université de Provence, p.113-125.

Raimbault M. (ed.) 2009. *Le Sahara il y a 7 000 ans : des lacs, des rivières et des hommes.* Les Eyzies-de-Tayac, Musée National de Préhistoire, RMN.

Roset J.-P. 1988. Iwelen, un site archéologique de l'époque des chars dans l'Aïr septentrional, au Niger. *Presses Universitaires de France, Études et documents UNESCO* 11, p.121-155.

Roset J.-P. 2007. La culture d'Iwelen et les débuts de la métallurgie du cuivre dans l'Aïr, au Niger, *In:* J. Guilaine (ed.) *Le Chalcolithique et la construction des inégalités. Proche et Moyen-Orient, Amérique, Afrique* (Tome II). Paris, Éditions Errance, p.107-136.

Smith A.B. 1974. Preliminary report of excavations at Karkarichinkat Nord and Sud, Tilemsi Valley, Mali, spring 1972. *West African Journal of Archaeology* 4, p.33-55.

Smith A.B. 1978. Terracottas from the Tilemsi valley, Mali. *Bulletin de l'IFAN* 40 (B), p.223-228.

Spruytte J. 1977. *Études expérimentales sur l'attelage.* Paris, Crépin-Leblond.

Spruytte J. 1996. *Attelages antiques libyens.* Paris, Éditions de la Maison des Sciences de l'Homme.

Takezawa S., Cissé M. (eds) 2017. *Sur les traces des grands empires. Recherches archéologiques au Mali.* Paris, L'Harmattan.

Tostain S. 1998. Le mil, une longue histoire : hypothèses sur sa domestication et ses migrations. *In:* M. Chastanet (ed.) *Plantes et paysages d'Afrique. Une histoire à explorer.* Paris, Karthala-CRA, p.461-490.

Van Albada A., Van Albada A.-M. 2000. *La montagne des hommes-chiens.* Paris, Editions du Seuil.

Vogelsang R., Albert K.-D., Kahlheber S. 1999. Le sable savant : les cordons dunaires sahéliens au Burkina Faso comme archive archéologique et paléoécologique de l'Holocène. *Sahara* 11, p. 51-68.

Troupeaux Holocène au Sahara

Une nouvelle représentation de la relation homme-animal dans l'économie et l'art

Barbara E. Barich[1]

Abstract

The establishment of the Holocene climate in the Sahara facilitated many new innovations, including the introduction of domesticated animal species from the Near East. This is the most evident proof of contacts between North Africa and the Southwest Levant, a corridor that was probably also used earlier, but which from that date onwards was traveled more and more intensively. Exploitation of domesticated animals, first and foremost goats/sheep and oxen, facilitated significant transformations in the economic and social structure of human groups. These groups continued to practice hunting,evidenced by the considerable presence of wild species (gazelle, hartebeest, hares) alongside the domesticated ones. This article deals with some relevant case-studies in Libya (Tadrart Akakus), in Algeria (Tassili and Hoggar/Ahaggar) and in Niger (Aïr) that document the presence of the first goats and pose the problem of ox domestication in the Sahara. At the Tenerian complexes of the Adrar Bous, eastern Aïr, the so-called 'Tenerian meals' have yielded cattle bones burned during ritual meals that provide evidence for a culturally structured practice. The importance of cattle is testified not only by their inhumations, found in many regions from the Sahara to Sudan, but also by the pictorial representations of large herds, or isolated figures, well known in the Tassili and Tadrart Akakus. Compared to the numerous attestations of these domestic species, the ones of other less important species, such as the dog, the donkey and, in more recent times, the dromedary are also present in the Saharan rock art repertoire.

Résumé

Le réchauffement climatique de l'Holocène au Sahara voit comme une innovation extraordinaire l'introduction d'espèces animales domestiquées venant du Proche-Orient. Il s'agit de la preuve la plus évidente des contacts entre l'Afrique du Nord et le Sud-Ouest du Levant, un couloir qui était probablement aussi utilisé à des époques antérieures, mais qui, à partir de ce moment, sera parcouru de manière de plus en plus intensive. Les animaux domestiqués, en premier lieu les chèvres/moutons et les bœufs, ont produit des transformations significatives dans la structure économique et sociale des groupes humains, même si la chasse a continué à être pratiquée, étant

1 Association Internationale pour les Études Méditerranéennes et Orientales (ISMEO) et Fondation de l'Université de Rome La Sapienza

donnée la présence considérable d'espèces sauvages (gazelle, Alcelaphus, lièvre) à côté des espèces domestiquées. L'article considère des exemples pertinentes en Libye (Tadrart Akakus), en Algérie (Tassili et Hoggar/Ahaggar) et au Niger (Aïr) qui documentent la présence des premières chèvres et posent la question de la domestication du bœuf au Sahara. Les complexes ténéréens de l'Adrar Bous, à l'Est de l'Aïr, dits «repas ténéréens», ont révélé des ossements de bovins brûlés lors de repas rituels qui témoignent d'une pratique culturellement structurée. L'importance culturelle du bétail est soulignée non seulement par les inhumations, attestées dans de nombreuses régions du Sahara au Soudan, mais aussi par les représentations picturales de grands troupeaux ou d'un animal isolé bien connues dans le Tassili et le Tadrart Akakus. Par rapport aux nombreuses attestations de ces espèces domestiques, d'autres moins importantes, comme le chien, l'âne et, à une époque plus récente, le dromadaire sont également présentes dans le répertoire de l'art rupestre saharien.

Introduction

Durant la 'Période Humide Africaine' de l'Holocène ancien (de Menocal, Tierney 2012), le Sahara a connu des phases d'humidité alternant avec des périodes d'aridité. Les recherches archéologiques conduites depuis la moitié du 20ème siècle ont défini la séquence paléo-climatique de ce vaste territoire (Hassan *et al.* 2001; Kuper, Kröpelin 2006; Wendorf *et al.* 2001 ; Gatto, Zerboni 2015) (Figure 1). Les régions sahariennes désertifiées par l'extrême sécheresse du Pléistocène supérieur-final (*Last Glacial Maximum*) furent à nouveau accessibles à l'occupation humaine à partir de 11 500 cal BP, grâce au rétablissement du cycle des pluies de mousson (Kuper, Kröpelin 2006; Barich 2019).

Les groupes humains ont pratiqué une économie largement ouverte sur l'environnement (*broad spectrum*), dans laquelle parallèlement à la chasse, qui a continué à jouer un rôle important (gazelles, lièvres, *Addax*, *Alcelaphus*, oiseaux, petits rongeurs), se sont développées des collectes de plantes sauvages, de mollusques et, dans certains cas, des activités de pêche. En outre, des tentatives de domestication du mouflon et de l'aurochs ont été observées dès la phase initiale de l'Holocène. Ces tentatives d'apprivoisement pourraient avoir été entreprises et pourraient résulter d'une longue familiarité et de pratiques de management et de soin. Les sites de Uan Afuda (Di Lernia 1999) et de Uan Tabu (Garcea 2001) dans la Tadrart Akakus, semblent avoir servi de camps spécialisés réservés au parcage d'*Ammotragus lervia*. Ainsi l'horizon supérieur d'Uan Afuda (Unité 1) suggère une pratique constante de gestion et de contrôle du mouflon. D'autres exemples presque contemporains situés en Egypte dans la région de Nabta Playa, à El Ghorab et à El Nabta-Al Jerar semblent correspondre à des sites semi-permanents d'activités d'apprivoisement et de gestion des aurochs (Gautier 1987, 2001).

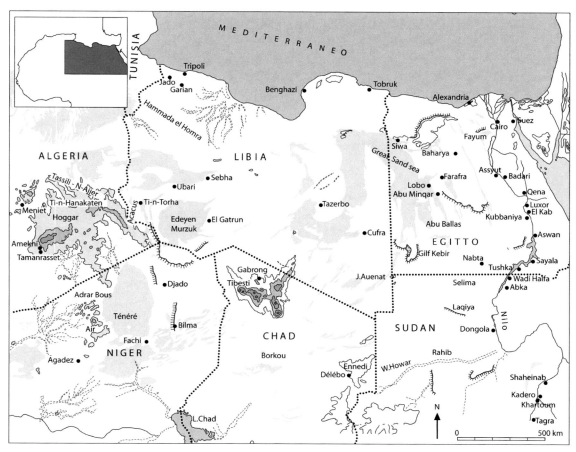

Figure 1. Sahara central et oriental et les principaux lieux mentionnés dans le texte.

Plus personne ne considère désormais la chèvre domestique comme résultant d'une hybridation possible avec l'*Ammotragus lervia*. Par rapport à l'Algérie littorale, on se souvient que Saxon (Saxon 1974) avait avancé l'hypothèse d'une tentative de domestication de l'*Ammotragus* dans la grotte ibéromaurusienne de Tamar Hat, cette hypothèse désormais écartée ne pourrait-elle pas déjà évoquer une pratique de contrôle, comme dans l'Akakus ? (C. Roubet *in litt.* 2020).

Rappelons que les moutons et les chèvres ont été initialement domestiqués au Proche-Orient entre la Mésopotamie et l'Asie centrale et se sont répandus dans la vallée du Nil vers 6500-6000 cal BC (Espérandieu 1994). De récentes datations indiquent que la chèvre a été le premier animal domestique introduit en Afrique depuis le Levant méridional. Les résultats chronologiques actuellement disponibles montrent que les sites de Sodmein sur la côte de la Mer Rouge (Veermeersch *et al.* 2015), de Farafra dans le Désert Occidental égyptien (Barich 2016; Gautier 2014), et de Cyrénaïque en Libye (de Faucamberge 2014), situent l'introduction d'animaux domestiques tantôt un peu avant, tantôt un peu après l'oscillation aride de 8200 cal BP / 6200 cal BC généralisée en Méditerranée. La diffusion des chèvres fut ensuite rapide d'Est en Ouest, le long de la côte Sud de la Méditerranée et du Nord au Sud le long la vallée du Nil. En Egypte, à Nabta Playa, des moutons et chèvres ont été identifiés dans les sites Al Jerar-El Ghanam dès la fin 7ème millénaire cal BC (entre 6200-5400 cal BC, Schild, Wendorf 2001; Linseele *et al.* 2014). En Libye, dans la Tadrart Akakus et au Niger, des caprins ont été identifiés vers 6100 cal BC, ils présentent une certaine contemporanéité avec ceux de la région de El Nabta, située à peu près à la même latitude. Comme le montrent les peintures rupestres les premiers moutons portaient une queue mince (*Ovis longipes*), les moutons à queue épaisse subsahariens sont entrés en Afrique plus tard, peut-être par le Nord-est et la corne de l'Afrique. L'analyse de l'ADN mitochondrial soutient une origine maternelle ancestrale commune pour tous les moutons africains. Tandis que l'analyse de l'ADN du chromosome Y, indique une origine génétique distincte pour les moutons africains à queue mince et à queue épaisse subsaharienne (Muigai, Hanotte 2013).

La reconstruction de l'origine et de la diffusion des bovins domestiques de l'espèce *Bos taurus* en Afrique, et en particulier au Sahara, reste compliquée, incertaine et non encore résolue. L'idée que l'Afrique pourrait avoir été un centre de domestication du boeuf, indépendant du Proche-Orient et de la vallée de l'Indus, a été longuement discutée. L'hypothèse d'un «putative domestic *Bos*» au Sahara égyptien entre le 9ème et le 8ème millénaire cal BC (Gautier 1984, 2001), reste très vague encore. Ces résultats et le statut du *Bos* continuent de faire l'objet de débats (Wendorf, Schild 1994, 2001; Marshall, Weissbrod 2011; Di Lernia 2013. Voir Brass 2017 pour une revue complète du problème).

Une de plus récente étude génétique conduite à l'échelle mondiale sur un échantillon de 1 500 bovins modernes, montre que les boeufs taurins africains actuels descendent de géniteurs domestiques du Proche-Orient, mais avec une forte hybridation faite avec des aurochs africains locaux (Decker *et al.* 2014). Ces derniers ayant disparu, il n'est plus possible de rétablir leur génome. On retiendra qu'entre 8300-7300 cal BP / 6200-5400 cal BC, les restes de boeuf reconnus dans la séquence de Nabta Playa (unités Al Jerar-El Ghanam) se trouvaient associés à ceux de mouton / chèvre domestiques. Ce constat n'entraîne pas de changement significatif dans le modèle économique des groupes humains et conserve un caractère exceptionnel (Linseele *et al.* 2014). Dans le Désert Occidental égyptien, et seulement à Dakhla pour le moment, on signale d'autres preuves de la présence d'un bétail avec boeuf (McDonald 2016).

La diffusion en Afrique des boeufs domestiques entre les 7ème-6ème millénaires cal BC reste un problème en suspens. Pour la chèvre, la route du Nord reste la plus vraisemblable depuis le Levant du Sud-ouest, le Delta du Nil et le Sinaï méridional (Close 2002; Barich 2016), mais elle n'a pas été retenue pour *Bos taurus*. Une des plus anciennes références de restes squelettiques de boeuf provient du Fayum (site QS IX / 81), la datation est de 6380 ± 60 BP (Gd-1499, Linseele *et al.* 2014).

On peut conclure que l'introduction des animaux domestiques en Afrique et leur diffusion progressive vers de nouveaux territoires, se sont produites par étapes successives (Linseele *et al.* 2016 ; Barich 2016). Selon les dates ^{14}C, une première diffusion, liée aux mouvements rapides et intermittents de petits groupes de troupeaux à la recherche d'eau et de pâturages, aurait pu se produire d'Est en Ouest à partir du désert Egyptien depuis Nabta (Hassan 2002). A ce type de mouvements peuvent être rattachés les restes d'Enneri Bardagué au Tchad (Hv2775–7455 ± 180 BP, Gabriel 1972); ceux de la Tadrart Akakus, en Libye à Takarkori (LTL914A–7327 ± 65 BP, Cremaschi *et al.* 2014) et d'Uan Muhuggiag (LTL17843A–6987± 45 BP, non publié[2]).

Entre le 5ème et 4ème millénaire cal BC, on constate que la région du Delta du Nil présente un modèle néolithique et une sorte de «*package*» néolithique proche-oriental presque complet: avec orge, blé, lentilles mais aussi *Bos taurus*, chèvres, porc (Barich 2021). En Égypte, la première apparition du chien en tant qu'animal domestique de troupeau est également enregistrée, il se pourrait qu'il ait été domestiqué localement (Linseele et al. 2014: 17). Au contraire Gautier, signalant la présence de restes de chien dans les niveaux de l'Holocène moyen-tardif à Nabta Playa (Site E-75-8: *Canis aureus* et *Canis aureus lupaster*), pensait plutôt que le chien aurait pu parvenir en Afrique depuis l'Asie, au cours du même événement ayant introduit le mouton / chèvre (Gautier 2001, p.620). Dans le désert égyptien, ce n'est qu'à partir de cette date qu'il y a eu dans les sites étudiés une augmentation significative de la présence d'animaux domestiques (Brass 2017, Tableau 4).

La mise en évidence de troupeaux au Sahara provient de travaux approfondis en stratigraphie, ayant fourni des données archéologiques multidisciplinaires, au cours des deux dernières décennies. La synthèse suivante repose sur les données les plus fiables, de valeur inégale d'une région à l'autre.

Moutons / chèvres et boeufs domestiques d'après les fouilles (Tableau 1).

Tadrart Akakus, Libye

La Tadrart Akakus en Libye a livré les plus anciennes données de la présence de bétail domestique associées à la culture pastorale (Barich 1974, 1987; Biagetti, Di Lernia 2013; Cremaschi, Di Lernia 1998; Garcea 2001). Des dizaines d'abris rocheux ont été utilisés par les bergers pendant des périodes plus ou moins longues. Les restes osseux de bovins (de l'espèce *Bos taurus*) ont été collectés dans la plupart des sites caractérisés par d'importants dépôts anthropiques (voir Uan Muhuggiag, Ti-n-Torha Nord, Takarkori, Uan Telokat). Une chronologie en trois phases a été établie pour la Tadrart, qui

---

2   Datation d'un échantillon venant des fouilles Barich 1981-82: Site UM A layer 2D (CEDAD 2018).

Tableau 1. Datations des témoins osseux de mouton / chèvre et bœuf domestiques mentionnés dans le texte. De bas en haut les dates sont listées à partir des plus anciennes (Holocène ancien) aux plus récentes (Holocène moyen et tardif). L'étalonnage à 2 sigmas a été effectué avec INTCAL13 (Reimer et al. 2013).

Lab code	Region	Site	Materiel	Date ¹⁴C non calibré bp	Date cal BP (95%)	Date cal BC (95%)	Bibliographie
inconnu	Niger	Adrar Bous, Agorass n'Tast	Em	4145±45	4828-4568	2878-2618	Clark et al.2008b
Geo	Libya	U.Muhuggiag, T.A.	Ch	4739±310	6184-4788	4234-2838	Mori 1965
Gif-1725	Niger	Arlit	Ch	5200±140	6224-5662	4274-3712	Vernet 1998
Gif-1380	Algérie	Adrar Tiouyine	Ch	5320±130	6325-5881	4375-3931	Camps 1974
Sa59	Algérie	Baguena V	Sol	5410±300	6893-5582	4943-3632	Delibrias et al.1957
inconnu	Algérie	In Relidjen	____	5420±130	6451-5918	4501-3968	Aumassip, Delibrias 1982-83
R-1165a	Libya	Oued Athal, T.A.	Ch	5630±50	6504-6301	4554-4351	Barich, Mori 1970
Ucla 1658	Niger	Adrar Bous, Agorass n'Tast	Coll	5760±500	7164-6173	5214-4223	Clark et al.1973
Mc-483	Algérie	Timidouine	Ch	6050±100	7015-6777	5065-4827	Maître 1974
PA0330	Niger	Adrar Bous 1	Bov	6325±300	7736-6498	5786-4548	Roset 1987
Gd-1499	Egypt	QS IX/81 Fayum	Ch(?)	6380±60	7426-7239	5476-5227	Linseele et al.2014
LTL17843A	Libya	U.Muhuggiag, T.A.	Ch	6987±45	7933-7707	5983-5757	Cedad 2018
Hv-14022	Niger	Dogomboulo	Veg	7125±290	8202-7672	6252-5722	Roset 1987
SMU-2745	Egypt	Nabta E-75-8	Ch	7220±75	8185-7931	6235-5981	Schild, Wendof 2001
Gif-5419	Algérie	Tin Hanakaten	Ch	7220±100	8073-7962	6123-6012	Aumassip, Delibrias 1982-83
R-2456	Egypt	H.Valley, Farafra	Ch	7251±67	8186-7954	6236-6000	Barich, Lucarini 2014
LTL914A	Libya	Takarkori,T.A.	Ch	7327±65	8320-8010	6370-6060	Cremaschi et al.2014
Hv-2775	Tchad	Bardagué	Girafe	7455±180	8604-7932	6654-5982	Gabriel 1972
Gd-6509	Egypt	Nabta E-75-6	Ch	7480±110	8459-8029	6509-6079	Schild, Wendof 2001

Légende: Ch=charbon; Bov=bovidé; Em= émail dentaire; Sol=sédiment; Veg= reste végétal.

reste une référence également pour les régions voisines: Pastoral ancien 8300-7200 cal BP / 6300-5200 cal BC; Pastoral moyen 7100-5600 cal BP / 5100-3600 cal BC et Pastoral final 6900-4400 cal BP / 4900-2000 cal BC (Biagetti, Di Lernia 2013; Di Lernia 2019). Les premiers bergers et leurs troupeaux installés entre le 7ème et le 6ème millénaire BC montrent une mobilité réduite et séjournent longtemps à l'intérieur du massif; pendant le Pastoral moyen, les groupes commencent à utiliser une grande variété de niches écologiques s'étendant à l'Est à l'Erg Uan Kasa (Di Lernia 2019, p.192-193). Les bovins (presque toujours associés aux caprins) deviennent de plus en plus fréquents et acquièrent une plus grande importance dans l'économie et la vie sociale. Durant cette période, on enregistre pour la première fois l'utilisation du lait, en s'appuyant sur des résidus d'acide alcanoïque de la matière grasse du lait, identifiés dans des fragments de poterie provenant de Takarkori, 5ème millénaire cal BC (Dunne *et al.* 2012).

Tassili n'Ajjer et Hoggar(Ahaggar), Algérie

Grâce aux mouvements saisonniers des éleveurs de bovins qui peuvent également être reconstruits à partir de leurs haltes (*steinplatze*) (Gabriel 1987), un réseau d'échanges entre territoires voisins s'est mis en place. Dans le massif du Tassili n'Ajjer, le phénomène pastoral - appelé ici «bovidien» (Camps 1974; Aumassip 1996) - a eu un impact tout aussi important que dans la Tadrart libyenne. Les vestiges archéologiques appartiennent à l'Holocène moyen, mais à Tin-Hanakaten, l'un des gisements les plus importants, l'horizon «bovidien» commence vers le 7ème millénaire cal BC (Gif-5419–7220 ± 100 BP, Aumassip, Delibrias 1982-83; Aumassip 1996), comme le Pastoral ancien de la Tadrart. Dans un site en plaine, et une station de surface située à In Relidjem, dans l'Erg d'Admer (versant ouest du Tassili), des dents et des os des bovins domestiques associés à des tessons céramiques décorés par impression ont été datés 5420 ± 130 BP (Aumassip, Delibrias 1982-83; Gautier 1987, p.171). Dans les sites d'Anou Oua Lelioua et de Tahort de l'Erg d'Admer, bien qu'aucun résidu squelettique de bœuf n'ait été collecté dans les dépôts, on admet qu'il s'agit de sites de pasteurs.

Dans le Hoggar, l'un des sites les plus importants de la region de Méniet – le site de Baguena V - a livré une séquence importante dans laquelle les bovins et les caprins sont datés de 5410 ± 300 BP (Delibrias *et al.* 1957; Gautier 1987, p.171). Dans le gisement de Timidouine daté de 6050 ± 100 BP (Mc483, Maître 1974), il se peut qu'ait existé une phase plus ancienne comparable à celle de Tin-Hanakaten, Séquence 4. D'autres restes de bovins, pour lesquels des informations précises font défaut, sont signalés dans l'Adrar Tiouyine (Smith 1980, Tab. 20.2).

Aïr, Niger

La culture ténéréenne pratiquée par des éleveurs de bovins semble avoir été mise en place entre l'Adrar Bous et Areschima, Aïr oriental, vers 4000 cal BP. Cependant, dans les sites de Dogomboulo, du secteur de Fachi (Hv14022 – 7125 ± 290 BP, Schulz 1987) et de l'Adrar Bous 1 (PA0330 - 6325 ± 300 BP, Roset 1987), on a récolté des restes brûlés de bœuf sous un monticule de pierres : premiers exemples d'une occupation pastorale au début de l'Holocène moyen. À Arlit, dans l'Aïr occidental, d'autres restes de boeuf récoltés par Lhote ont été datés de 5200 ± 140 BP (Gif-1725, Vernet 1998).

La tendance aride reconnue en Libye vers la fin du 6ème millénaire cal BC, fut peut-être la cause d'une migration vers les régions du Sud (Barich 2010, p.215). Pour Roset également, l'impact des fortes conditions d'aridité sur l'environnement pourrait avoir conduit les pasteurs à se déplacer en direction des zones montagneuses, où l'eau était encore suffisante (Roset 1987). On doit la reconnaissance de la culture ténéréenne aux premières investigations de l'Expédition Berliet Ténéré sur le site d'Adrar Bous III (Tixier 1962). Le Ténéréen présente une importante industrie lithique avec des instruments à retouche bifaciale et des céramiques hémisphériques décorées par impression (Roset 1987). La chasse et la cueillette de végétaux sauvages étaient toujours importantes; la présence de *Brachiaria* et de *Sorghum* sont mises en évidence par des impressions conservées dans les parois des récipients céramiques.

Dans les années 1970, l'Adrar Bous et, en particulier, le secteur Agorass n'Tast, ont fait l'objet d'autres recherches conduites par la *British Expedition* dirigée par J.D. Clark ; les données n'ont été que récemment publiées dans leur intégralité (Clark *et al.* 2008a). La découverte d'un squelette presque complet de *Bos taurus* représente une importante découverte. L'âge de ce *Bos* a été établi à partir de deux datations : l'une 5760 ± 500 BP sur le collagène (UCLA 1658, Clark *et al.* 1973, p.291), et l'autre 4145 ± 45 BP, date AMS, sur une molaire de l'animal (Clark *et al.* 2008b). Les auteurs estiment que cette dernière date est beaucoup plus fiable que la première, car elle établit une relation convaincante entre la mort naturelle de l'animal et un épisode aride de l'Holocène final, confirmant ainsi l'hypothèse d'une mort probablement due à la déshydratation (Clark *et al.* 2008b, p. 391).

Toujours en relation avec l'occupation ténéréenne, l'Adrar Bous a livré un grand nombre de restes domestiques de *Bos*, considérés comme les plus importants du Sahara. Pas un seul reste d'ovin ou de caprin ou bien de faune aquatique n'a été collecté, malgré la présence importante de lagunes en eau. Le bœuf d'Agorass n'Tast est en réalité une vache au squelette complet, dont les caractères sont similaires à ceux des spécimens d'Afrique de l'Ouest (Carter, Clark 1976, pl. 2). Les caractères du frontal et des cornes ressemblent au type de Kobadi. La taille au garrot de l'individu de 104 cm est conforme à celle d'animaux de petites races actuelles d'Afrique de l'Ouest (Clark *et al.* 2008b).

La vache de l'Adrar Bous, comme les spécimens provenant d'Uan Muhuggiag entre le 5ème et le 4ème millénaire cal BC, est donc de petite taille avec des cornes courtes et pourrait représenter une femelle d'un type africain commun (Grigson 2000). Cependant, parmi les autres spécimens de l'Adrar Bous, il y a des éléments osseux beaucoup plus grands, témoins d'une diversité dimensionnelle considérable qui ne peut s'être produite que pendant une période d'au moins deux millénaires. Au contraire, dans la Vallée du Nil, un large échantillon d'individus (y compris ceux de Badari) est caractérisé par un type de bœuf (dimension inconnue) à longues cornes, caractère présent chez le bœuf sauvage, localement connu (Brunton, Caton-Thompson 1924; Grigson 2000).

À l'Adrar Bous, la plupart des restes de bovins ont été collectés dans des *Tenerian meals*: concentrations issues de consommations, constituées d'os de bœufs brûlés. Mais ce ne sont pas de simples déchets de cuisine, il s'agirait peut-être de regroupements et de recompositions délibérées d'os de bovins, après consommation, destinés ensuite à être brûlés à l'intérieur de fosses (Gifford-Gonzalez, Parham 2008, p.360). Les auteurs estiment que les fosses ont été initialement utilisées pour la cuisson complète des pièces à consommer, puis pour récupérer à nouveau les restes destinés à une forme d'incinération. Tout cela semble être la preuve d'une pratique «cultuellement structurée» (Gifford-Gonzalez, Parham 2008, p.360). Une évaluation de la quantité de viande qui a pu être consommée –souvent plus d'un individu par fosse - suggère que les repas ténéréens pourraient avoir représentés des épisodes d'abattage récurrents, comprenant division et consommation de jeunes animaux, dans un contexte social et symbolique de fêtes et de célébrations tout au long de l'année.

Gobero, Niger

La culture ténéréenne est présente plus au sud, au bord du paléolac appelé Gobero (Sereno *et al.* 2008). Les mollusques (*Mutela*), le poisson-chat (*Clarias*) et le *Tilapia* dominent la faune, qui comprend également des ossements et des dents d'hippopotame, une petite antilope, de petits carnivores, des tortues et des crocodiles. Contrairement aux sites de l'Adrar Bous, les taureaux domestiques ne représentent qu'une petite composante alimentaire, compte tenu de la rareté des collectes d'os de bovins. Les auteurs semblent pouvoir reconstruire un modèle économique basé sur la pêche en eau peu profonde et sur la chasse aux espèces de savane, ainsi que sur la collecte de plantes sauvages.

Importance culturelle du boeuf: les enfouissements cultuels

L'un des indicateurs les plus surprenants de l'importance des bovins dans la subsistance saharienne est l'inhumation des bœufs répandue à travers le Sahara central, en Algérie, en Libye et au Niger (Paris 1997; Jelinek 2003; Tauveron *et al.* 2009; Di Lernia *et al.* 2013). Mais aussi à Kerma, au Soudan, où les sites d'enfouissement peuvent compter des centaines de crânes, y compris les exemples de regroupement de crânes aux cornes intentionnellement déformées (Chaix, Hansen 2003).

Herskovitz (1928) s'était penché sur ce puissant phénomène d'attachement social des pasteurs à leurs troupeaux en tant qu'élément constitutif d'une organisation pastorale en Afrique orientale et méridionale. L'univers mythique de ces populations pastorales parut centré sur ce bétail, et un ordre social en aurait découlé. La pratique des inhumations de bovins a été mise en évidence dans plusieurs sites au Niger, comme par exemple à Iwelen, Chin Tafidet, Adrar Bous 1, Talak-Timersoi, In Tuduf (Paris 2000). En Libye, on connait l'exemple d'In Habeter IIIa et les restes osseux de bovins dans les monuments funéraires en corbeilles du Messak Settafet (Zampetti 2019). En Egypte, dans la région de Nabta Playa la découverte d'un centre cérémoniel près du site E-75-8 a conduit les chercheurs à suggérer une structuration de la société et l'hypothèse de l'adoption de valeurs ayant inspiré les communautés du Néolithique tardif. Le site aurait

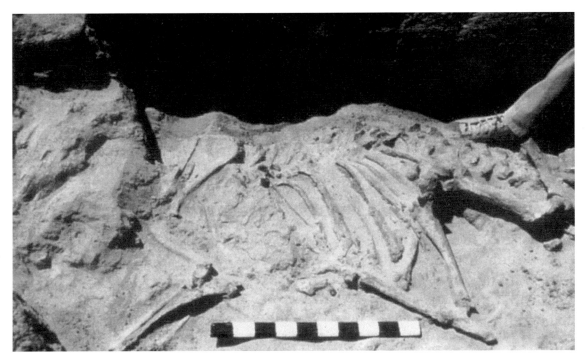

Figure 2.
Inhumation d'un
bovin dans le site
E-94-1n, Nabta
Playa, Egypte
(d'après Barich
2010, fig.5.5b).

alors joué un rôle important en tant que lieu de rassemblement, pour des groupes célébrant des cérémonies en relation avec le retour des pluies (naissance possible d'un culte des pluies ?). Deux tumulus (Sites E-91-1n et E-94-1n) (Applegate *et al.* 2001), ont été édifiés pour des sépultures bovines (Figure 2) durant le 6ème millénaire cal BC. La pratique du sacrifice et de l'enfouissement des ces animaux souligne l'importance attribuée au bœuf dans ces contextes sahariens et la place qui leur a été reconnue au centre de significations symboliques qui, d'une certaine manière, pourraient anticiper l'apparition du domaine religieux dans le monde égyptien. Cette originalité africaine méritait d'être soulignée.

Les animaux domestiques dans les représentations pariétales : un art animalier éblouissant

Une forme de ritualisation de la figure du *Bos,* exprimée par les sépultures bovines, se trouve également exprimée à travers les peintures et gravures des premiers bergers sahariens. Dans l'art pariétal du Sahara central, il semble que le troupeau ait constitué le principal moteur des manifestations sociales les plus importantes: naissance, initiation, mariage, mort. Chaque individu, en fonction de son âge, pourrait avoir traversé différentes formes de relations avec son troupeau et les œuvres d'art pourraient l'avoir exprimé à travers des codes figuratifs nourris de structures symboliques issues de leur culture. À côté d'une lecture immédiate de la scène, basée sur une analyse formelle sous laquelle l'oeuvre apparaît, se cache une autre représentation profonde et symbolique à percevoir qui devrait tendre à comprendre le sens que l'auteur a voulu communiquer en s'appuyant sur la construction et l'usage de divers détails. Il faut également admettre que les dimensions dans lesquelles les animaux des troupeaux sont représentés et les caractéristiques anatomiques signalées s'éloignent de la précision qu'exige un expert naturaliste. Malgré cela, la valeur de ces représentations est d'autant plus grande qu'il ne subsiste plus rien de cette réalité et que seuls ces témoignages restent vivants, dynamiques et mystérieux.

Les documents artistiques marquant le passage et le séjour de groupes de bergers sont connus dans tous les massifs sahariens, du Hoggar (Hugot 1963), au Tibesti et Ennedi (Huard, Massip 1963), jusqu'au Jebel Uweinat (Van Noten 1978) et Gilf El Kebir (Kuper 2013; Zboray 2011), jusque dans la vallée du Nil. Ils sont le signe de l'expansion du pastoralisme Holocène et de l'ampleur des mouvements de transhumance.

Les répertoires les plus importants proviennent de deux massifs contigus celui du Tassili n'Ajjer et celui de la Tadrart Akakus; plusieurs missions scientifiques ont identifié, enregistré et sauvegardé des œuvres constituant désormais un vaste corpus documentaire (Lajoux 1964; Lhote 1973; Hachid 2000a; Mori 1965, 1998; Barich *et al.* 1986; Di Lernia, Zampetti 2008). Celui-ci reste cependant difficile à mettre en relation avec les occupations au sol des bergers. Cette possibilité n'existe qu'en quelques endroits où le niveau archéologique est en contact avec la paroi ornée et avec des fragments de celle-ci incorporés au niveau archéologique. Cependant, il est vrai qu'une séquence d'occupation humaine du territoire, établie avec précision, peut offrir un cadre chronologique aux représentations rupestres.

Pour le Tassili n'Ajjer, une documentation très riche a été réunie à la suite des recherches pionnières de H. Lhote; on lui doit la reproduction des plus importants scènes de Iheren, Ta-n-Zoumaitek, Tamrit, Jabbaren, Aouanrhet, Sefar et Ti-n-Aboteka. Le panneau de Ti-n-Tazarift (Figure 3), dans lequel les hommes et les bœufs avancent sur plusieurs rangées montrant les têtes qui alternent, on distingue leurs membres inférieurs comme coupés par le niveau de l'eau, cette scène a été interprétée par Lhote comme une représentation symbolique de la cérémonie du Lootori, celle de la lustration sacrée typique de la tradition peule (Lhote 1973). De plus, dans cette scène le motif digité évoquant une main à cinq doigts a été associé à la main de Kikala, premier berger fondateur légendaire de l'ethnie Peule; ce motif est également récurrent dans l'univers symbolique de ce peuple. Cette interprétation ne faisant pas l'unanimité, d'autres auteurs ont proposé d'autres interprétations de la scène. Pour Smith (Smith 1993), diverses formes d'images communes au répertoire saharien ("tettiformes", spirales, motifs géométriques) peuvent faire penser à une tentative d'exécution produite dans un état de conscience altérée. D'après cet auteur cette scène serait chargée de réminiscence peule et des signes de transe.

Le panneau de Ti-n-Tazarift offre un exemple de la complexité sémantique que présente de nombreuses peintures pastorales et des problèmes qui

Figure 3. Art pastoral de Ti-n-Tazarift, Tassili n'Ajjer (Algérie), interprété comme une représentation de cérémonie lustrale (Lootori) par H. Lhote (d'après Barich 2010, fig.8.10).

surgissent aussi dans l'esprit des autochtones. On retrouve ces difficultés d'interprétation sur les représentations les plus réalistes du bétail: par exemple le bœuf de Tadjelamine a été dessiné avec des motifs en spirale élégants sur tout le corps, motifs restés chargés d'interrogation (Lajoux 1964, p.106); le beau dessin d'une vache n'ayant qu'une seule corne de Jabbaren reste énigmatique (Lajoux *idem*, p. 97 et 99); tout comme l'individu à deux têtes de Sefar (Lajoux *idem*, p.102). Quant à la signification de la gravure de Tegharghart, près de Djanet, qui montre des bœufs assoiffés buvant dans une flaque d'eau presque asséchéee, nul doute qu'il s'agisse d'une référence à une sécheresse (Hachid 2000a, p.230, fig. 346).

Ce qui rend ces représentations extraordinaires, ce sont les troupeaux composés de nombreux individus, aux robes colorées de différentes taches en blanc, noir, marron, rougeâtre, les animaux étant placés sur plusieurs rangées comme l'exemple de Jabbaren (Lajoux 1964, p.103), ou bien les troupeaux de Teshuinat II (Di Lernia, Zampetti 2008, p.140-141), et ceux de Maharalgeli, les deux derniers dans l'Akakus. Si à Maharalgeli on décompte au moins 30 bœufs aux robes colorées en blanc et rouge de la Tadrart (Di Lernia, Zampetti 2008, p.101), dans la grande fresque qui orne la paroi de Uan Tabu, il y en a au moins 150 bœufs (Garcea 2001; Di Lernia, Zampetti 2008, p.120). Représentés dans un style naturaliste, ces peintures ont un caractère réaliste très vivant même s'il est difficile de penser que le climat semi-aride et le territoire déjà appauvri, ne pouvaient déjà plus soutenir autant de troupeaux

À une courte distance d'Uan Tabu, l'abri d'Uan Muhuggiag est l'un des rares sites sahariens où une relation directe entre l'occupation pastorale et la peinture a été établie. Deux figures de bœuf, également en blanc et rouge, sont représentées sur l'une des surfaces d'un grand bloc qui s'est détaché de la paroi, puis est tombé dans un dépôt anthropique recouvert à son tour par une nouvelle couche archéologique. La datation de 4739 ± 310 BP (Geo: 4234-

2838 cal BC, Mori 1965) de cette couche, représente le *terminus ante quem* de l'exécution des figures bovines. Celles-ci peuvent donc être associées à la phase pastorale moyenne-finale de la Tadrart.À Tin-Hanakaten, au Tassili n'Ajjer, la peinture de deux taureaux accompagnés d'un berger et la fresque de l'abri (Aumassip 1978: fig.14), étudiée par S. Hachi (Hachi 1998) peuvent être connectées à une période de l'occupation pastorale (7ème-4ème millénaire cal BC) qui reste encore imprécise.

Dans de nombreuses scènes le troupeau est conduit par son berger. À Sefar et Titeghast n'Elias, Tassili n'Ajjer, les bœufs sont conduits par des bergers et les scènes s'ouvrent sur le foyer et la vie familiale, à l'intérieur de la hutte, avec des ustensiles soigneusement disposés sur une étagère (Hachid 2000a, p.224, figures 332, 333). À l'Oued Kessi, Tadrart Akakus (Di Lernia, Zampetti 2008, p. 108), une rangée de six bœufs dirigés par un berger est dessinée dans le style appelé de «Ti-n-Lalan», de la phase pastorale tardive (Mori 1965). Une autre peinture pastorale tardive, à Afozzigiar V présente un berger qui dirige son troupeau de bœufs pommelés en blanc et rouge, qui comprend des individus sans cornes (Di Lernia, Zampetti 2008, p.223).

Les représentations de bovins sont si nombreuses et si attractives qu'on en oublierait presque d'évoquer les autres animaux domestiques, certes plus rares. Ainsi les **chèvres**, dont les restes osseux ont été recueillis en grand nombre dans les gisements de cette période, sont mal représentées dans les scènes. Dans l'Akakus, le troupeau de bœufs en mouvement observé à In Eidi I (Di Lernia, Zampetti 2008, p. 97) est également accompagné de chèvres; le beau fragment d'In Farden présente côte à côte le berger et sa chèvre (Zampetti 2019, p.177). Au Tassili, une image de Sefar représente un grand troupeau de moutons avec quelques bovins (Hachid 2000a, p.254, fig. 395); une belle scène représentant une chèvre et son chevreau a été découverte à Amguid (Balout, Espérandieu 1954); tandis qu'une scène d'Iheren présente de nombreuses têtes de chèvres et de moutons groupées vers l'arrosage (Hachid 2000a, p. 249, fig.383). En Egypte occidentale une image de chèvre est connue dans la grotte de Wadi el Obeiyid, Oasis de Farafra du désert égyptien (Barich 1998 ; 2014, fig.16.9a). L'image, gravée sur la paroi de la première salle de la grotte (*front gallery*), est l'une des nombreuses représentations animalières de la grotte. Les restes de chèvre collectés *in situ* dans le village de Hidden Valley, voisin de la grotte, sont datés entre 7200 et 6900 BP / ca 6200-5900 cal BC (Barich, Lucarini 2014). Ces datations fournissent une référence chronologique approximative pour la gravure de la chèvre, on peut penser que la grotte fut probablement visitée lors de l'occupation de ce village.

Le **chien** est un autre animal domestique également lié aux activités pastorales avec un rôle précis en rapport avec les mobilités des troupeaux bovins (Espérandieu 1994). Au Tassili n'Ajjer, dans la scène de Ti-n-Aboteka, un chien apparaît accompagnant un berger portant une curieuse coiffure. Son museau n'est pas reconnaissable, les pattes sont minces et la queue est tournée vers le haut (Lajoux 1964, p.145). Dans d'autres scènes, au Tassili n'Ajjer et dans l'Akakus, des groupes de chiens sont engagés dans la chasse aux antilopes ou aux mouflons. Par exemple, nous pouvons citer la scène de Teshuinat V et celle de Ti-n-Teghaghit situées dans la Tadrart, où un groupe de chiens peints en blanc s'est lancé à la poursuite d'un mouflon (Di Lernia, Zampetti 2008, p.125, 134 et 165). Un autre exemple propose la même situation à Eberer, dans le Tassili du Nord (Gauthier *et al.* 1996, fig.40). Dans ces représentations de chiens on reconnaît le Tessem (le sloughi), avec un corps puissant et élancé, un museau allongé, des oreilles bien droites et une queue assez longue. Il s'agit d'un «lévrier africain» (Coppe *et al.* 1994), espèce qui apparaît avec une certaine fréquence également dans l'art de la phase pastorale tardive du Tassili (Hachid 2000b, p. 148). Enfin, je veux mentionner une représentation différente du chien qui nous est offerte par la figurine en argile modelée et cuite trouvée dans le gisement pastoral de l'Oued Athal (Barich, Mori 1970; Barich 2017, p. 111, fig. 6.2). La figurine fut découverte dans le niveau supérieur du gisement. La datation du site (R-1165α, 5630 ± 50 BP / 4554-4351 cal BC), offre une référence chronologique précise à cet objet. Le chien représenté ici est un molosse (Cesarino 1997, p. 106) donc d'un type différent du Tessem et qui, compte tenu du contexte pastoral joue le rôle de chien de berger au 5ème millénaire cal BC. L'origine du chien européen est attribuée à des géniteurs nordiques datant d'une période comprise entre 18 800 et 32 000 cal BC (Thalmann *et al.* 2013). Mais en Afrique nous avons peu de données sur les espèces de chiens domestiques africains, une domestication locale ne doit pas être exclue (voir Boyko *et al.* 2009; Linseele *et al.* 2014).

Dans les contextes pastoraux d'époques finales (<IIème mill. BC.), le **cheval** devient le sujet principal des représentations graphiques et les séquences culturelles prennent le nom d'«équidiennes» ou «caballines». On sait qu'en Anatolie, au sud du Caucase, le cheval est attelé depuis le 4ème millénaire cal BC, et qu'entre le 3ème et le 2ème millénaire le cheval se propage à l'Est et à l'Ouest. En Afrique, le cheval apparaît pour la première fois à Tell Heboua, dans le Sinaï (Chaix 2000), au cours de la Deuxième Période Intermédiaire (1786-1552 BC), date qui apporte une indication précise *post-quem* transférable, avec prudence, à la présence du cheval dans l'art rupestre du Sahara. Depuis le milieu du 2ème millénaire, il y a eu une véritable explosion des représentations dans lesquelles le cheval est associé au char, reconnus le long des piémonts des principaux massifs sahariens du Tibesti, de la Tadrart Akakus (ici, les scènes de Ti-n-Teghaghit et Ti-n-Abrukin, Di Lernia, Zampetti 2008, p. 167-168 et 171), au Tassili (Oued Djerat, Amguid et Tamadjert, Hachid 2000b, p. 118-123, fig. 180, 182) (Figure 4) et au Hoggar (En Daladj, Lhote 1982, fig.1), et au Maroc méridional, pour ne citer que quelques exemples. A propos de la diffusion rapide des scènes de chars et de chevaux au «galop volant», Lhote (Lhote, 1982) s'appuie sur une «Route des Chars», itinéraire qui peut être suivi depuis la côte des Syrtes (Tunisie) jusqu'au cœur du Sahara, pour aller jusqu'au Niger et au Mali.

Figure 4. Art équidien d'Amguid, Tassili n'Ajjer (Algérie), représentant un cheval et un char au galop volant (d'après Hachid 2000b).

Ainsi, suppose-t-il effectuée la pénétration des groupes méditerranéens, probablement proto-berbères, jusqu'au Sud du Sahara ?

Pour d'autres auteurs (Camps, Dupuy 1996), la berbérisation des régions méridionales du Sahara serait plutôt à considérer comme un phénomène graduel, basé sur des échanges et des relations sociales à distance, qui se seraient développées en plusieurs phases successives. Ainsi l'avènement de groupes méditerranéens (berbères) au Sud du Sahara se serait accompagné de structures funéraires monumentales, se situant vers le premier millénaire BC. Selon ce modèle, il se pourrait que les scènes reproduisant cavaliers et chars soient des œuvres de bergers sahariens, produites lors de leurs transhumances récurrentes dans les massifs. La combinaison cheval-char - connue plus au Nord - aurait pu exercer sur eux une attraction extraordinaire et devenir symbole de puissance et de statut supérieur.

Le rôle attribué au cheval comme moyen de transport et de communication entre les différentes régions et les communautés du Sahara, à été reconnu et exploité de façon décisive par **l'âne,** dont les sociétés ont toujours admiré la force et la capacité de résister aux situations environnementales difficiles. Les études génétiques des spécimens modernes des ânes, d'une part ont exclu que leur domestication ait eu lieu en Asie (Beja Pereira *et al.* 2004) et, d'autre part, ont révélé la présence de deux haplogroupes d'ADN mitochondriaux distincts, conséquence probable de deux événements séparés de domestication, survenus tous les deux en Afrique du Nord

(Kimura *et al.* 2011). Les résultats de l'étude de Kimura *et Alii* ont montré que l'âne sauvage de Nubie (*Equus africanus africanus*) est l'ancêtre du premier des deux haplogroupes (Clade 1), tandis que pour l'autre haplogroupe aucune conclusion convaincante n'a été tirée.

Les premiers exemples domestiques connus en Afrique proviennent des sites égyptiens de Maadi, daté d'environ 4000 cal BC (Boessneck *et al.* 1989) et d'Abydos, ceux-ci concernant un complexe funéraire daté d'environ 3000 cal BC (Rossel *et al.* 2008). Peu de restes d'ânes domestiques sont connus à partir de fouilles du Sahara central. Aussi les témoins recueillis dans le site d'Uan Muhuggiag, dans la Tadrart Akakus libyenne, sont-ils importants. Deux échantillons provenant de la section supérieure de la séquence (fouilles Pasa 1960-62) et datés ca.1200-1000 cal BC, ont été inclus dans l'étude de Kimura cité. La reproduction et l'isolement de l'ADNmt ont établi le statut domestique des restes et leur appartenance au Clade 1, dérivé de l'ancêtre *nubian wild ass* (Kimura *et al.* 2011). Toujours en Libye, des restes osseux de l'âne -beaucoup plus récents - entre 50 BC et 150 AD - ont été collectés dans les fouilles du site d'Aghram Nadharif, citadelle de la période Garamantique de l'Oued Tanezzuft, dans le sud-ouest de la Libye (Alhaique 2005, p. 351, Table 29. XIV ; Liverani 2005, p. 365).

Rares sont les représentations connues de l'âne. En Algérie atlasique une tête d'âne a été peinte dans les monts des Ksour et la gravure d'une ânesse avec son petit a été gravée à el Richa, tous deux dans l'Atlas saharien (Vaufrey 1939 ; Muzzolini 1995). Toujours en Libye, dans le plateau du Messak, une gravure incisée d'un âne est représentée schématiquement (Lutz, Lutz 1995); tandis qu'une autre incision est connue de l'Oued Mathendush dans laquelle des individus thériomorphes à tête d'âne sont représentés dans une scène d'accouplement (Muzzolini 1995).

L'autre grande innovation qui introduisit de grands changements au Sahara a été l'adoption du **dromadaire.** Présent au Maghreb durant le Pléistocène il s'est ensuite éteint; il fut réintroduit en Égypte depuis la péninsule arabique, où il est domestiqué autour 3000 BC (Uerpmann, Uerpmann 2002). Sa présence en Afrique du Nord est datée des 4ème-3ème siècle BC (Wilson 1984). Cet animal, abondamment représenté dans l'art rupestre (phase «cameline»), a grandement contribué à faciliter les communications entre les différentes régions du Sahara, il aurait permis d'anticiper l'installation des premières routes caravanières, grâce à sa capacité à supporter de lourdes charges et de rudes conditions climatiques.

Remerciements

Je remercie le Président de l'UISPP, Prof. François Djindjian, pour l'invitation à participer à la publication de ce volume en tant que contribution de la Commission: «Art et civilisations au Sahara en temps préhistoriques». Merci aussi à Colette Roubet pour les échanges d'idées et de suggestions et pour une première lecture de l'article.

Bibliographie

Alhaique F. 2005. The faunal remains. In: M. Liverani (ed.), *Aghram Nadharif – The Barkat Oasis (Sha'Abiya of Ghat, Libyan Sahara) in Garamantian Times. The Archaeology of Libyan Sahara*, Vol. II. Florence, All'Insegna del Giglio, p.349-360.

Applegate A., Gautier A., Duncan S. 2001. The North Tumuli of the Nabta Late Neolithic Ceremonial Complex. In: F.Wendorf, R. Schild and Associates (eds.), *Holocene Settlement of the Egyptian Sahara. Volume 1: The Archaeology of Nabta Playa*. New York, Kluwer Academic/Plenum Publishers, p.468-488.

Aumassip G. 1978. Ti-n- Hanakaten— Bilder einer Ausgrabung, In: *Sahara. 10.000 Jahre zwischen Weide und Wüste*. Köln, Museen der Stadt, p.208-213.

Aumassip G. 1996. Propos sur le Bovidien. In: G. Aumassip, J.D. Clark, F. Mori (eds.), *The Prehistory of Africa*. XIIIème Congrès UISPP, Forlì 1996, Colloquium XXX. Forlì, ABACO, p.209-218.

Aumassip G., Delibrias G. 1982-83. Ages des dépôts néolithiques du gisement de Ti-n-Hanakaten (Tassili-n-Ajjer, Algérie). *Libyca* XXX-XXXI, p.207-211.

Balout L., Esperandieu G. 1954. La chèvre peinte d'Amguid. *Libyca* II, p. 155-162.

Barich B.E. 1974. La serie stratigrafica dell'Uadi Ti-n-Torha (Tadrart Acacus, Libia) - Per una interpretazione delle facies a ceramica saharo-sudanesi. *Origini* VIII, p.7-184.

Barich B.E. (ed.) 1987. *Archaeology and Environment in the Libyan Sahara: The Excavations in the Tadrart Acacus 1978-1983*. Oxford, British Archaeological Reports, International Series 368.

Barich B.E. 1998. The Wadi el-Obeiyd Cave, Farafra Oasis: A new pictorial complex in the Libyan-Egyptian Sahara. *Libya Antiqua* N.S. 4, p.9-19.

Barich B.E.2010. *Antica Africa - Alle Origini delle Società*. Roma, L'Erma di Bretschneider.

Barich B.E. 2014. The Wadi el Obeiyid Cave 1: the rock art archive. In: B.E. Barich, G. Lucarini, M.A. Hamdan, F.A. Hassan (eds.), *From Lake to Sand: The Archaeology of Farafra Oasis (Western Desert, Egypt)*. Florence, All'Insegna del Giglio, p.367– 387.

Barich B.E. 2016. The introduction of Neolithic resources to North Africa: A discussion in light of the Holocene research between Egypt and Libya. *Quaternary International* 410 (part A), p.198–216.

Barich B.E. 2017. The Sahara. In: T. Insoll (ed.), *The Oxford Handbook of Prehistoric Figurines*. Oxford, Oxford University Press, p.105-127.

Barich B.E. 2019. Herder-Foragers and Low-Level Food Producers. Some Insights into the Early Food Production in Northern Africa. In: M. Baldi, R.Dan, M. Delle Donne, G.Lucarini, G. Mutri (eds.), *Archaeology of Food – New Data from International Missions in Africa and Asia*. Rome, Serie Orientale Roma n.s. 17, ISMEO and Scienze e lettere, p.75-106.

Barich B.E. 2021. Rethinking the North African Neolithic – The multifaceted aspects of a long-lasting revolution. In: J. Rowland, G.Lucarini, G.J. Tassie (eds.), *Revolutions. The Neolithisation of the Mediterranean Basin.* Berlin, Berlin Studies of the Ancient World vol.68, Edition Topoi.

Barich B.E., Garcea E.A.A., Lupacciolu M., Sebastiani R. (eds.) 1986. *Arte Preistorica del Sahara.* Roma-Milano, De Luca-Mondadori.

Barich B.E., Lucarini G. 2014. Social dynamics in northern Farafra from the middle to late Holocene: changing life under uncertainty. In: B.E.Barich, G. Lucarini, M.A. Hamdan, F.A. Hassan (eds.), *From Lake to Sand: The Archaeology of Farafra Oasis* (*Western Desert, Egypt*). Florence, All'Insegna del Giglio, p.467-484.

Barich B.E., Mori F. 1970. Missione Paletnologica Italiana nel Sahara Libico. Risultati della Campagna 1969. *Origini* IV, p.79-144.

Beja Pereira A., England P.R., Ferrand N., Jordan S., Bakhiet A.O., Abdalla M.A. *et al.* 2004. African origins of the domestic donkey. *Science* 304 (5678), p.1781.

Biagetti S., Di Lernia S. 2013. Holocene deposits of Saharan rock shelters: the case of Takarkori and other sites from the Tadrart Acacus Mountains (Southwest Libya). *African Archaeological Review* 30, p.305– 338.

Boessneck J., Von Den Driesch A., Ziegler R. 1989. Die Tierreste von Maadi und Wadi Digla [The animal remains of Maadi and Wadi Digla]. In: I. Rizkana and J. Seeher (eds.), *Maadi III.* Mainz, Phillipp von Zabern, p.87-128.

Boyko A.R., Boyko R.H., Boyko C.M., Parker H.G., Castelhano M., Corey L., Degenhardt J.D. *et al.* 2009. Complex population structure in African village dogs and its implications for inferring dog domestication history. *PNAS,* 106, p.13903–13908.

Brass M. 2017. Early North African Cattle Domestication and its Ecological Setting: A Reassessment. *Journal of World Prehistory*, published on line 14 December 2017, doi:10.1007/s10963-017-9112-9.

Brunton G., Caton-Thompson G. 1924. *The Badarian Civilisation.* London, British School of Archaeology in Egypt.

Camps G. 1974. *Les Civilisations Préhistoriques de l'Afrique du Nord et du Sahara.* Paris, Doin.

Camps G., Dupuy Ch. 1996. Equidiens. *Encyclopédie Berbère*, fasc.17, Editions Peeters, en ligne. http://encyclopedieberbere.revues.org/2665-2677.

Carter P.L., Clark J.D. 1976. Adrar Bous and African cattle. In: J.E.G. Sutton (ed.), *Proceedings of the Panafrican Congress of Prehistory and Quaternary studies VIIth Session, 1971.* Addis Ababa, Ministry of Culture, p.487-493.

Cesarino F.1997. I cani del Sahara. *Sahara* 9, p. 93-113.

Chaix L. 2000. An Hyksos horse from Tell Heboua (Sinaï, Egypt). In: M. Mashkour (ed.), *Archaeozoology of the Near East IV.* Paris – Groningen, ARC-Publicatie 32, p.177-186.

Chaix L., Hansen J.W. 2003. Cattle with 'forward-pointing horns: archaeozoological and cultural aspects. In: L. Krzyżaniak, K. Kroeper, M.

Kobusiewicz (eds.), *Cultural Markers in the Later Prehistory of Northeastern Africa and Recent Research*. Poznan, Poznan Archaeological Museum, p.269-281.   https://en.wikipedia.org/wiki/PMID.

Clark J.D., Williams M.A.J., Smith A.B. 1973. The geomorphology and archeology of Adrar Bous, central Sahara: a preliminary report. *Quaternaria* 17: 245-297.

Clark J.D., Agrilla E.J., Crader D.C., Galloway A., Garcea E.A.A., Gifford-Gonzales D., Hall D.N., Smith A.B., Williams M.A.J. 2008a. *Adrar Bous. Archaeology of a Central Saharan Granitic Ring Complex in Niger*. Tervuren, Royal Museum for Central Africa.

Clark J.D., Carter P.L., Gifford-Gonzalez D., Smith A.B. 2008b. The Adrar Bous Cow and African Cattle. In: J.D.Clark *et al., Adrar Bous. Archaeology of a Central Saharan Granitic Ring Complex in Niger*. Tervuren, Royal Museum for Central Africa, p.389-403.

Close A.E. 2002. Sinai, Sahara, Sahel: The Introduction of Domestic Caprines to Africa. In: Jennerstrasse 8 (ed.), *Tides of the Desert – Gezeiten der Wüste, Contributions to the Archaeology and Environmental History of Africa in Honour of Rudolph Kuper*. Köln, Heinrich Barth Institut, p.459-469.

Coppe G. 1994. Le lévrier de l'Azawak. *Encyclopédie Berbère*, fasc.13, p.1919-1924.

Cremaschi M., Di Lernia S. (eds.) 1998. *Wadi Teshuinat. Palaeoenvironment and Prehistory in South Western Fezzan (Libyan Sahara)*. Milano, Quaderni di Geodinamica Alpina e del Quaternario n°7.

Cremaschi M., Zerboni A., Mercuri A.M., Olmi L., Biagetti S., Di Lernia S. 2014. Takarkori rock shelter (SW Libya): an archive of Holocene climate and environmental changes in the central Sahara. *Quaternary Science Reviews* 101, p.36-60.

Decker J.E., McKay S.D., Rolf M.M., Kim J., Molina Alcalá A., Sonstegard T.S., Hanotte O. *et al.* 2014. Worldwide Patterns of Ancestry. Divergence, and Admixture in Domesticated Cattle. *PLoS Genet*, 10(3): e1004254. Doi:10.1371/ journal. P gen.1004254.

De Faucamberge E. 2014. *Le site néolithique d'Abou Tamsa (Cyrénaïque, Libye). Apport à la préhistoire du nord-est de l'Afrique*. Paris, Collection Études Libyennes n°2, Riveneuve éditions.

Delibrias G., Hugot H.J., Quezel P. 1957. Trois datations de sédiments sahariens récents par le radiocarbone. *Libyca* V, p.267-270.

De Menocal P.B., Tierney J.E. 2012. Green Sahara: African Humid Periods Paced by Earth's Orbital Changes. *Nature Education Knowledge* 3 (10), p.12-18.

Di Lernia S. (ed.) 1999. *The Uan Afuda Cave: Hunter- Gatherer Societies of Central Sahara*. Florence, All'Insegna del Giglio.

Di Lernia S. 2013. The emergence and spread of herding in Northern Africa - A critical reappraisal. In: P. Mitchell and P.Lane P. (eds.), *The Oxford Handbook of African Archaeology*. Oxford, Oxford University Press, p.527–40.

Di Lernia S. 2019. From "Green" to "Brown" – The archaeology of the Holocene Central Sahara. In: E. Chiotis (ed.), *Climate Changes in the Holocene. Impacts and Human Adaptation.* London, Taylor & Francis Group, p.183-200.

Di Lernia S., Tafuri M.A., Gallinaro M., Alhaique F., Balasse M., Cavorsi L., Fullagar P.D., Mercuri A.M. *et al.* 2013. Inside the "African cattle complex": Animal burials in the Holocene central Sahara. *PLoS One*, 8(2), p.e 56879

Di Lernia S., Zampetti D. 2008. *La memoria dell'Arte – Le pitture rupestri dell'Acacus tra passato e futuro.* Florence, All'Insegna del Giglio.

Dunne J., Evershed R.P., Salque M., Cramp L., Bruni S., Ryan K., Biagetti S.,Di Lernia S. 2012. First dairying in green Saharan Africa in the fifth millennium BC. *Nature* 486, p.390-394.

Esperandieu G. 1994. Chèvre. *Encyclopédie Berbère* fasc.13, p.1913-1918.

Gabriel B. 1972. Neuere Ergebnisse der Vorgeschichtsforschung in östlichen Zentralsahara. *Berliner Geogr. Abh.* 16, p.153-156.

Gabriel B. 1987. Palaeocological evidence from Neolithic fireplaces in the Sahara. *African Archaeological Review* 5, p.93-103.

Garcea E.A.A. (ed.) 2001. *Uan Tabu in the Settlement History of the Libyan Sahara.* Florence, All'Insegna del Giglio.

Gatto M.C., Zerboni A. 2015. Holocene Supra-Regional Environmental Changes as Trigger for Major Socio-Cultural Processes in Northeastern Africa and the Sahara. *African Archaeological Review* 32, p.301-333. http://doi.org/10.1007/s10437-015-9191-x.

Gautier A. 1984. Archaeozoology of the Bir Kiseiba region, Eastern Sahara. In: F. Wendorf, R. Schild (assemblers), A.E. Close (ed.), *Cattle- Keepers of the Eastern Sahara.* Dallas, Department of Anthropology, Southern Methodist University, p.49-72.

Gautier A. 1987. Prehistoric Men and Cattle in North Africa: A Dearth of Data and a Surfeit of Models. In: A.E. Close (ed.), *Prehistory of Arid North Africa, Essay in Honor of Fred Wendorf.* Dallas, Southern Methodist University Press, p.163-187.

Gautier A. 2001. The Early to Late Neolithic Archeofaunas from Nabta and Bir Kiseiba. In: F. Wendorf, R. Schild and Associates (eds.), *Holocene Settlement of the Egyptian Sahara. Volume 1: The Archaeology of Nabta Playa.* New York, Kluwer Academic Plenum Publishers, p.609-635.

Gautier A. 2014. Animal remains from the Hidden Valley Neolithic site, Farafra Oasis, Egypt. In: B.E. Barich, G. Lucarini, M.A. Hamdan, F.A. Hassan, (eds.), *From Lake to Sand: The Archaeology of Farafra Oasis (Western Desert, Egypt).* Florence, All'Insegna del Giglio, p.369-374.

Gauthier Y., Gauthier C., Morel A., Tillet T. 1996. *L'Art du Sahara-Archives des Sables.* Paris, Seuil.

Gifford Gonzales D., Parham J.F. 2008. Fauna from Adrar Bous and surrounding Areas. In: J.D. Clark et al., *Adrar Bous. Archaeology of a Central Saharan Granitic Ring Complex in Niger.* Tervuren, Royal Museum for Central Africa, p.337-388.

Grigson C. 2000. *Bos africanus* (Brehm) ? Notes on the archaeozoology of the native cattle of Africa. In: K.C. MacDonald (ed.), *The Origins and Development of African Livestock. Archaeology, Genetics, Linguistics, and Ethnography*. London, UCL Press, p.38-60.

Jelinek J. 2003. Pastoralism, Burials and Social Stratification in Central Sahara. *Les Cahiers de l'AARS* 8, p.41-44.

Hachi S. 1998. Une Approche anthropologique de l'art figuratif préhistorique d'Afrique du Nord - Analyse d'une fresque de Tin Hanakaten (Tassili-n-Ajjer). *Etudes et Documents Berbères* 15-16, p.163-184.

Hachid M. 2000a. *Le Tassili des Ajjer. Aux sources de l'Afrique 50 siècles avant les Pyramides*. Paris, Éditions Paris Méditerranée.

Hachid M. 2000b. *Les Premiers Berbères - Entre Méditerranée, Tassili et Nil*. Aix-en-Provence, Édisud.

Hassan F.A. (ed.) 2002. *Droughts, Food and Culture: Ecological Change and Food Security in Africa's Later Prehistory*. New York, Kluwer Academic/Plenum Publishers.

Hassan F.A., Barich B.E., Mahmoud A.M., Hamdan M.A. 2001. Holocene playa deposits of Farafra Oasis, Egypt, and their palaeoclimatic and geoarchaeological significance. *Geoarchaeology: An International Journal* 16 (1), p.29-46.

Herskovitz M.J. 1928. The cattle complex in East Africa. *American Anthropologist* 28, p.230-272.

Holl A. 2004. *Saharan Rock Art: Archaeology of Tassilian Pastoralist Iconography*. Walnut Creek, Altamira Press.

Huard P., Massip J. 1963. Gravures rupestres du Tibesti méridional et du Borkou. *B.S.P.F.* 60/7-8, p.468-481.

Hugot H.J. 1963. *Recherches Préhistoriques dans l'Ahaggar Nord-Occidental (1950-1957)*. Paris, Mémoires du CRAPE (I).

Kimura B., Marshall F.B., Chen S., Rosenbom S., Moehlman P.D., Tuross N. *et al.* 2011. Ancient DNA from Nubian and Somali wild ass provides insights into donkey ancestry and domestication. *Proc Biol Sci,* 278(1702), p.50–57.

Kuper R. (ed.) 2013. *Wadi Sura : The cave of Beasts - A rock art site in the Gilf Kebir (SW- Egypt)*. Köln, Africa Praehistorica 26, Heinrich Barth Institute.

Kuper R., Kröpelin S. 2006. Climate-controlled Holocene occupation in the Sahara: motor of Africa's evolution. *Science* 313, p.803-807.

Lajoux J.D. 1964 (éd.Italienne). *Le Meraviglie Del Tassili N'ajjer - L'arte Preistorica Del Sahara*. Bergamo, Istituto Italiano d'Arti Grafiche.

Linseele V., Van Neer W., Thys S., Phillipps R., Cappers R., Wendrich W., Holdaway S. 2014. New Archaeozoological Data from the Fayum "neolithic" with a Critical assessment of the Evidence for Early Stock Keeping in Egypt. *Plos One* 9 (10): e108517.

Linseele V., Holdaway S.J., Wendrich W. 2016. The Earliest phase of introduction of Southwest Asian domesticated animals into Africa. New evidence from the Fayum Oasis in Egypt and its implications. *Quaternary International* 412, p.11-21.

Liverani M. 2005. *Aghram Nadharif - The Barkat Oasis (Sha'Abiya of Ghat, Libyan Sahara) in Garamantian Times. The Archaeology of Libyan Sahara*, Vol. II. Florence, All'Insegna del Giglio.

Lhote H. 1973. The *search for the Tassili frescoes - The story of the prehistoric rock-paintings of theSahara.* Hutchinson, Penguin Random House UK.

Lhote H. 1982. *Les chars rupestres sahariens (des Syrtes au Niger, par le pays des Garamantes et des Atlantes).* Paris, éd. Hespérides.

Lutz R., Lutz G. 1995. Spears and ovoids in the rock art of Messak Sattafet and Mellet. *Sahara* 7, p.89-96

Maitre J.P. 1974. Perspectives nouvelles sur la préhistoire récente de l'Ahaggar. *Libyca* 22, p.93-143.

Marshall F., Weissbrod L. 2011. Domestication Processes and Morphological Change. Through the Lens of the Donkey and African Pastoralism. *Current Anthropology* 52, Supplement 4, p.S397- S413.

McDonald M.M.A. 2016. The pattern of Neolithization in Dakhleh Oasis in the Eastern Sahara. *Quaternary International* 410 (A), p.181-197.

Mori F. 1965. *Tadrart Acacus. Arte rupestre e Culture del Sahara preistorico.* Torino, Einaudi.

Mori F. 1998. *The Great Civilisations of the Ancient Sahara.* Roma, L'Erma di Bretschneider.

Muigai A.W.T., Hanotte O. 2013. The Origin of African Sheep: Archaeological and Genetic Perspectives. *African Archaeological Review* 30(1), p.39-50.

Muzzolini A. 1995. *Les images rupestres du Sahara.* Toulouse, Préhistoire du Sahara.

Paris F. 1997. Les inhumations de Bos au Sahara méridional au Néolithique. *Archaeozoologia* IX, p.113-122.

Paris F. 2000. African Livestock Remains from Saharan Mortuary Contexts. In: R.M. Blench and K.C. MacDonald (eds.), *The Origins and Development of African Livestock: Archaeology, Genetics, Linguistics and Ethnography.* London, UCL Press, p.111-126.

Reimer P.J., Bard E., Bayliss A., Beck J.W., Blackwell P.G., Bronk Ramsey C. *et al.* 2013. IntCal13 and MARINE13 radiocarbon age calibration curves 0-50000 yr calBP. *Radiocarbon* 55(4). doi: 10.2458/azu_js_rc.55.16947.

Roset G.-P. 1987. Néolithisation, néolithique et post-néolithique au Niger nord-oriental. *Bulletin de l'Association Française pour l'Etude du Quaternaire* 24(32), p.203-214.

Rossel S., Marshall F., Peter J., Pilgram T., Adams M.D., O'Connor D. 2008. Domestication of the donkey: Timing, processes, and indicators. *PNAS* 105(10), p.3715-3720

Saxon E. C. (with contribution of A.Close, C. Cluzel, V. Morse, N.V. Shackleton) 1974. Results of recent investigations at Tamar Hat. *Libyca* 22, p.49-82.

Schild R., Wendorf F. 2001. Combined Prehistoric Expedition's Radiocarbon Dates associated with Neolihic Ooccupations in the Southern Western Desert of Egypt. In: F. Wendorf, R. Schild and Associates (eds.), *Holocene*

Settlement of the Egyptian Sahara. Volume 1: The Archaeology of Nabta Playa. New York, Kluwer Academic/Plenum Publishers, p.51-56.

Schulz E. 1987. Die holozäne vegetation der zentralen Sahara. *Palaeoecology of Africa* 18, p.143-162.

Sereno P.C., Garcea E.A.A., Jousse H., Stojanowski C.M., Saliège J.-F., Maga A., Ide O.A., Kelly J., Knudson K.J., Mercuri A.M. *et al.* 2008. Lakeside Cemeteries in the Sahara: 5000 Years of Holocene Population and Environmental Change. *PLoS ONE* 3(8): e2995. doi:10.1371/journal.pone.000299.

Smith A.B. 1980. Domesticated cattle in the Sahara and their introduction into West Africa. In: M.A.J. Williams and H. Faure (eds.), *The Sahara and the Nile*. Rotterdam, Balkema, p.451-465.

Smith A.B. 1993. New approaches to Saharan rock art. In: G. Calegari (ed.), *L'Arte e l'ambiente del Sahara preistorico: dati e interpretazioni*. Milano, Museo Civico di Storia Naturale, p.467-477.

Tauveron M., Striedter K.H., Ferhat N. 2009. Neolithic domestication and pastoralism in central Sahara: The cattle necropolis of Mankhor (Tadrart algérien). In: R. Baumhauer, J. Runge (eds.), *Holocene palaeoenvironmental history of the central Sahara*. Boca Raton, CRC Press, p.201-208.

Thalmann O., Shapiro B., Cui P., Schuenemann V.J., Sawyer S.K., Greenfield D.L., Germonpré M.P. *et al.* 2013. Complete mitochondrial genomes of ancient canids suggest a European origin of domestic dogs. *Science* 342, p.871– 874.

Tixier J. 1962. Le «Ténéréen» de l'Adrar Bous III. In: H.J. Hugot (ed.), *Mission Berliet Ténéré-Tchad*. Paris, AMG, p.333-348.

Uerpmann H.-P., Uerpmann M. 2002. The appearance of the domestic camel in South-east Arabia. *J Oman Stud.* 12, p.235–260.

Van Noten F. 1978. *Rock art of the Jebel Uweinat (Libyan Sahara)*. Graz, Akademische Druck u.Verlagsanstalt.

Vaufrey R. 1939. *L'art rupestre nord-africain*. Paris, Archives de l'Institut de Paléontologie Humaine, Mémoire 20, Masson.

Vermeersch P.M., Linseele V., Marinova E., Van Neer W., Moeyersons J., Rethemeyer J. 2015. Early and Middle Holocene Human Occupation of the Egyptian Eastern Desert: Sodmein Cave. *African Archaeological Review* 32, p.465-503.

Vernet R. 1998. *Le Sahara et le Sahel. Paléoenvironnements et occupation humaine à la fin du Pléistocène et à l'Holocène. Inventaire des datations 14C, 2e édition: jusqu'en 1997*. Nouakchott, Université de Nouakchott, Centre de Recherches inter-africain en Archéologie.

Wendorf F., Schild R. 1994. Are the Early Holocene cattle in the Eastern Sahara domestic or wild? *Evolutionary Anthropology* 1994, p.118-128.

Wendorf F., Schild R. 2001. Conclusion. In: F. Wendorf, R. Schild and Associates (eds.), *Holocene Settlement of the Egyptian Sahara. Volume 1: The Archaeology of Nabta Playa*. New York, Kluwer Academic/Plenum Publishers, p.648-675.

Wendorf F., Schild R., and Associates (eds.) 2001. *Holocene Settlement of the Egyptian Sahara. Volume 1: The Archaeology of Nabta Playa.* New York, Kluwer Academic/Plenum Publishers.

Wilson R.T. 1984. *The Camel.* London, Longmans.

Zampetti D. 2019. Subsistence Strategies and Food in North African Rock Art: The Central Saharan Massifs. In: M. Baldi, R. Dan, M. Delle Donne, G. Lucarini, G. Mutri (eds.), *Archaeology of Food –New Data from International Missions in Africa and Asia.* Rome, Serie Orientale Roma, n.s. 17, ISMEO and Scienze e Lettere, p.165-185.

Zboray A. 2011. The prehistoric cultures of the Gilf Kebir and Jebel Uweinat - The evidence from rock art. In: *XXIV Valcamonica Symposium 2011 Papers.* Milano, Jaca Book.

L'évènement climatique 4.2 ka BP et la transition du Néolithique à l'âge du Bronze dans le Sud-est de la France dans son contexte euro-méditerranéen

Olivier Lemercier[1]

Résumé

Parmi les évènements climatiques ayant entrainé des conséquences pour les sociétés humaines, l'évènement 4.2 Ka BP est aujourd'hui largement étudié et reste utilisé par de nombreux chercheurs pour expliquer le déclin ou l'effondrement de certaines des premières sociétés avancées de l'Indus à la Mer Égée et à l'Égypte. Pourtant de nombreuses questions demeurent. Les interprétations possibles de mêmes observations divergent considérablement lorsque les observations elles-mêmes ne sont pas contestées d'une étude à l'autre. Quelle est la réalité de cet évènement, son intensité et ses répercutions climatiques dans les diverses régions du monde ? Quel a pu être son impact sur l'environnement et sur les activités humaines ; impact qui semble différent, voire opposé, selon les régions ? Après un rappel de la nature de l'évènement lui-même et des questionnements sur ses conséquences environnementales et socio-économiques, cet article propose un panorama global de ces questions pour quelques régions du monde avant de se focaliser sur une approche régionale dans le nord-ouest de la Méditerranée (la France méditerranéenne) afin d'examiner les données archéologiques concernant les évolutions environnementales, démographiques et socio-économiques.

Abstract

Among the climatic events that have had consequences for human societies, the 4.2 Ka BP event is now widely studied and is still used by many researchers to explain the decline or collapse of some of the first advanced societies of the Indus to the Aegean Sea and to Egypt. Yet many questions remain. The possible interpretations of the same observations diverge considerably when the observations themselves are not disputed from one study to another. What is the reality of this event, its intensity and its climatic repercussions in the various regions of the world? What may have been its impact on the environment and on human activities; impact that seems different, or even opposite, depending on the region? After a reminder of the nature of the event itself and the questions about its environmental and socio-economic consequences, this article offers a global overview of these questions for some regions of the world before focusing on a regional approach in the north-western Mediterranean (Mediterranean France) in order to examine archaeological data concerning environmental, demographic and socio-economic evolutions.

1 Commission Civilisations néolithiques de la Méditerranée et de l'Europe de l'UISPP; Université Paul Valéry – Montpellier 3 – UMR 5140 ASM. olivier.lemercier@univ-montp3.fr

Introduction

En Europe et en Méditerranée occidentales, l'évènement 4.2 Ka BP, généralement daté entre 2300 et 1900 BCE, coïncide avec la fin du Néolithique et les premiers temps de l'âge du Bronze. Il est donc tentant d'établir un lien direct entre le changement climatique et l'évolution des sociétés à cette époque, dans un contexte où l'évènement 4.2 ka BP est utilisé depuis une cinquantaine d'années pour expliquer un effondrement généralisé des premières sociétés avancées de l'Indus au Nil, en passant par le monde égéen. Que nous disent les données environnementales et archéologiques dans le Midi de la France ? Comment interprète-t-on réellement cet évènement climatique dans les autres régions aujourd'hui ? Et comment aborder les relations entre changements climatiques et sociétés humaines, d'une façon plus générale ?

L'évènement 4.2 ka BP et la question de ses conséquences : données générales

L'évènement climatique 4.2 ka BP

Un évènement climatique, au sens de « *Climatic Event* » ou « *Rapid Climatic Change* » en anglais, décrit un phénomène climatique brutal et inattendu, sur un temps court. Plusieurs de ces phénomènes sont reconnus pour l'Holocène et seraient liés à un refroidissement de l'atmosphère (Bond *et al.* 1997 ; Mayewsky *et al,* 2004) à l'échelle de l'hémisphère nord vers 7200, 6200, 4300, 2200 et 1200 BCE. Les causes de ces changements climatiques sont en réalité toujours discutées aujourd'hui (Bradley, Bakke, 2019 par exemple). L'évènement qui nous intéresse ici, celui de 2200 BCE, généralement appelé « *4.2 ka BP event* » ou plus rarement « *Bond Holocene IRD Event 3* » est considéré comme le principal refroidissement depuis l'autre évènement majeur de l'Holocène (8.2 ka BP). Températures plus basses en raison d'un blocage anticyclonique sur l'Atlantique nord et précipitations inférieures à la normale caractériseraient principalement cet épisode (Jalali, Sicre, 2019). Il s'étend selon les études à partir de 2400 ou 2300 BCE pour une période de 400 à 600 ans au maximum mais pourrait être plus brutal et court selon de nombreuses études. Le plus souvent les chercheurs évoquent des changements rapides entre 2200 et 1900 BCE. 2200 BCE est généralement reconnu comme un point de changement, proposition défendue par le Comité international de stratigraphie selon laquelle 2200 BCE est la limite entre l'Holocène moyen, plus chaud et plus humide, et l'Holocène récent, plus frais et sec (Ran, Chen, 2019). La précision chronologique de ces évènements s'améliore et une étude récente place un évènement d'aridité important au Moyen Orient autour de 4,26 ka BP soit 2260 BCE, pour une durée de 290 ans (Carolina *et al.* 2019). Cependant, en Méditerranée, les études les plus récentes tout en montrant d'importants indices d'un épisode d'aridité mettent aussi en évidence d'importantes différences dans les enregistrements selon les régions, qui correspondent sans doute à un impact localement très varié (Bini *et al.* 2019) ou ne permettent même pas de faire apparaitre cet épisode (Finné *et al,* 2019).

Les conséquences environnementales

Même à l'échelle globale, les conséquences de cet évènement 4.2 ka BP ne semblent pas uniformes et, tout au contraire, se manifestent selon les régions par la forte augmentation de la fréquence des sécheresses, des conditions plus humides ou des inondations, des tempêtes de poussière majeures, des réactivations de dunes, des incendies et peut-être d'autres anomalies climatiques (Wang *et al.* 2016). En Chine, on évoque une sécheresse dans le nord et des inondations dans le sud alors qu'en Europe, les relevés montreraient plutôt des conditions plus humides ou plus fraiches dans le nord et une aridité prononcée dans les régions méditerranéennes comme en Asie occidentale (Wang *et al.* 2016). En Méditerranée centrale, le scénario pourrait être encore plus compliqué selon certaines études et légèrement décalé dans le temps par rapport à d'autres régions, avec un épisode effectivement sec entre 2100 et 2000 BCE encadré par deux épisodes plus humides entre 2350 et 2100 puis entre 1950 et 1850 BCE (Magny *et al.,* 2009). Cette période de changements climatiques correspondrait peu ou prou à la mise en place du climat méditerranéen (Magny *et al.* 2013). Aujourd'hui, la période comprise entre 2,3 et 2,1 ka BC est souvent considérée comme une phase de « seuil environnemental global » dans la circulation atmosphérique sud-nord au début de l'Holocène récent, provoquée par un forçage orbital (Lespez *et al.* 2016).

La question de l'impact socio-économique

Ces évènements qui ne durent que d'un à quelques siècles sont aujourd'hui souvent considérés comme à l'origine d'importants changements dans les sociétés. Ces changements, aussi bien d'ordre culturel que socio-économique, provoqueraient parfois migrations voire effondrements (Lespez *et al.* 2016). Ce lien est envisagé comme une causalité depuis au moins 50 ans (Bell, 1971), mais les deux dernières décennies ont vu un regain d'intérêt pour l'étude des relations climats/sociétés dans le passé avec quelques auteurs particulièrement prolifiques comme B. Weninger (Weninger, 2012 ; Weninger, Harper, 2015 ; Weninger *et al.* 2006, 2009, 2014 etc.) ou H. Weiss par exemple (Weiss, 2015, 2016). L'idée d'un lien simple et direct entre grands évènements climatiques planétaires, changements environnementaux et transformations des sociétés a donc souvent été discutée (Dalfes *et al.* 1997 ; Abate, 1994 ; Kuzucuoğlu, Marro, 2007), nuancée voire contestée et même totalement rejetée, en avançant le plus souvent des arguments concernant la qualité et la précision des données régionales ou locales et la complexité et de la variété des situations en fonction d'un grand nombre de paramètres (Kuzucuoğlu, 2012 ; Kuzucuoğlu, Tsirtsoni, 2016 ; Middleton, 2012 ; 2018 ; Kleijne *et al.,* 2020 par exemple) qui mèneraient à une aussi grande variété d'effondrements, transformations, adaptations propres à chaque région et chaque situation (Höflmayer, 2017). Mais certains commentateurs de la « méga-sécheresse » et théoriciens de l'effondrement ne désarment pas en décrédibilisant les arguments de réelles adaptations et en niant même les sites non abandonnés (à part dans les zones « refuges ») comme H. Weiss pour l'Égée, l'Anatolie et le Proche Orient, qui évoque effondrements politiques, migrations et retour à une vie nomade ou regroupements dans des zones refuges (Weiss, 2015). Dans d'autres régions, comme la Chine, on insiste sur les évolutions diverses montrant selon les régions des effondrements culturels (par exemple, les cultures Liangzhu et Shijiahe dans le Sud ou les cultures Qijia et Shandong Longshan dans le Nord) ou au contraire des formes de dépassements de la crise ou de diverses formes de résilience (culture Henan Longshan dans les plaines centrales) (Ran, Chen, 2019).

Par ailleurs, le lien de causalité simple entre évènement climatique et transformations socio-économiques est aujourd'hui remis en cause pour envisager des scénarios plus complexes intégrant de multiples paramètres combinés. L'évolution des sociétés a longtemps été envisagée par l'archéologie historico-culturelle sous la forme de longues périodes de stabilité socio-économique, séparées par de courts moments de perturbation causés par des facteurs externes tels que les mouvements de population, la conquête territoriale ou les invasions avant que l'archéologie processuelle ne vienne mettre l'accent sur les facteurs internes d'évolution dans les systèmes économiques, sociaux et politiques (Risch *et al.* 2015). La mise en évidence de perturbations climatiques, aujourd'hui bien calées en chronologie, amène à s'interroger de nouveau sur la part des facteurs externes d'évolution potentiels, et bien sûr de l'évolution et des évènements environnementaux, comme la génétique et les études sur la violence préhistorique remettent sur le devant de la scène les notions de migrations et de guerres pour ces périodes.

L'évènement 4.2 ka BP et ses conséquences : un aperçu pour quelques régions

Quelques auteurs ont tentés de dresser des panoramas généraux ou réunit de nombreux chercheurs (Höflmayer 2017), depuis déjà plusieurs décennies, pour décrire les âges sombres de l'âge du Bronze (*Dark Ages*), commençant autour de 2200 BCE par une crise climatique ayant entraîné un déclin politique et économique (Chew, 2005), souvent dans des tentatives d'histoire globale, parfois simplificatrices mais pas toujours simplistes. D'autres ont tenté des approches régionales plus modestes mais plus précises. En 2014, l'évènement 4.2 ka BP fait même l'objet d'un très important colloque spécifique dont la publication, de plus de 800 pages, est parue dès l'année suivante (Meller *et al.* 2015).

L'Orient et l'Afrique

À l'extrême Est de la région que Sing C. Chew considère, dans le secteur de l'Indus, la cause première de l'effondrement serait à rechercher dans le système économique lui-même avec une exploitation intensive de la nature à des fins de commerce, dans un contexte d'accumulation, d'urbanisation et de croissance démographique provoquant une vulnérabilité dans le système économique vis-à-vis des changements climatiques et calamités naturelles. Il en serait de même en Mésopotamie où l'agriculture serait avant 2200 BCE poussée aux limites de sa productivité (Chew, 2005). Les études portant sur la zone levantine du Proche Orient sont plus nuancées. Si l'évènement 2200 BCE y est généralement reconnu, il est peut-être moins abrupt ou émergerait dans le Levant centre-sud, deux phases arides encadrant une période plus humide (Kaniewsky *et al.* 2018). Archéologiquement, la situation est encore moins nette. Ainsi, si la Jordanie montrerait l'abandon de sites urbains autour de cette date (Höflmayer, 2017), au Liban, l'impact de la crise de 2200 BCE est sans doute très différent selon les régions. Dans la zone côtière, sans doute moins affectée par une aridification plus sensible à l'intérieur des terres, le long de la zone désertique, son impact n'est pas observable et aucun changement majeur ne peut être détecté autour de 2200 BCE (Genz, 2015).

En Anatolie, l'évènement 4.2 ka BP aurait eu un impact significatif mais dont les conséquences se feraient sentir sur la longue période entre 2300/2200 et 1900 BCE (Massa, Sahoglu, 2015). Diverses réponses environnementales et sociales sont probables. D'un côté, principalement sur la côte égéenne, le processus d'urbanisation commencé antérieurement s'interrompt. Des destructions et abandons de sites sont observés en lien peut-être avec une plus grande mobilité pastorale. De l'autre côté, en Anatolie centrale on observe une augmentation de la complexité sociale avec des épisodes généralisés de violence organisée et de nombreux autres aspects (apparition de temples, formalisation du panthéon religieux, augmentation de la taille et la richesse des bâtiments publics, contrôle plus strict des ressources - stockage centralisé, pratiques administratives, contrôle militaire etc.). Ici aussi, une vulnérabilité initiale vis-à-vis des aléas climatiques de ces sociétés serait envisageable en raison de leur dépendance excessive à une agriculture intensive dans un contexte de croissance démographique et d'urbanisation. En Asie Mineure, la période est celle de la transition du Bronze ancien (Early Helladic) au Bronze moyen (Middle Helladic). Sur le site de Troie

(Hisarlik Tepe, Turquie), la période serait marquée par un gap observable dans l'évolution de la céramique et le corpus de datations radiocarbone (Jung, Weninger, 2015 ; Weninger, Easton, 2017), mais selon d'autres, si la région est marquée par des modifications du couvert végétal (diminution des espèces forestières, extension de la garrigue, caractéristiques de la mise en place du climat méditerranéen) et des différences dans la production agricole, elle ne montrerait pas d'effondrement réel du site toujours occupé, ni de transformations majeures même dans les contacts et réseaux à longue distance (Blum, Riehl, 2015). Les mêmes données semblent donc pouvoir être interprétées de diverses manières.

En Égypte, l'aridification autour de 2200 BCE correspondrait selon certains chercheurs à l'effondrement du premier Empire et au début de la première Période Intermédiaire même si la précision chronologique de ces évènements climatiques n'est pas encore calée précisément avec les évènements politiques connus en Égypte (Dee, 2017). Pour d'autres (Moreno Garcia, 2015), la fin du 3e millénaire avant notre ère ne serait pas une période de crises économiques, d'invasions étrangères ni d'effondrement social. La fin de l'Ancien Empire et la première Période Intermédiaire auraient été une période de profonds changements dans la société pharaonique plutôt en raison d'une crise politique interne, alors que l'Égypte s'intégrait de plus en plus aux réseaux commerciaux internationaux qui fleurissaient à la fin du Bronze ancien et à l'âge du Bronze moyen.

L'Europe

En Europe orientale, la période 2300-2100 BCE constituerait un moment de bouleversements culturels interprétés en termes de migrations de populations. La culture d'Abashevo qui se met en place sur la moyenne Volga, correspond à de très nettes influences d'Europe centrale, de traditions campaniformes (Mimokhod, 2018).

En Allemagne, dans le secteur du lac de Constance, il semble que si l'économie agricole campaniforme est en rupture avec celle du Cordé antérieur, quelque part autour de 2500 BCE, il n'y a pas de rupture majeure de ces traditions au début du Bronze ancien (Lechterbeck *et al.* 2014). À partir d'une étude multi-paramètres de la culture matérielle, il en serait de même en Allemagne centrale où l'événement 2200 BCE ne constitue pas une rupture (Müller, 2015). Dans le sud de l'Allemagne, grâce à une révision des datations radiocarbone et de nouvelles séries chronométriques, le début du Bronze ancien a été replacé autour de 2150 BCE et non plus entre 2300 et 2200 BCE, ce qui place ce changement consécutivement à l'évènement 2200 BCE (Stockhammer *et al.* 2015). Dans la région du Danube, la transition de l'âge du Cuivre au Bronze ancien n'est pas considérée comme un tournant majeur de civilisation, mais montre au contraire une remarquable continuité alors que dans les régions au nord du Danube, une rupture culturelle plus remarquable avec de nouvelles traditions funéraires (culture d'Únětice rejetant les distinctions claires entre les sexes et introduisant le bronze à l'étain par exemple) (Bertemes, Heyd, 2015). Dans le bassin des Carpates, et en particulier dans sa zone centrale en Hongrie, 2200 BCE concorde avec la fin du Campaniforme, l'apparition de traditions céramiques plus restreintes et un retour aux habitats en tells. Mais ces changements ne sont pas considérés localement comme une crise mais plutôt comme le « point de départ » d'une croissance qui va durer jusque vers 1600-1400 BCE (Pusztainé Fischl *et al.* 2015).

En Europe du Nord, les changements observés à la fin du 3ᵉ millénaire BCE ne sont ni identiques ni simultanés dans les différentes régions. Au Pays-Bas, l'événement 2200 BCE ne semble pas non plus constituer la date d'un changement important dans les sociétés (Kliejne *et al.* 2020). Il en serait de même dans la région de Schleswig-Holstein, en Allemagne du nord (Müller, 2015). Au Danemark, l'impact aurait été plus important et marqué par des changements plus évidents (Müller, 2015 ; Kliejne *et al.* 2020). Selon une étude assez récente (Roland *et al.* 2014), l'évènement 4.2 ka BP serait insensible dans les îles britanniques et 2200 BCE ne semble pas le moment d'un bouleversement sociétal majeur dans cette région (Fitzpatrick, 2015), où le passage du Campaniforme au Bronze ancien se fait sous la forme d'une continuité assez remarquable. De nouvelles études (Jordan *et al.* 2017) montreraient finalement un possible impact de cet évènement sur les îles britanniques, mais encore assez mal évalué.

La Méditerranée

Les études archéobotaniques à large échelle de temps et d'espace sur la Méditerranée nord et l'ensemble du Néolithique et de l'âge du Bronze ne semblent pas faire ressortir l'évènement 4.2 ka BP du point de vue de l'usage de terres (Mercuri *et al.* 2019). Pourtant, selon certains, l'épisode d'aridité de 2200 BCE bien documenté en Méditerranée orientale s'étend bien à la Méditerranée occidentale (Wienelt *et al.* 2015). Selon eux, la frontière entre les conditions sèches et humides se situe au niveau du 45ᵉ parallèle en Méditerranée centrale. Ils évoquent des hivers rigoureux avec une augmentation de l'activité des tempêtes en Méditerranée occidentale autour de 2200 BCE. Ils notent aussi une augmentation de la population de l'âge du Bronze ancien rapportée conjointement dans toutes les régions considérées, à l'exception de la France méditerranéenne, ce qui en fait une « caractéristique supra régionale très robuste ».

En Crète, la période 2200 BC serait celle d'un arrêt dans les relations entre l'île et l'Anatolie et la Méditerranée orientale et de destructions ou abandons de sites de la fin du Minoen ancien II, mais il ne s'agirait pas ici d'un réel effondrement mais des conditions qui ont permis le « décollage de la société crétoise palatiale » remarquable dans la période suivante avec le Minoen ancien III (Manning, 2017).

En Méditerranée centrale, une étude récente sur un transect du nord de l'Italie à la Tunisie, a montré qu'entre le 43e et le 45e parallèles, il n'y a pas de changement significatif de végétation autour de 2200 BCE, en revanche entre le 39e et le 43e parallèles, l'ouverture de la forêt est systématique et entre le 36e et le 39e, le déclin forestier est parfois très important en lien avec l'évènement 4.2 ka BP.

En Italie plus précisément, l'évènement 4.2 ka BP a pu être enregistré dans des analyses multi-proxy (Drysdale *et al.* 2006 ; Isola *et al.* 2019). Plusieurs études ont mis en évidence des différences entre les secteurs avec, en Italie du nord, des conditions plus fraiches et humides, alors qu'en Italie du sud l'évènement 4.2 ka BP correspond à une aridification qui culmine autour de 2200 BCE (Zanchetta *et al.* 2016). En Italie du nord, la transition au Bronze ancien est datée autour de 2200 avant notre ère, avec une phase formative de la culture de Polada, peut-être contemporaine des dernières manifestations du Campaniforme et avant le développement de l'habitat sur pilotis vers 2050 BCE. C'est à ce moment, vers 2050-2000 BCE que les mutations socio-économiques sont les plus importantes sans qu'il soit possible de les corréler avec un changement climatique (Leonardi *et al.* 2015). En Italie du sud et en Sicile, la situation est probablement assez différente. En Italie péninsulaire méridionale, l'impact du Campaniforme a été assez faible et la transition à l'âge du Bronze entre 2200 et 2100 BCE environ se marque d'une discontinuité culturelle et d'un dépeuplement. En Sicile, l'évolution du Chalcolithique final au Bronze ancien serait progressive et différentes selon les régions. Le lien avec l'aridification supposée autour de 2200 BCE demeure peu clair (Pacciarelli *et al.* 2015). Une récente étude sur l'Italie centrale tyrrhénienne qui croise les données chronologiques, démographiques et paléoenvironnementales ne relève rien de très particulier autour de 2200 avant notre ère se contentant de mentionner une baisse de population au début de l'âge du Bronze (Stoddart *et al.* 2019).

En Sardaigne et en Corse, aucun changement net de végétation n'est observé à ce moment (Di Rita *et al.* 2019).

Dans le Nord-ouest de la péninsule Ibérique, le 3e millénaire est considéré comme un tournant dans l'intensité de l'impact humain sur la dynamique forestière où la déforestation conduit à des paysages ouverts ce qui a accéléré l'érosion des sols et amené à la genèse de nouvelles formes de reliefs. L'augmentation des indices d'incendies ou de brulis irait dans le même sens d'une anthropisation généralisée et de plus en plus forte au cours de la période (Costa Casais *et al.* 2015), sans accident brutal relevé autour de 2200 BCE ou après. Le changement climatique reconnu comme une baisse de 2 degrés autour de 2200 BCE étant envisagé comme « conjugué à une pression humaine ». Sur un plan culturel, 2200 BCE n'est pas vu comme un moment de changement important mais plutôt de continuité du Chalcolithique au Bronze ancien (Fabregas Valarce *et al.* 2003). C'est aussi en termes de continuité et peut-être de résilience que la région est vue dans le cadre de la vaste synthèse réalisée en 2018 (Blanco Gonzalez *et al.* 2018).

Dans l'Ouest, au Portugal, la fin du 3ᵉ millénaire marque un profond changement dans l'organisation socio-économique politique et culturelle des communautés, en particulier dans le centre-sud, les données du Nord montrant quelques continuités d'occupations de sites qui sont néanmoins exceptionnelles. Cependant, la part de la crise climatique demeure très incertaine dans la mesure où ces changements s'amorcent dès le milieu du 3ᵉ millénaire pour certaines et selon un processus qui semble plutôt continu (Valera, 2015).

Sur la Meseta Nord, les données environnementales encore peu nombreuses conduisent les chercheurs à un peu de prudence. Une phase d'aridité est remarquée dans les diagrammes palynologiques entre 2150 et 2050 BCE, mais celle-ci voit plutôt une reprise d'activité céréalière et pastorale ainsi que l'intensification de l'exploitation du sel, alors que les siècles précédents étaient plutôt marqués par une déprise. Culturellement, les changements observés au 3ᵉ millénaire sont plus importants avec le début du Campaniforme qu'en 2200 BCE (Delibes de Castro *et al.* 2015). Dans la Mancha, le virage est plus sensible avec la culture des Motillas, parallèle au développement argarique du Sud-est (Blanco Gonzalez *et al.* 2018). Les motillas, ces puits permettant de collecter l'eau souterraine seraient synchrones de l'évènement 4.2 Ka BP et auraient constitué une solution efficace dans une période de sécheresse sévère et prolongée permettant la poursuite de l'évolution socio-économique des cultures concernées (Mejias Moreno *et al.* 2014).

Dans l'Est de la péninsule, l'épisode 4.2 ka BP qui marque la transition du Chalcolithique de Los Millares au Bronze ancien argarique intervient au moment de l'une des transformations les plus importantes de Méditerranée occidentale à cette époque (Blanco-Gonzalez *et al.* 2018), qui ne peut être vue comme un effondrement mais plutôt le point de départ de quelque chose, dont la part climatique demeure à déterminer. D'autres réfutent pour l'essentiel cette part, soulignant que la transition entre le Chalcolithique final et la culture d'El Argar dure probablement seulement quelques années et non quelques décennies et trouve son origine dans des processus sociaux et politiques (Lull *et al.* 2015). L'évènement 4.2 ka BP n'aurait ici pas de conséquences réelles ni sur l'évolution démographique qui marque un pic juste après 2200 BC dans les estimations, ni pour la végétation (Fyfe *et al.* 2019). Mais l'évènement climatique lui-même pose encore des problèmes de détermination dans le Sud-est de la péninsule Ibérique, puisqu'une étude de l'évolution des températures moyennes réalisée sur des coquilles de Glycymeris d'un site stratifié et bien daté ne montre pas d'accident particulier autour de 2200 BCE mais une baisse plutôt régulière entre 2500 et 1500 BCE (Kölling *et al.* 2015). Parmi les études les plus récentes, une analyse multi-proxy des sédiments marins des côtes sud-ibériques montre une phase sèche entre 2400 et 2300 BCE (± 100 ans), suivie d'un changement rapide vers des conditions plus humides révélant un modèle plus complexe en termes de moment et de durée que celui décrit pour l'événement de 4.2 ka BP dans d'autres régions (Schirrmacher *et al.* 2019).

Dans la région du Guadalquivir, d'importants développements socio-économiques et culturels ont eu lieu entre 3000 et 2500 BCE probablement en lien avec l'exploitation du cuivre, et dans une période où le sud de la péninsule Ibérique est l'aboutissement de réseaux d'échanges à l'échelle de la Méditerranée et de l'Afrique du nord. L'effondrement de cet ensemble ne date pas de 2200 BCE mais le précède nettement entre 2500 et 2300 BCE. À partir de 2200 BCE, sur le haut bassin du Guadalquivir, un nouvel ensemble culturel et politique se met en place qui ne semble plus fondé sur la métallurgie (Nocete *et al.* 2010).

Dans le Sud-ouest de la péninsule, les archéologues observent un effondrement culturel marqué, mais qui débute clairement avant l'évènement 4.2 ka BP et se prolonge bien au-delà dans l'âge du Bronze (Blanco-Gonzalez *et al.* 2018). Une autre étude, plus récente (Hinz *et al.* 2019) observe une corrélation entre crise climatique (sécheresse),végétation, effondrement démographique et changements culturels, tout en soulignant

Figure 1. Europe, Afrique du nord et Asie entre 2400 et 2200 BCE. Principales entités culturelles mentionnées (O. Lemercier).

que les facteurs potentiels ne se limitent pas à un accident climatique et peuvent intégrer les inégalités sociales, la surexploitation et la concurrence pour les ressources, les changements démographiques dus à une mobilité accrue et éventuellement à d'importants changements de population, et les conflits violents qui ont pu contribuer aux changements en se conjuguant à un moment donné (Figure 1).

Le cas du Sud-est de la France

Évolutions climatiques et environnementales

Régionalement, en France méditerranéenne, un changement environnemental important autour de 2200 BCE est envisagé depuis longtemps, mais demeure assez mal caractérisé sur le plan végétal par exemple. La plupart des études réalisées n'ont pas la précision chronologique nécessaire pour corréler évènement climatique et évolution précise du couvert végétal ou sont inclues dans des synthèses sur de vastes chronologies et géographies (Fyfe *et al.* 2018 par exemple). Les études se contentent le plus souvent d'observer la mise en place de la végétation méditerranéenne à partir de 2500 BCE environ (Jalut *et al.* 2009 par exemple) marquée par la baisse du chêne à feuilles caduques et la progression des paysages de garrigues avec le buis et le chêne vert (Vernet, 1999). De grands changements sont néanmoins perceptibles, dans certains secteurs mieux étudiés et plus souvent sur un plan géo-archéologique et géomorphologique que paléobotanique. Dans la moyenne vallée du Rhône, Jean-François Berger a mis en évidence après un siècle de reconquête forestière entre 2200 et 2100 BCE une crise climato-anthropique générale entre 2100 et 1900 avant notre ère, qui se caractérise par une phase érosive importante, couplée à des incendies et une importante activité anthropique (Berger *et al.* 2000). Si cet épisode est encore reconnu aujourd'hui (Walsh *et al.*, 2019), il est aussi généralement réinséré dans une séquence plus longue d'observations qui montre tout d'abord à une échelle plus large que l'ouverture du milieu date probablement du Néolithique final et s'accentue dans le premier millénaire avant notre ère (Roberts *et al.* 2019). Surtout, la récente et vaste synthèse multi-proxy sous la direction de J.-F. Berger (Berger *et al.* 2019) montrerait une grande dépression post-néolithique de près de mille ans entre 2300/2000 et 1300/1200 BCE. L'étude enregistre une forte baisse du nombre de sites, de la démographie radiocarbone et envisage même un hiatus à la fin du Campaniforme récent rhodano-provençal d'environ un siècle avant le début du Bronze ancien épicampaniforme autour de 2150 BCE. Les auteurs de l'étude mettent ce phénomène en lien avec l'évènement 4.2 Ka BP, tout en indiquant que le nombre de sites du premier Bronze ancien demeure faible face aux maximums atteint dans la période du Néolithique final Fontbouisse-campaniforme précédent (Figure 2). Dans le même temps,

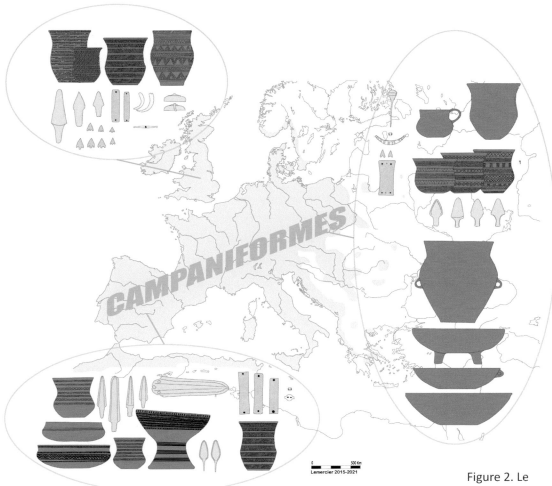

Figure 2. Le Campaniforme : relative uniformisation des cultures et pratiques en Europe entre 2400 et 2200 BCE (O. Lemercier).

les indicateurs anthropiques sont tous bas alors qu'une légère reprise des bois s'observe dans certains secteurs. De plus forts reboisements sont d'ailleurs observables entre le Bronze ancien 2 et le début du Bronze final au moment des plus bas indicateurs démographiques vers 1500 BCE (Berger *et al.* 2019). Dans la vallée du Rhône, la présence de nombreux charbons de bois dans les alluvions datés entre 2150 et 2100 BCE est interprétée comme un important défrichement par le feu du couvert végétal qui avait regagné le milieu entre la période campaniforme et la période épicampaniforme (Carozza *et al.* 2015).

Évolutions démographiques

Depuis les années 1970, en France, l'idée d'une crise démographique à l'échelle générale du pays actuel, au début de l'âge du Bronze est fréquemment évoquée. Raymond Riquet mentionne qu'après le Chalcolithique, un effondrement démographique extraordinaire car inexplicable (épidémies?)

laisse les lieux vides (Riquet, 1976). Encore une fois, dans les années 1990, Jean Guilaine évoque la théorie d'une déprise humaine ou agricole (avec un mode de vie plus mobile) émise par certains chercheurs concernant le Bronze ancien (Guilaine, 1996). André D'Anna, pour la Provence, évoque le fait que l'augmentation de la population à la fin du Néolithique en Provence fait partie d'une crise dont la conséquence est probablement le déclin du nombre de sites au début de l'âge du Bronze (D'Anna, 1995a). Et, d'une manière générale, de nombreux chercheurs (Gateau, 1996 ; D'Anna, 1992 par exemple) ont noté le très faible nombre de sites archéologiques connus pour l'âge du Bronze ancien (et parfois l'âge du Bronze moyen) souvent en comparaison avec le nombre d'occupations attribuées au Bronze final). La déprise envisagée ne peut être réellement envisagée qu'en comparaison avec le boom démographique du Néolithique final, qui atteint son paroxysme vers le milieu du 3e millénaire BCE. Le nombre de sites connus augmente alors considérablement. Un petit comptage de sites réalisé pour le département du Vaucluse à partir des cartes archéologiques pour comparer environ 1000 ans de durée radiocarbone de Néolithique moyen et 1000 ans de durée de Néolithique final avait permis de montrer un quadruplement du nombre de sites archéologiques (Lemercier, 2020). Dans les deux cas, des sites de tous types et de toutes superficies sont observables et des essais d'analyses micro-régionales ont montré qu'il ne s'agit pas du produit d'évènements érosifs ayant pu faire disparaitre certains sites de façon différentielle (Jallot, 2011). Les environnementalistes arrivent à des conclusions similaires, qu'ils envisagent les indices d'anthropisation du milieu ou utilisent des artéfacts mathématiques tels les sommes de probabilité des distributions des datations radiocarbone (SCPD) (Blanchemanche *et al.* 2003 ; Berger *et al.* 2007 ; Roberts *et al.* 2019 ; Walsh *et al.* 2019) : maximum du nombre de sites au milieu du 3e millénaire puis fort recul au début du Bronze ancien.

Ce déclin du Bronze ancien a pu être interprété soit en termes de démographie réelle (une perte de populations), soit en termes de transformations des activités ou d'abandon agricole, c'est-à-dire la mise en place d'un mode de vie plus pastoral qu'agraire (Berger *et al.* 2000 ; Berger, 2003 ; Vital, 2001). André D'Anna indiquait que presque partout en Provence, on assistait au début de l'âge du Bronze, quelques centaines d'années après le boom démographique du Néolithique final, à une forte diminution du nombre de sites. Ce phénomène était alors mal expliqué mais pouvait indiquer un déclin démographique important dont les causes restaient à déterminer (D'Anna, 1992). Cette idée a été progressivement nuancée à partir des années 2000 notamment avec d'autres chercheurs qui envisagent que, pour le Bronze ancien du Midi français, l'agriculture et l'élevage se sont stabilisés, même si on évoque localement un essor du pastoralisme, voire son rôle déterminant dans les massifs (Pyrénées et Massif Central) (Gascó, 2004) ou relativisent cet abandon en fonction des résultats sur certaines régions géographiques (Vital, 2004a ; Vital *et al.* 2007).

Les études pluridisciplinaires organisées autour de Jean-François Berger, sur la moyenne vallée du Rhône, ont aussi permis de prendre en compte le biais taphonomique par une étude systématique de l'évolution des différentes unités paysagères et ainsi de proposer un modèle de correction des cartes archéologiques (Berger *et al.* 2000). La baisse apparente du nombre de sites archéologiques doit aussi être analysée en fonction de ces possibilités de destructions / recouvrements différentiels en fonction des choix d'implantation propres à chaque période et des évènements érosifs consécutifs.

Joël Vital avait, de son côté, constaté au début des années 2000 une absence ou une grande rareté des datations radiocarbone entre 2250/2200 et 2100/2050 (Vital, 2004a, 2004b), phénomène encore mis en évidence récemment (Berger *et al.* 2019). Cependant, la question de la continuité de l'occupation d'un certain nombre de sites et celle de la filiation culturelle entre le Campaniforme rhodano-provençal et les groupes à céramiques à décor incisé et barbelé (Epicampaniforme) qui apparaissent vers 2100 avant notre ère ne sont pas univoques (Vital *et al.* 2012).

Si on considère le moment même de la transition, le cycle campaniforme entre 2600/2500 et 1950/1900 BCE, qui a fait l'objet de recherches spécifiques sur la périodisation (Lemercier, 2012, 2018) et de nombreux inventaires (Lemercier, 2004, 2014), l'effondrement du nombre de sites n'est peut-être pas aussi spectaculaire. Dans un grand Sud-est de la France, étendu de la région toulousaine à la région lyonnaise, on dénombre 170 sites de la céramique à décor incisé et barbelé de l'âge du Bronze ancien sur 732 sites campaniformes et épicampaniformes (tous styles et phases confondus) (Lemercier, sous presse). Un examen plus précis des seules implantations domestiques de la région de l'arc méditerranéen défini ici permet de dénombrer 299 sites totalisant 358 occupations (Lemercier *et al.* 2019). Parmi ceux-ci, si 54 ne peuvent être attribués à une phase typo-chronologique, 65 correspondent à la phase ancienne (styles international et pointillé-géométrique), 146 à la phase récente (styles rhodano-provençal et pyrénéen) et 93 à la phase tardive (style épicampaniforme incisé et barbelé). Les sites domestiques du premier Bronze ancien représentent ainsi 26% des sites d'habitat connus pour le Campaniforme et 30% des sites attribués à l'une des trois phases typo-chronologiques. S'il est possible d'évoquer un déclin significatif, il ne s'agit peut-être pas d'un véritable effondrement (Figure 3).

La question reste donc ici sans réponse : il y a moins de sites mais cela peut correspondre pour partie à des problèmes d'ordre taphonomique, et à l'inverse l'activité humaine serait particulièrement importante à ce moment. Il pourrait alors s'agir d'un changement de mode de vie, d'une réorganisation socio-économique plutôt que d'une réelle crise démographique (Lemercier, 2020).

Figure 3. Évolution de la distribution des sites en France méditerranéenne entre 2400-2200 BCE (Campaniforme récent) et 2150-1950 BCE (Bronze ancien épicampaniforme) (O. Lemercier).

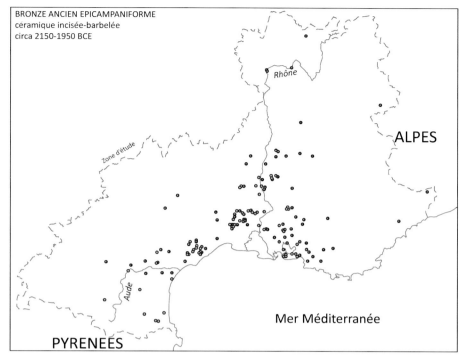

BRONZE ANCIEN EPICAMPANIFORME
céramique incisée-barbelée
circa 2150-1950 BCE

Rhône

ALPES

Zone d'étude

Aude

Mer Méditerranée

PYRENEES

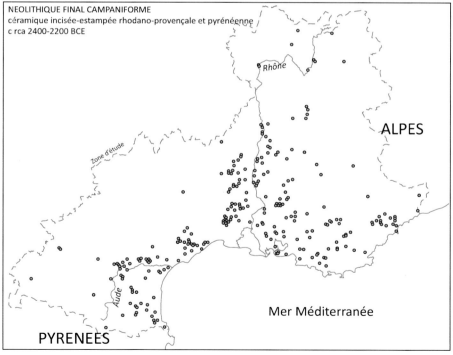

NEOLITHIQUE FINAL CAMPANIFORME
céramique incisée-estampée rhodano-provençale et pyrénéenne
c rca 2400-2200 BCE

Rhône

ALPES

Zone d'étude

Aude

Mer Méditerranée

PYRENEES

Évolutions culturelles et socio-économiques

Dans le Sud-est de la France, le 3e millénaire est un moment de forte croissance démographique mais aussi d'évolutions culturelles rapides avec des changements nettement observables dans la culture matérielle. Selon les terminologies et les périodisations des différents chercheurs, on distingue trois à quatre phases culturelles pendant le millénaire (Gutherz, Jallot, 1995 ; D'Anna, 1995a, 1995b ; Carozza *et al.* 2005 ; Lemercier, 2007 ; Cauliez, 2011 ; Jallot, Gutherz, 2014, par exemple).

La région, très diversifiée d'un point de vue géologique, topographique, hydrologique et environnemental, est aussi marquée par un morcellement culturel important, malgré quelques grandes tendances évolutives. A partir du milieu du 3e millénaire, la présence campaniforme tend à réduire les différences micro-régionales. Même si les cultures locales, telle celle de Fontbouisse, en Languedoc oriental, semble survivre quelques temps, peut-être même jusqu'au-delà de 2300 BCE dans certains secteurs, et que la culture de la phase récente du Campaniforme n'est pas totalement uniforme (du point de vue de l'habitat par exemple), les temps sont à une unification culturelle bien perceptible à travers la céramique commune. La céramique décorée campaniforme montre l'existence de groupes distincts en France méditerranéenne. Géographiquement, l'ouest de la région, des Pyrénées à la moyenne Garonne jusqu'au Languedoc central est occupé par le groupe pyrénéen, alors que l'est, du Languedoc oriental aux Alpes, est occupé par le groupe rhodano-provençal.

La céramique de ces deux groupes présente une grande variété de morphologies (gobelets, bouteilles ou pseudo-bouteilles et grands pots ou pichets ansés, bols, écuelles, jattes, coupes) où les gobelets ne dominent plus comme au moment de la première diffusion. La principale différence réside dans la technique de décor (essentiellement pointillée dans le groupe pyrénéen, généralement incisée et estampée dans le groupe rhodano-provençal) pour des dispositions en registres horizontaux, en bandes, séparées ou non de bandes réservées. Une autre différence importante avec la phase précédente est l'existence d'une céramique commune, non décorée, composée de récipients de préparation, cuisson et stockage, très similaire pour les deux groupes régionaux.

L'industrie lithique, mieux connue pour le groupe rhodano-provençal, témoigne de fortes liaisons avec les traditions techniques des cultures du Néolithique final local. L'approvisionnement en matières premières est majoritairement local afin de produire le plus grand nombre possible de petits éclats. Les pièces esquillées et les grattoirs (très souvent unguiformes) dominent l'outillage domestique, et on observe une réintégration des produits de spécialistes du Néolithique final indigène tels que les grandes lames et poignards, mais aussi l'apparition de nouveaux outils (segments de cercle et micro-denticulés). Les armatures sont moins nombreuses et le type à pédoncule et ailerons équarris fréquent dans la phase ancienne n'est plus présent (Furestier, 2007). Les outillages métalliques en cuivre issus d'un contexte fiable sont peu nombreux (alênes et poignards). Concernant la parure, les boutons en os à perforation en V et les pendeloques grossièrement arciformes, non décorés, voisinent avec tous les types de parures présents à la fin du Néolithique. Les brassards d'archer, généralement en pierre (calcaire, grès) sont bien présents.

Les habitats montrent une très grande variété et une régionalisation évidente avec des cabanes ovalaires dallées en Languedoc oriental, des constructions sur poteaux dans la vallée du Rhône et probablement des constructions mixtes pierre et bois en Provence.

Les sépultures sont collectives en grottes et monuments mégalithiques, à l'exception de rares sépultures individuelles d'enfants. Les sites se répartissent sur l'ensemble de la région considérée, y compris dans les zones marginales et montagneuses (Lemercier, 2012 ; Lemercier *et al.* 2014a).

Autour de 2150-2100 BCE, un nouvel ensemble apparait, généralement appelé Bronze ancien 1 ou Campaniforme tardif ou Epicampaniforme (Lemercier, sous presse). Il est principalement caractérisé par une céramique d'influences diverses selon les secteurs. Dans la moyenne vallée du Rhône, apparaît entre 2150 et 2100 BCE un ensemble caractérisé selon Joël Vital (Vital *et al.* 2012) par des influx orientaux correspondant à des morphologies des phases récentes et terminales du Campaniforme d'Europe centrale, du Cordé et du Proto-Únětice, donc des régions de Bavière, de Bohème et de Moravie. Des éléments d'origine nord-alpine, mais aussi du Midi méditerranéen sont notables. L'Auvergne présente une situation plus complexe, avec les mêmes influences notables, mais aussi une composante méridionale plus importante de tradition Campaniforme tardif à décor incisé et barbelé en provenance de Provence ou du Languedoc oriental. En Provence et en Languedoc oriental, l'héritage campaniforme est très important aussi bien dans les supports que dans les thématiques et les techniques décoratives. La céramique décorée est marquée par les fameux décors barbelés synchrones, réalisés au moyen d'un outil généralement composite. Le décor barbelé diachrone, composée d'une ligne repiquée, présent dans le Campaniforme récent rhodano-provençal n'est pas représenté dans les séries du premier Bronze ancien. Si une composante campaniforme locale est possible, une composante italique forte est aussi remarquable. Parmi les éléments très spécifiques de ces productions, le décor en panneau organisé autour d'une anse pourrait renvoyer au Campaniforme d'Italie centrale, dans une phase « évoluée » mais considérée comme « pré-épicampaniforme ». Cette région est sans doute aussi très importante concernant les comparaisons morphologiques des récipients. Mais toutes les caractéristiques épicampaniformes ne sont pas présentes dans ces séries, en particulier le décor barbelé lui-même. Ce dernier est en revanche présent en Italie septentrionale. Il est probable au Monte Covolo (province de Brescia) (Poggiani-Keller *et al.* 2006), bien qu'initialement identifié comme un décor réalisé au peigne, et attesté à Bernardine di

Coriano (province de Vérone) (Gilli *et al.* 2000, 2005). Ce dernier site est mis en relation – avec d'autres éléments d'Italie septentrionale – avec les groupes péricampaniformes de l'Est-adriatique et de l'Ouest des Balkans, comme la culture de Ljubljana, qui utilise le décor barbelé au peigne fileté, et la culture de Cetina (Dimitrijevič, 1967 ; Della Casa, 1995 ; Nicolis, 1998 ; Forenbaher, 2018). L'industrie lithique ne témoigne pas d'une rupture importante mais plutôt d'une continuité (Furestier, 2007 ; Lemercier sous presse). Les objets métalliques demeurent peu nombreux et méconnus (Lemercier, sous presse). Concernant la pratique métallurgique elle-même, si le site d'Al Claus dans le Tarn et Garonne (Carozza *et al.* 1999) a livré des dates qui correspondent plutôt à l'extrême fin du Néolithique qu'au Bronze ancien I (2500-2175 BCE) et que le secteur des Rousses en Oisans (Moulin *et al.* 2012) avec des dates qui s'étalent de 2100 à 1600 BCE semble exploité surtout plus tard, nous disposons tout de même des datations dans les Hautes-Alpes du site des Clausis (Carozza *et al.* 2011) qui s'étalent entre 2550 et 2030 BCE environ, et celles du site proche au vallon du Longet (Carozza *et al.* 2010), qui a fourni une datation entre 2280 et 2030 BCE, soit la fin du Campaniforme récent et le Bronze ancien I (Figure 4).

Figure 4. Évolution des traditions céramiques en France méditerranéenne entre 2400-2200 BCE (Campaniforme récent) et 2150-1950 BCE (Bronze ancien épicampaniforme) (d'après Lemercier, 2004 et Vital *et al*. 2012).

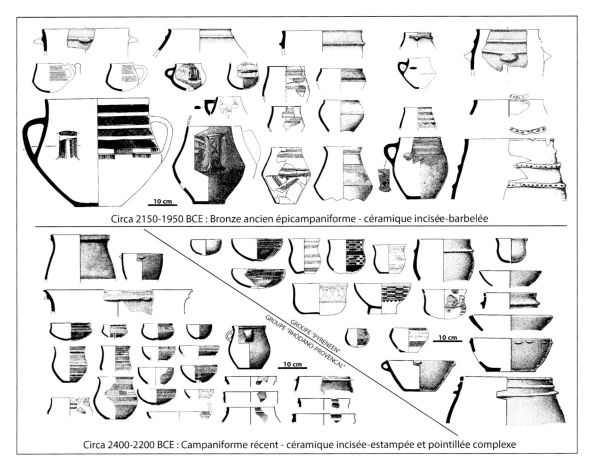

Circa 2150-1950 BCE : Bronze ancien épicampaniforme - céramique incisée-barbelée

GROUPE "PYRÉNÉEN"
GROUPE "RHODANO-PROVENÇAL"

Circa 2400-2200 BCE : Campaniforme récent - céramique incisée-estampée et pointillée complexe

L'habitat est majoritairement en plein air (110 sites de plein air pour 10 occupations supposées domestiques en cavités (Lemercier, sous presse). Toutes les topographies semblent concernées, des sites de plaine aux sites les plus perchés. Les formes de l'habitat et de l'habitation demeurent en revanche encore méconnues, comme c'est le cas pour le Campaniforme (Lemercier, Strahm, 2018 ; Lemercier *et al.* 2019), malgré l'augmentation des grands décapages avec l'archéologie préventive. Quelques plans de maisons attribués au Bronze ancien sont connus dans le Var et les Bouches-du-Rhône, mais ne permettent pas une attribution précise ou font référence à un Bronze ancien avancé. La petite structure de la terrasse XI du site du Col Sainte-Anne dans les Bouches-du-Rhône qui a été attribuée à l'Épicampaniforme, est très atypique et de dimensions restreintes. Un ensemble de trous de poteaux sur une longueur de 8 m a été observé sur le site de Châteaublanc dans les Bouches-du-Rhône, mais la forme de l'édifice dont l'attribution chronologique au Bronze ancien reste hypothétique, demeure inconnue. Le site des Juilléras dans le Vaucluse avait montré une nette organisation des structures préservées mais pas d'indices sur les formes des constructions. Plus au nord, dans la vallée du Rhône, les choses sont plus intéressantes. Au Serre I dans la Drôme, une occupation assez étendue avec des constructions quadrangulaires sur poteaux. Il en est de même sur le site du boulevard périphérique nord à Lyon. Ces architectures sur poteaux correspondent régionalement à ce qui est connu pour le Campaniforme récent à Montélimar, au Serre 1 pour l'occupation précédente ou probablement aux Vignarets à Upie, toujours dans la Drôme. Dernier élément, concernant les régions les plus méridionales : la présence d'enceintes, de natures diverses : levées de terre, murs de pierre, fossés. Au moins cinq à six cas sont aujourd'hui connus dans les Bouches-du-Rhône, l'Hérault et l'Aude. Il s'agit du Camp de Laure au Rove, le Collet-Redon à Martigues, le Clos Marie-Louise à Aix-en-Provence dans les Bouches-du-Rhône, le Mas de Garric à Mèze (Hérault) et le Roc d'en Gabit à Carcassonne (Aude) et peut-être Montredon (Saint-Pons-de-Mauchiens (Hérault) (fouilles en cours). Aucune enceinte ne peut en revanche être clairement attribuée au Campaniforme récent dans ces mêmes régions.

Dans le domaine funéraire, Les sépultures épicampaniformes ou du tout début du Bronze ancien sont représentées seulement par environ 33 sites mais présentent en même temps une remarquable diversité. Dans un large Sud-est de la France, l'usage de la sépulture collective, principalement en cavité, demeure très important. Une petite vingtaine de sépultures en cavité est ainsi dénombrée, sachant qu'un certain nombre de céramiques barbelées découvertes en grotte correspondent sans doute à des sépultures démantelées non répertoriées comme telles. Nous avons aussi ce type de mobilier dans 5 monuments mégalithiques de type néolithique final. Il reste donc quelques sites à sépultures individuelles en fosse, ciste ou coffre. Ces dernières formes ne sont pas toujours faciles à dater précisément, il est possible d'évoquer la ciste des Gouberts (Gigondas, Vaucluse) et probablement celle de Georges Besse II-5 (Nîmes, Gard). Deux secteurs s'isolent. Il s'agit de l'Auvergne, avec dans le secteur de Clermont-Ferrand une série de nécropoles à tombes en fosses, plus ou moins complexes, qui débute généralement dans l'Épicampaniforme ou le BzA1 pour se prolonger par la suite, comme à Chantemerle à Gerzat ou au Petit Beaulieu à Clermont-Ferrand. Ce phénomène de nécropole semble beaucoup plus rare dans les régions méridionales, à l'exception des petites cellules d'inhumations des Juilléras dans le Vaucluse et de Rec de Ligno 2 dans l'Hérault et peut-être de celle de la Céreirède (Lattes, Hérault) qui associent systématiquement quelques tombes en fosses parfois appareillées de pierres et un unique petit coffre (ou ciste) de pierre. Pour le Campaniforme antérieur les nécropoles n'existent pas dans le Midi méditerranéen mais se concentrent dans le nord-est de la France. Dans la basse vallée du Rhône (en Vaucluse et dans le Gard) avec une prolongation dans l'Hérault et les Pyrénées Orientales, on observe aussi la présence de quelques sépultures en cistes ou coffres, parfois associées à des sépultures en fosse et attribuables à l'Épicampaniforme ou au tout premier Bronze ancien. Si des structures de ce type préexistent au Bronze ancien, elles semblent bien se développer en nombre à ce moment pour se prolonger par la suite (Lemercier, sous presse).

Concernant finalement les données économiques, La grande synthèse collective récemment réalisée sur l'alimentation à l'âge du Bronze en France (Toulemonde *et al.* 2018) s'avère inutilisable dans le détail du temps et de l'espace pour synthétiser les données de la France méditerranéenne au début du Bronze ancien. Pour les données de l'économie végétale tout d'abord, et malgré la récente synthèse collective réalisée dans le cadre de l'enquête Bronze de l'Inrap (Carozza *et al.* 2017), il faut bien avouer que les données demeurent indigentes concernant le premier Bronze ancien, plus encore face à l'absence quasi-totale de données actuellement disponible concernant le Campaniforme récent dans le Sud-est de la France. Tout au plus noterons-nous que les traditions néolithiques semblent encore très fortes au Bronze ancien, les principaux changements observés n'intervenant qu'à partir du Bronze final (Bouby *et al.* 2017).

Cette idée s'oppose néanmoins aux observations réalisées en vallée du Rhône où Joël Vital note à l'inverse (Vital, 2004a) :

- l'importance accrue de la production fourragère,
- l'apparition de structures pouvant être interprétées en termes de greniers,
- le remplacement de l'orge nue par l'orge vêtue,
- la diminution de l'engrain,
- la part des légumineuses avec la fève
- la fréquence du millet commun et de l'épeautre.

Pour les données de l'économie animale, on regrettera l'absence d'une réelle synthèse dans la même enquête Bronze (Auxiette, 2017), alors que les données sont à l'inverse beaucoup plus abondantes pour la fin du Néolithique et le Campaniforme dans le grand Sud-est de la France

(Blaise, 2010 ; Blaise *et al.* 2010, 2014). Peu de données sont généralement disponibles pour le premier Bronze ancien. Les quelques études dont les données sont mentionnées dans la littérature montrent un élevage dominé par les caprinés mais aussi une part de chasse, même si elle ne concerne probablement que faiblement l'alimentation, ce qui est conforme à ce qui est connu pour le Campaniforme antérieur dans le Sud-est de la France, et s'oppose aux pratiques du Néolithique final où la chasse était pour ainsi dire inexistante. Deux sites du Gard montrent une situation différente – sur des corpus certes assez restreints – avec une économie animale fondée exclusivement sur l'élevage et dominée par les caprinés, sans présence de chasse.

Éléments de discussions

Chronologie, continuités et ruptures

Les données chronologiques de la transition du Néolithique à l'âge du Bronze demeurent délicates à interpréter. Le Campaniforme récent demeure difficile à caler en chronologie (Lemercier *et al.* 2014b). Le résultat des modélisations pour le Midi de la France en place le début entre 2500 et 2300 BCE et la fin entre 2200 et 2000 BCE. La phase tardive, avec les ensembles à céramiques à décor incisé et barbelé, disposait de 34 datations dont 19 à faible écart-type car elle a bénéficié de nombreuses datations récemment effectuées dans le cadre du Projet Collectif de Recherche sur les premières productions céramiques du Bronze ancien dans le Sud-est de la France, coordonné par J. Vital (Vital *et al.* 2012). Malgré un étalement toujours important, l'essentiel des dates semble se concentrer entre 2150 et 1800 BCE et même sans doute entre 2100 et 1900 BCE.

La question d'un hiatus entre le Campaniforme récent et le premier Bronze ancien ne semble pas réglée. La baisse des datations radiocarbone entre le Campaniforme récent et le début de l'âge du Bronze est un fait qui a été remarqué à plusieurs reprises (Vital *et al.* 2007 ; Carozza *et al.* 2015) mais, il ne semble pas se placer au même moment selon les régions envisagées, plutôt entre 2200 et 2100 BCE dans l'extrême sud-est, entre 2300 et 2200 BCE ou 2250 et 2150 BCE dans l'arrière-pays languedocien et la vallée du Rhône. Pourtant, la part campaniforme locale semble réelle dans la constitution du premier Bronze ancien du Sud de la France, particulièrement dans les régions les plus méditerranéennes à partir du groupe rhodano-provençal (Vital *et al.* 2012 ; Lemercier, sous presse). La part pouvant provenir du groupe pyrénéen du Campaniforme dans le premier Bronze ancien, n'a pas encore fait l'objet d'une recherche spécifique. Le premier Bronze ancien des régions les plus méridionales, même s'il subit des influences orientales évidentes, semble de fait bien plus ancré dans les traditions campaniformes que le secteur rhodanien qui évolue plus vite sous un très fort influx d'origine centre-européenne. En revanche, l'idée d'une participation des groupes locaux du Néolithique final dans la genèse de ces ensembles du Bronze ancien peut sans doute être aujourd'hui écartée. Même si certains groupes de la fin du Néolithique montrent aussi bien en Languedoc qu'en Provence, une durée relativement longue dans la seconde moitié du 3ᵉ millénaire, en synchronie avec les phases récentes du Campaniforme parfois, les datations semblent disparaître entre 2300 et 2200 BCE. Enfin, tous les cas d'associations suspectées entre objets du Néolithique final et du Bronze ancien ont pu être écartés (Lemercier, sous presse).

Il ne faut donc probablement pas envisager une disparition totale de la population du Sud de la France entre 2300 et 2100 avant notre ère, même si une réduction très importante du nombre de sites est très probable.

Les continuités observables sont évidentes concernant certains traits de la céramique, mais à l'inverse, les ruptures sont donc assez importantes entre Campaniforme récent régional et la phase épicampaniforme. Cet étrange constat pourrait correspondre à une communauté d'origine et, d'héritage campaniforme au sens large. Il semble bien que les ensembles du premier Bronze ancien découlent en grande partie de ce qui s'est mis en place avec le Campaniforme dans les différentes régions que le phénomène a touchées, en Europe centrale et en Italie comme dans le Sud-est de la France (Lemercier, sous presse). Finalement, en l'état des connaissances, la coupure entre Néolithique et âge du Bronze pour cette région pourrait plus facilement être placée au début du Campaniforme ou après l'Epicampaniforme qu'au moment de l'apparition des rares premiers objets de bronze (Lemercier, 2019).

L'évolution sur le temps long, tendances et évènements

Selon les résultats des études récentes, il faut pour comprendre l'impact de l'évènement 4.2 Ka BP, observer les changements sur le temps long mais aussi spécifiquement dans chaque région, car les transformations ne semblent ni identiques ni totalement synchrones.

Du point de vue archéologique (Carozza *et al.* 2015), l'optimum d'occupations en Languedoc semble être atteint vers 2500 BCE, la diminution commencerait dès 2400/2300 BCE. Même si les données montrent un large éventail dans la forme des habitats comme dans les choix d'implantation dans la zone côtière, le nombre de sites diminue drastiquement autour de 2200 BCE. Dans les zones de collines, les garrigues, une diminution du nombre de sites est observée autour de 2350-2250 BCE suivie par des réoccupations au début de l'âge du Bronze ancien épicampaniforme affectant la surface des effondrements des constructions antérieures. Dans ce secteur de l'arrière-pays, un hiatus est possible entre 2300 et 2200 BCE. En Languedoc, le forçage climatique (aridification) serait à dater entre 2300 et 2200 BCE. À partir de 2100 BCE avec l'Epicampaniforme il s'agirait d'une période de réorganisation et de réoccupation des anciens sites.

La basse vallée du Rhône, à la hauteur de la Drôme et du Vaucluse, montre une densification des sites dans les plaines inondables et les bassins humides vers 2600/2300 BCE. Puis la majorité de ces sites sont rapidement ensevelis sous les alluvions dès 2300/2200 BCE (Carozza *et al.* 2015). Les auteurs notent une disparition des sites entre 2250 et 2150 BCE avant l'apparition dans les mêmes secteurs de sites épicampaniformes vers 2150-2100 BCE, soit une interruption de 100 à 150 ans, selon la chronologie radiocarbone.

La situation dans les zones de montagnes serait encore différente. Dans les Pyrénées, le rythme des changements de fréquentation humaine, sans doute principalement lié aux activités pastorales, ne cadre pas avec l'idée d'un impact important de l'évènement 4.2 Ka BP (Carozza *et al.* 2015). Dans les Alpes, entre 2200 et 1800 BCE, on observerait une réorganisation de l'habitat autour d'activités pastorales et cynégétiques ainsi que le développement de l'exploitation des ressources métallurgiques (Carozza *et al.* 2015) à une époque qui correspond à un climat plus doux en haute altitude, pendant la phase de régression glaciaire identifiée, qui atteint son maximum entre 2200 et 1900 BCE (Joerin *et al.*, 2006 ; 2008).

Selon Laurent Carozza et ses collègues, l'évènement 4.2 ka BP doit donc être considéré dans le Sud-est de la France, sur le temps long entre 2300 et 1850 BCE couvrant trois phases environnementales distinctes. Dans cette période la mise en place du climat méditerranéen montre deux gradients sud-nord et est-ouest. Le Néolithique final se marque par une forte croissance démographique et une spécialisation économique grandissante. L'économie, selon les secteurs, serait de plus en plus orientée vers les activités pastorales. La superficie et le nombre des sites réduisent vers 2300-2200 BCE, avec une véritable crise socio-environnementale entre 2300 et 2100 BCE. Celle-ci voit l'abandon de vastes zones qui étaient densément occupées jusqu'à cette période. Dans le même temps, une crise hydro-climatique avec une fréquence élevée d'événements extrêmes est enregistrée de la vallée du Rhône au pied des Alpes du sud avec des effets importants sur les systèmes de peuplement dans les plaines inondables (Carozza *et al.* 2015). Le modèle économique du Néolithique final, fondé sur une forte spécialisation (pastoralisme, métallurgie) a été confronté à une aridité croissante entraînant une faible résilience environnementale (rareté de l'eau et de la végétation, érosion des sols, etc., dépassant les seuils de résistance et la capacité de régénération des écosystèmes), en particulier dans les zones de garrigue, qui ont probablement cristallisé les tensions sociales (aménagement des enceintes et des sites fortifiés). À partir de 2400-2300 BCE les sociétés auraient subi les effets conjoints d'une crise de croissance et d'une crise climatique rapide amenant à leur effondrement (Carozza *et al.* 2015).

Le Sud-est de la France, la Méditerranée et l'Europe

En Méditerranée et en Europe, le 3e millénaire est celui des réseaux d'échanges à longues ou très longues distances finissant par donner une relative uniformité de certaines pratiques sociales ou de certains usages, même si elle cache une diversité de pratiques économiques et ne s'apparente en rien à l'uniformité de fond des grandes cultures du Néolithique ancien où beaucoup de choses étaient réellement normées. Ce sont ces réseaux d'échanges, de circulations sociaux, économiques et symboliques suprarégionaux qui sont désarticulés autour de 2200 avant notre ère (Risch *et al.* 2015). Mais il s'agit sans doute moins d'un effondrement que d'une réorganisation qui voit émerger de nouveaux centres et de nouveaux réseaux.

Le Sud-est de la France semble subir un double influx, au moment de la transition au Bronze ancien entre 2150 et 2100 BCE. Le premier est d'origine nord-orientale (nord-est des Alpes, région danubienne) et va toucher la vallée du Rhône et s'étendre jusqu'à l'Auvergne. Son impact s'observe à la fois dans la culture matérielle (céramique, rares vestiges métalliques) mais aussi dans l'organisation de l'habitat, le développement des nécropoles et l'économie agricole qui semble évoluer à ce moment. Mais il influence aussi de façon secondaire les régions les plus méridionales, comme en témoigne certains objets céramiques et métalliques mais aussi les nécropoles et probablement les sépultures en ciste. L'autre est d'origine sud-orientale et orientale (domaine italique, est-adriatique et nord-ouest des Balkans). Il va surtout toucher les régions les plus méridionales du Sud-est de la France, et aura un impact en Languedoc et jusqu'en Auvergne. C'est lui qui est à l'origine des céramiques à décor barbelé dont les influences probablement adriatiques sont très sensibles. L'extrême Sud-est de la France (est-Varois, Alpes-Maritimes) demeure en revanche dépourvu d'ensembles relatifs à cette période et à ces influences pour une raison inconnue (Lemercier, sous presse).

Dans le Sud-est de la France, aucune relation n'est actuellement reconnue avec les centres émergeants de Méditerranée occidentale comme la culture d'El Argar dans le Sud-est de la péninsule Ibérique, mais plutôt avec ce qui se passe dans le bassin adriatique qui connait à ce moment une remarquable dynamique dans toutes les directions, comme le mentionnent plusieurs chercheurs. La Grèce, en particulier la partie occidentale, a subi un fort impact de la zone adriatique et de la Méditerranée centrale, comme en témoigne la céramique, certains petits objets et les brassards de pierre à partir de 2300 BCE surtout qui pointent vers l'ouest et le nord-ouest, et en fait vers le monde campaniforme (Rahmstorf, 2015). Les relations entre la côte dalmate et le monde Égéen s'étendent sans doute jusqu'à la Sicile et à Malte où certains objets partagent aussi des caractères communs (Recchia, Fiorentino, 2015). Le Sud de la France semble, au début du Bronze ancien, participer à des réseaux de contacts et d'échanges assez différents de ceux qui existaient à la fin du Néolithique et jusqu'au Campaniforme récent.

Si le début de l'âge du Bronze voit, dans un certain nombre de régions, les courbes démographiques s'accentuer, le modèle du Midi français avec son maximum au milieu du 3ᵉ millénaire suivi d'une baisse assez spectaculaire, qui étonne certains grands synthétiseurs (Wienelt *et al.* 2015) montre l'existence de trajectoires différentes, des histoires, dont les processus dépendent de facteurs très nombreux aujourd'hui reconnus comme interférant les uns avec les autres. Malcolm H. Weiner envisage les changements de la fin du 3ᵉ millénaire BCE en Grèce et dans les Cyclades non seulement en fonction d'un évènement climatique qui a pu effectivement être un facteur déclenchant, mais aussi en fonction des évolutions locales socio-économiques et des innovations dans les transports en particulier car il propose de tenir compte d'évènements et de développements de toutes sortes qui ont pu avoir lieu dans d'autres régions plus ou moins éloignées (Wiener, 2013, 2014). Mais pour la Méditerranée occidentale aussi, cette période a vu la combinaison complexe de variations climatiques, ou d'imprévisibilité, coïncider avec l'émergence de systèmes socio-économiques complexes et variés et ce n'est peut-être pas le climat en soi qui a causé l'érosion, mais plutôt une intersection de processus socio-écologiques complexes (Walsh *et al.* 2019).

Conclusions

Au terme de ce court panorama concernant l'évènement climatique 4.2 ka BP, trois constats peuvent être faits.

Le premier concerne les connaissances elles-mêmes, qu'elles soient climatiques, environnementales ou archéologiques, qui demeurent extrêmement lacunaires d'une région à l'autre, reposant parfois sur des données extrêmement réduites ou peu sûres, difficilement comparables et de fait interprétables assez aisément dans un sens ou dans un autre. Dans le cas de la France méditerranéenne, on observe un déficit relatif concernant l'archéologie de la transition du Néolithique à l'âge du Bronze et du début de l'âge du Bronze mais surtout concernant les approches économiques de ces périodes, par rapport à ce qui est connu pour la fin du Néolithique antérieure ou la fin de l'âge du Bronze. Ces lacunes s'ajoutent à nos carences en termes de chronologie absolue qui, malgré l'usage de corrections et de probabilités de plus en plus sophistiquées, demeurent incapables de nous offrir ni la certitude ni la précision nécessaires à certaines observations. Dans ce contexte, observer en l'état des continuités et des ruptures et les caractériser demeure difficile et détecter d'éventuels ou potentiels liens de causalité relève de la gageure.

Le second concerne la dimension régionale de l'histoire, qui ne saurait être simplement globale. Il semble qu'un même évènement climatique supposé global au départ se traduise de façon extrêmement différente d'une région à une autre selon des gradients à la fois latitudinaux mais aussi longitudinaux, selon les conditions locales préexistantes aussi.

Un même évènement peut ainsi avoir des conséquences écologiques ou environnementales extrêmement différentes, voire opposées d'une région à une autre. Et ces conséquences écologiques et environnementales peuvent avoir à leur tour différentes conséquences économiques, sociales et politiques diverses selon les sociétés, leur organisation et leur production etc. Enfin, les réponses de ces sociétés à ces potentiels dérèglements économiques, sociaux ou politiques peuvent elles-mêmes être différentes selon les sociétés en question autant que selon la pression économique, environnementale, climatique etc. : entre effondrement, adaptation, résilience, métamorphose, point de départ... Une approche globale semble totalement vaine. Il faut au contraire multiplier les approches régionales. En même temps, chaque région ne peut sans doute être comprise que dans un contexte plus général à une période où des réseaux de contacts et d'échanges semblent développés parfois sur de très longues distances, où les moyens de transport ont fait de notables progrès et où les études génétiques et isotopiques nous montrent des déplacements importants d'individus, de groupes et parfois de populations. Il convient d'identifier des trajectoires régionales originales dans un contexte où toutes les régions sont liées entre-elles.

Enfin, la recherche d'une causalité simple et unique des évènements climatiques sur l'évolution des sociétés n'est clairement plus de mise. Au-delà des querelles d'écoles archéologiques ou d'époques privilégiant tantôt des facteurs externes, tantôt des évolutions internes pour les transformations socio-économiques, la plupart des chercheurs privilégient aujourd'hui des combinaisons de facteurs parmi lesquels les changements climatiques ne sont que l'un des éléments potentiels. Il n'y a pas de causalité simple entre un évènement climatique et un changement social. Il est nécessaire de prendre en compte la réalité de l'évènement au niveau régional, son impact sur l'environnement régional (hydrologique, végétal etc.), mais aussi le type de société concernée, son système économique, sa capacité à évoluer ou sa capacité de résilience etc. Cependant, le rôle d'un évènement climatique comme élément déclencheur des changements, des évolutions et des éventuels effondrements – même s'ils dépendant aussi effectivement d'autres paramètres, demeure très probable et ne doit donc pas être négligé mais analysé.

Bibliographie

Abate T. 1994. Climate and the Collapse of Civilization, *BioScience* 44, p.516-519.

Auxiette G. 2017. Les consommations carnées à l'âge du Bronze : bilan et perspectives. *In:* L. Carozza, C. Marcigny, M. Talon (dir.), *L'habitat et l'occupation des sols à l'âge du Bronze et au début du premier âge du Fer*, Recherches archéologiques 12, Paris, CNRS/Inrap, p.327-336.

Bell B., 1971. The Dark Ages in Ancient History. I. The First Dark Age in Egypt, *American Journal of Archaeology* 75, p.1-26.

Berger J.-F. 2003. La « dégradation des sols » à l'Holocène dans la moyenne vallée du Rhône : contexte morpho-climatique, paléobotanique et culturel. *In:* S. Van der Leeuw, F. Favory, J.-L. Fiches (dir.), *Archéologie et systèmes socio-environnementaux. Études multiscalaires sur la vallée du Rhône dans le programme Archeomedes*, Paris, CNRS (monographies du CRA 27), p.45-167.

Berger J.-F., Magnin F., Thiébault S., Vital J. 2000. Emprise et déprise culturelle à l'Age du Bronze : l'exemple du Bassin Valdainais (Drôme) et de la moyenne vallée du Rhône, *Bulletin de la Société préhistorique française* 97, p.95-119.

Berger J.-F., Brochier J.-L., Vital J., Delhon C., Thiébault S. 2007. Nouveau regard sur la dynamique des paysages et l'occupation humaine à l'Âge du Bronze en moyenne vallée du Rhône. *In:* C. Mordant, H. Richard, M. Magny (dir.), *Environnements et cultures à l'âge du Bronze en Europe occidentale*. Actes du 129e colloque du CTHS, (Besançon, avril 2004), Paris, Éditions du CTHS (Documents préhistoriques 21), p.260-283.

Berger J.-F., Shennan S., Woodbridge J., Palmisano A., Mazier F., Nuninger L., Guillon S., Doyen E., Begeot C., Andrieu-Ponel V., Azuara J., Bevan A., Fyfe R., Roberts C. N. 2019. Holocene land cover and population dynamics in Southern France, *The Holocene* 29, p.776-798.

Bertemes F., Heyd V. 2015. 2200 BC – Innovation or Evolution? The genesis of the Danubian Early Bronze Age. *In:* H. Meller, H.W. Arz, R. Jung, R. Risch (dir.), *2200 BC – Ein Klimasturz als Ursache für den Zerfall der Alten Welt? - 200 BC – A climatic breakdown as a cause for the collapse of the old world? 7.* Mitteldeutscher Archäologentag vom 23. bis 26. Oktober 2014 in Halle (Saale), Tagungen des Landesmuseums für Vorgeschichte Halle, 12/II, p.561-578.

Bini M., Zanchetta G., Persoiu A., Cartier R., Català A., Cacho I., Dean J.R., Di Rita F., Drysdale R. N., Finnè M., Isola I., Jalali B., Lirer F., Magri D., Masi A., Marks L., Mercuri A. M., Peyron O., Sadori L., Sicre M.-A., Welc F., Zielhofer C., Brisset E. 2019. The 4.2 ka BP Event in the Mediterranean region: an overview, *Climate of the Past* 15, p.555-577.

Blaise E. 2010. *Économie animale et gestion des troupeaux au Néolithique final en Provence : approche archéozoologique et contribution des analyses isotopiques de l'émail dentaire*, Oxford, John and Erica Hedges Ltd. (British Archaeological Reports, International Series 2080).

Blaise E., Bréhard S., Carrère I., Favrie T., Gourichon L., Helmer D., Rivière J., Tresset A., Vigne J.-D. 2010. L'élevage du Néolithique moyen 2 au Néolithique final dans le Midi méditerranéen de la France : état des données archéozoologiques. *In:* O. Lemercier, R. Furestier, E. Blaise (dir.), *4è Millénaire. La transition du Néolithique moyen au Néolithique final dans le sud-est de la France et les régions voisines*, Lattes, ADAL et UMR 154 CNRS (Monographies d'Archéologie Méditerranéenne, 27), p.261-284.

Blaise E., Helmer D. Convertini F., Furestier R., Lemercier O. 2014. Bell Beakers herding and hunting in south-eastern France: technical, historical and social implications. *In:* M. Besse (dir.), *Around the Petit-Chasseur Site in Sion (Valais, Switzerland) and New Approaches to the Bell Beaker Culture.* Proceedings of the International Conference held at Sion (Switzerland) October 27th–30th, 2011, Oxford, Archaeopress, p.163-180.

Blanchemanche P., Berger J.-F., Chabal L., Jorda C., Jung C., Raynaud C. 2003. Le littoral languedocien durant l'Holocène : milieu et peuplement entre le Lez et le Vidourle (Hérault, Gard). *In:* T. Muxart, F.-D. Vivien, B. Villalba, J. Burnouf (dir.), *Des milieux et des hommes : fragments d'histoires croisées.* Bilan du Programme PEVS/SEDD, Elsevier (coll. Environnement), p.79-92.

Blanco-González A., Lillios K.T., López-Sáez J.A., 2018. Cultural, demographic and environmental dynamics of the Copper and early Bronze Age in Iberia (3300–1500 BC): Towards and interregional multiproxy comparison at the time of the 4.2 ky BP event, *Journal of World Prehistory* 31, p.1-79.

Blum S.W.E., Riehl S. 2015. Troy in the 23rd century BC – environmental dynamics and cultural change. *In:* H. Meller, H.W. Arz, R. Jung, R. Risch (dir.), *2200 BC – Ein Klimastur zals Ursache für den Zerfall der Alten Welt? - 2200 BC – A climatic breakdown as a cause for the collapse of the old world ? 7.* Mitteldeutscher Archäologentag vom 23. bis 26. Oktober 2014 in Halle (Saale), Tagungen des Landesmuseums für Vorgeschichte Halle, 12/I, p.181-204.

Bond G.W., Showers M., Cheseby R., Lotti P., Almasi P., De Menocal P., Priore P., Cullen H., Hajdas I., Bonani G. 1997. A pervasive Millennial-Scale Cycle in North Atlantic Holocene and Glacial Climates, *Science* 278, p.1257-1266.

Bouby L., Zech-Matterne V., Bouchette A., Cabanis M., Derreumaux M., Dietsch-Sellami M.-F., Durand F., Figueiral I., Marinval P., Paradis L., Pradat B., Rousselet O., Rovira N., Schaal C., Toulemonde F., Wiethold J. 2017. Ressources et économie agricole en France à l'âge du Bronze et au Premier âge du Fer : les données carpologique. *In:* L. Carozza, C. Marcigny, M. Talon (dir.), *L'habitat et l'occupation des sols à l'âge du Bronze et au début du premier âge du Fer*, Recherches archéologiques 12, Paris, CNRS/Inrap, p.299-326.

Bradley R.S., Bakke J. 2019. Is there evidence for a 4.2 ka BP event in the northern North Atlantic region? *Climate of the Past* 15, p.1665-1676.

Carolina S.A., Walker R.T., Day C.C., Ersek V., Sloan R.A., Dee M.W., Talebian M., Henderson G.M., 2019. Precise timing of abrupt increase in dust activity in the Middle East coincident with 4.2 ka social change, *PNAS* 116, p.67-72.

Carozza L., Bourgarit D., Mille B., Burens A. 1999. L'habitat et l'atelier de métallurgiste d'Al Claus (Tarn et Garonne), *Archéologie en Languedoc*, 21, Actes du colloque de Cabrières (mai 1997), p.147-160.

Carozza L., Georjon C., Vignaud A. (dir.) 2005. *La fin du Néolithique et les débuts de la métallurgie en Languedoc central. Les habitats de la colline du Puech Haut à Paulhan*, Hérault, Toulouse, Archives d'Écologie Préhistorique.

Carozza L., Rostan P., Bourgarit D., Mille B., Coquinot Y., Burens A., Escanilla Artigas N. 2010. Un site métallurgique du Bronze ancien dans le vallon du Longet à Molines-en-Queyras (Hautes-Alpes) : caractérisation du contexte archéologique et des déchets liés aux activités de métallurgie extractive. *In:* S. Tzortzis, X. Delestre (dir.), *Archéologie de la montagne européenne.* Actes de la table ronde internationale de Gap, 29 septembre-1er octobre 2008, Paris-Aix-en-Provence, Éditions Errance / CCJ, Bibliothèque d'Archéologie Méditerranéenne et Africaine 4, p.261-281.

Carozza L., Mille B., Bourgarit D., Rostan P., Burens-Carozza A. 2011. Mine et métallurgie en haute montagne dès la fin du Néolithique et le début de l'âge du Bronze: l'exemple de Saint-Véran en Haut-Queyras (Hautes-Alpes, France). *In: L'Etàdel rame in Italia*, XLIII Riunione Scientifica IIPP in memoria di Gianni Bailo Modesti, Bologna, 26-29 novembre 2008, Firenze, IIPP, p.151-155.

Carozza L., Berger J.-F., Burens-Carozza A., Marcigny C. 2015. Society and environment in Southern France from the 3rd millennium BC to the beginning of the 2nd millennium BC: 2200 BC a tipping point? *In:* H. Meller, H.W. Arz, R. Jung, R. Risch (dir.), *2200 BC - Ein Klimasturz als Ursach efür den Zerfall der Alten Welt? - 2200 BC - A climatic breakdown as a cause for the collapse of the old world?* 7. Mitteldeutscher Archäologen tag vom 23. bis 26. Oktober 2014 in Halle (Saale), Tagungen des Landesmuseums für Vorgeschichte Halle, 12/II, p.335-362.

Carozza L., Marcigny C., Talon M. (dir.) 2017. *L'habitat et l'occupation des sols à l'âge du Bronze et au début du premier âge du Fer*, Paris, Recherches archéologiques 12, CNRS/Inrap.

Casa P. D. 1995. The Cetina group and the transition from Copper to Bronze Age in Dalmatia, *Antiquity* 69, p.565-576.

Cauliez J. 2011. Restitution des aires culturelles au Néolithique final dans le sud-est de la France. Dynamiques de formation et d'évolution des styles céramiques, *Gallia Préhistoire* 53, p.85-202.

Chew S.C. 2005. From Harappa to Mesopotamia and Egypt to Mycenae: Dark Ages, Political–Economic Declines, and Environmental/Climatic Changes 2200 B.C.–700 B.C. *In:* C. Chase-Dunn, E. N. Anderson (dir.), *The Historical Evolution of World-Systems*, New-York, Palgrave Macmillan, p.52-74.

Costa-Casais M., López-Merino L., Kaal J., Martínez Cortizas A. 2015. Environmental changes in north-western Iberia around the Bell Beaker period (2800–1400 cal. BC). *In:* M. P. Prieto Martinez, L. Salanova (dir.), *The Bell Beaker Transition in Europe. Mobility and local evolution during the 3rd millennium BC*, Oxford, Oxbow Books, p.150-158.

D'Anna A. 1992. Le peuplement préhistorique du massif de Sainte-Victoire, *Méditerranée* 75, p.59-68.

D'Anna A. 1995a. La fin du Néolithique dans le Sud-est de la France. *In:* R. Chenorkian (dir.), *L'Homme Méditerranéen, Mélanges offerts à Gabriel Camps*, Aix en Provence, Publications de l'Université de Provence, p.299-333.

D'Anna A. 1995b. Le Néolithique final en Provence. *In:* J.-L. Voruz (dir.), *Chronologies néolithiques : de 6000 à 2000 avant notre ère dans le Bassin Rhodanien*, Actes des Rencontres néolithiques Rhône-Alpes, Ambérieu-en-Bugey, septembre 1992, Ambérieu-en-Bugey, Université de Genève et Éditions de la Société Préhistorique Rhodanienne (Document du Département d'Anthropologie et d'Écologie de l'Université de Genève, 20), p.265-286.

Dalfes H.N., Kukla G., Weiss H. (dir.) 1997. *Third Millennium BC Climate Change and Old World Collapse.* Proceedings of the NATO Advanced Research Workshop on Third Millennium BC Abrupt Climate Change and Old World Social Collapse, held at Kemer, Turkey, September 19-24, 1994, Berlin, Heidelberg, Springer-Verlag.

Dee M.W. 2017. Absolutely Dating Climatic Evidence and the Decline of Old Kingdom Egypt. *In:* F. Höflmayer (dir.), *The Late Third Millennium in the Ancient Near East: Chronology, C14 and Climate Change.* Papers from the Oriental Institute Seminar Held at the Oriental Institute of the University of Chicago, Oriental Institute seminars 11, Chicago, Oriental Institute, p.323-332.

Delibes De Castro G., Abarqueromoras F.J., Crespo Diez M., Garcia Garcia M., Guerra Doce E., Lopez Saez J.A., Perez Diaz S., Rodriguez Marcos J.A. 2015. The archaeological and palynological record of the Northern Plateau of Spain during the second half of the 3rd millennium BC. *In:* H. Meller, H.W. Arz, R. Jung, R. Risch (dir.), *2200 BC – Ein Klimasturz als Ursache für den Zerfall der Alten Welt? - 2200 BC – A climatic breakdown as a cause for the collapse of the old world ?* 7. Mitteldeutscher Archäologen tag vom 23. bis 26. Oktober 2014 in Halle (Saale), Tagungen des Landesmuseums für Vorgeschichte Halle, 12/I, p.429-448.

Dimitrijevič S. 1967. Die Ljubljana – Kultur, *Archaeol. Iugoslavica* VIII, p.1-25.

Di Rita F., Magri D. 2019. The 4.2 ka event in the vegetation record of the central Mediterranean, *Climate of the Past* 15, p.237-251.

Drysdale R., Zanchetta G., Hellstrom J., Maas R., Fallick A., Pickett M., Cartwright I., Piccini L., 2006. Late Holocene drought responsible for the collapse of Old World civilizations is recorded in an Italian cave flowstone, *Geology* 34, p.101-104.

Fábregas Valcarce R., Martínez Cortizas A., Blanco Chao R., Chesworth W. 2003. Environmental change and social dynamics in the second–third millennium BC in NW Iberia, *Journal of Archaeological Science* 30, p.859-871.

Finné M., Woodbridge J., Labuhn I., Roberts N., 2019. Holocene hydro-climatic variability in the Mediterranean: A synthetic multi-proxy reconstruction, *The Holocene* 29, p.847-863.

Fitzpatrick A. P. 2015. Great Britain and Ireland in 2200 BC. *In:* H. Meller, H.W. Arz, R. Jung, R. Risch (dir.), *2200 BC – Ein Klimasturz als Ursachefür den Zerfall der Alten Welt? - 2200 BC – A climatic breakdown as a cause for the collapse of the old world?* 7. Mitteldeutscher Archäologen tag vom 23. bis 26. Oktober 2014 in Halle (Saale), Tagungen des Landesmuseums für Vorgeschichte Halle, 12/II, p.805-832.

Forenbaher S. 2018. Ljubljana i Cetina: lončarskistilovi 3. tisućljećaprije Krista na prostoruistočnoga, *Prilozi* 35, p.113-157.

Fyfe R.M., Woodbridge J., Roberts C.N. 2018. Trajectories of change in Mediterranean Holocene vegetation through classification of pollen data. *Vegetation History and Archaeobotany* 27, p.351-364.

Fyfe R. M., Woodbridge J., Palmisano A., Bevan A., Shennan S., Burjachs F., Legarra Herrero B., García Puchol O., Carrión J. S., Revelles J., Roberts N. 2019. Prehistoric palaeodemographics and regional land cover change in eastern Iberia, *The Holocene* 29, p.799-815.

Furestier R. 2007. *Les industries lithiques campaniformes du sud-est de la France*, Oxford, John and Erica Hedges Ltd. (British Archaeological Reports, International Series, 1684).

Gascó J. 2004. Les composantes de l'Age du Bronze, de La fin du Chalcolithique à l'Âge du Bronze ancien en France méridionale, *Cypsela* 15, p.39-72.

Gateau F. 1996. *L'Étang-de-Berre. Carte Archéologique de la Gaule* 13/1, Paris, Académie des Inscriptions et Belles Lettres.

Genz H. 2015. Beware of environmental determinism: the transition from the early to the Middle Bronze Age on the Lebanese coast and the 4.2 ka BP event, *In:* H. Meller, H.W. Arz, R. Jung, R. Risch (dir.), *2200 BC – Ein Klimasturz als Ursachefür den Zerfall der Alten Welt? - 2200 BC – A climatic breakdown as a cause for the collapse of the old world?* 7. Mitteldeutscher Archäologen tag vom 23. bis 26. Oktober 2014 in Halle (Saale), Tagungen des Landesmuseums für Vorgeschichte Halle, 12/I, p.97-111.

Gilli E., Petrucci G., Salzani L. 2000. L'abitato di Bernardine di Coriano-Albaredo d'Adige (materiali degli scavi 1987-1990), *Bolletino del Museo Civico di Storia Naturale di Verona*, 24, 2000, p. 99-154.

Gilli E., Salzani L., Salzani P. 2005. New evidence of Barbed-Wire pottery from the Verona area (Northern Italy). *In:* R. Laffineur, J. Driessen, E.Warmenbol (dir.), *L'âge du Bronze en Europe et en Méditerranée /The Bronze Age in Europe and the Mediterranean*, Actes du XIVe Congrès international de l'UISPP, Liège, 2001, Oxford, Archaeopress (British Archaeological Reports International Series 1337), p.91-98.

Guilaine J. 1996. Le Bronze ancien en Méditerranée occidentale. *In:* C. Mordant, O. Gaiffe (dir.), *Cultures et sociétés du Bronze ancien en Europe »,*

Actes du 117e Congrès des Société savantes, Clermont-Ferrand, 1992, Commission de Préhistoire et de Protohistoire, Paris, CTHS, p.37-68.

Gutherz X., Jallot L. 1995. Le Néolithique final du Languedoc méditerranéen. *In:* J.-L. Voruz (dir.), *Chronologies néolithiques : de 6000 à 2000 avant notre ère dans le Bassin Rhodanien*, Actes des Rencontres néolithiques Rhône-Alpes, Ambérieu-en-Bugey, septembre 1992, Ambérieu-en-Bugey, Université de Genève et Éditions de la Société Préhistorique Rhodanienne (Document du Département d'Anthropologie et d'Écologie de l'Université de Genève, 20), p.231-263.

Hinz M., Schirrmacher J., Kneisel J., Rinne C., Weinelt M. 2019. The Chalcolithic–Bronze Age transition in southern Iberia under the influence of the 4.2 ka BP event? A correlation of climatological and demographic proxies, *Journal of Neolithic Archaeology* 21, p.1-26.

Höflmayer F. (dir.) 2017. *The Late Third Millennium in the Ancient Near-East: Chronology, C14 and Climate Change.* Papers from the Oriental Institute Seminar Held at the Oriental Institute of the University of Chicago, Oriental Institute seminars 11, Chicago, Oriental Institute.

Höflmayer F. 2017. The Late Third Millennium B.C. in the Ancient Near East and Eastern Mediterranean: A Time of Collapse and Transformation. *In:* F. Höflmayer (dir.), *The Late Third Millennium in the Ancient Near East: Chronology, C14 and Climate Change.* Papers from the Oriental Institute Seminar Held at the Oriental Institute of the University of Chicago, Oriental Institute seminars 11, Chicago, Oriental Institute, p. 1-30.

Isola I., Zanchetta G., Drysdale R.N., Regattieri E., Bini M., Bajo P., Hellstrom J.C., Baneschi I., Lionello P., Woodhead J., Greig A. 2019. The 4.2 ka event in the central Mediterranean: new data from a Corchia speleothem (Apuan Alps, central Italy), *Climate of the Past* 15, p.135-151.

Jallot L. 2011. *Milieux, sociétés et peuplement en Languedoc méditerranéen au Néolithique final*, Thèse de Doctorat, Montpellier, Université Paul Valéry.

Jallot L., Gutherz L. 2014. Le Néolithique final en Languedoc oriental et ses marges : 20 ans après Ambérieu-en-Bugey. *In:* I. Sénépart, F. Leandri, J. Cauliez, T. Perrin, E. Thirault (dir.), *Chronologie de la Préhistoire récente dans le sud de la France : Acquis 1992-2012. Actualité de la recherche.* Actes des 10e Rencontres Méridionales de Préhistoire Récente (Porticcio, 18-20 octobre 2012), Toulouse, Archives d'Écologie Préhistorique, p.137-158.

Jalut G., Dedoubat J.J., Fontugne M., Otto T. 2009. Holocene circum-Mediterranean vegetation changes: Climate forcing and human impact. *Quaternary International* 200, p.4-18.

Jordan S.F., Murphy B., O'Reilly S.S., Doyle K.P., Williams M.D., Grey A., Lee S., Mccaul M.V., Kelleher B. P. 2017. Mid-Holocene climate change and landscape formation in Ireland: evidence from a geochemical investigation of a coastal peat bog, *Organic Geochemistry* 109, p.67-76.

Joerin U.E., Stocker T.F., Schlüchter C. 2006. Multi-century glacier fluctuations in the Swiss Alps during the Holocene, *The Holocene* 16, p.697-704.

Joerin U.E., Nicolussi K., Fischer A., Stocker T.F., Schlüchter C. 2008. Holocene optimum events inferred from subglacial sediments at Tschierva Glacier, Eastern Swiss Alps, *Quaternary Science Reviews* 27, p.337-350.

Jung R., Weninger B. 2015. Archaeological and environmental impact of the 4.2 kacal BP event in the central and eastern Mediterranean. *In:* H. Meller, H. W. Arz, R. Jung, R. Risch (dir.), *2200 BC – Ein Klimasturz als Ursache für den Zerfall der Alten Welt? - 2200 BC – A climatic breakdown as a cause for the collapse of the old world ?* 7. Mitteldeutscher Archäologen tag vom 23. bis 26. Oktober 2014 in Halle (Saale), Tagungen des Landesmuseums für Vorgeschichte Halle, 12/I, p. 205-234.

Kaniewski D., Marriner N., Cheddadi R., Guiot J., Van Campo E. 2018. The 4.2 ka BP event in the Levant, *Climate of the Past* 14, p.1529–1542.

Kleijne J., Weinelt M., Müller J. 2020. Late Neolithic and Chalcolithic maritime resilience? The 4.2 ka BP event and its implications for environments and societies in Northwest Europe, *Environmental Research Letters* 15, 125003.

Kölling M., Lull V., Micó R., Rihuetcherrada C., Risch R. 2015. No indication of increased temperatures around 2200 BC in the south-west Mediterranean derived from oxygen isotope ratios in marine clams (Glycymeris sp.) from the El Argar settlement of Gatas, south-east Iberia. *In:* H. Meller, H.W. Arz, R. Jung, R. Risch (dir.), *2200 BC – Ein Klimasturz als Ursache für den Zerfall der Alten Welt? - 2200 BC – A climatic breakdown as a cause for the collapse of the old world ?* 7. Mitteldeutscher Archäologen tag vom 23. bis 26. Oktober 2014 in Halle (Saale), Tagungen des Landesmuseums für Vorgeschichte Halle, 12/I, p.449-460.

Kuzucuoğlu C. 2012. Le rôle du climat dans les changements culturels, du 5e au 1er millénaire avant notre ère, en Méditerranée orientale. *In:* J.-F. Berger (dir.), *Des climats et des hommes*, Paris, La Découverte, Inrap, p.239-256.

Kuzucuoğlu C., Marro C. (dir.) 2007. *Sociétés humaines et changement climatique à la fin du troisième millénaire : Une crise a-t-elle eu lieu en Haute Mésopotamie ?* Actes du Colloque de Lyon (5–8 décembre 2005), Istanbul, Institut Français d'Études Anatoliennes-Georges Dumézil (Varia Anatolica, 19).

Kuzucuoğlu C., Tsirtsoni Z. 2016. Changements climatiques et comportements sociaux dans le passé : quelles corrélations ? *Les nouvelles de l'archéologie*, 142, p.49-55.

Lechterbeck J., Kerig T., Kleinmann A., Sillmann M., Wick L., Rösch M. 2014. How was Bell Beaker economy related to Corded Ware and Early Bronze Age lifestyles? Archaeological, botanical and palynological evidence from the Hegau, Western Lake Constance region, *Environmental Archaeology* 19, p.95-113

Lemercier O. 2004. *Les Campaniformes dans le sud-est de la France*, Lattes, Publications de l'UMR 154 du CNRS / ADAL (Monographies d'Archéologie Méditerranéenne, n°18).

Lemercier O. 2007. La fin du Néolithique dans le sud-est de la France. Concepts techniques, culturels et chronologiques de 1954 à 2004. *In:* J. Evin (dir.),

Un siècle de construction du discours scientifique en Préhistoire, Actes du XXVIe Congrès Préhistorique de France, Avignon, 21-25 septembre 2004, Volume I, Paris, Société Préhistorique Française, p.485-500.

Lemercier O. 2012. The Mediterranean France Beakers Transition. *In:* H. Fokkens, F. Nicolis (dir.), *Background to Beakers. Inquiries into the regional cultural background to the Bell Beaker complex"*, Leiden, Sidestone Press, p.81-119.

Lemercier O. 2014. Bell Beakers in Eastern France and the Rhone-Saone-Rhine axis question, *In:* M. Besse (dir.), *Around the Petit-Chasseur Site in Sion (Valais, Switzerland) and New Approaches to the Bell Beaker Culture.* Proceedings of the International Conference held at Sion (Switzerland) October 27th – 30th, 2011, Oxford, Archaeopress, p.181-204.

Lemercier O. 2018. La question campaniforme, *In:* J. Guilaine, D. Garcia (dir.), *La Protohistoire de la France*, Paris, Hermann (Histoire et Archéologie), p.205-217.

Lemercier O. 2019. Campaniforme : fin du Néolithique et/ou début de l'âge du Bronze ? *In:* N. Buchez, O. Lemercier, I. Praud, M. Talon (dir.), *La fin du Néolithique et la genèse du Bronze ancien dans l'Europe du nord-ouest, Actes de la session 5 du XXVIIIe Congrès Préhistorique de* France (Amiens, 29 mai-3 juin 2016), Paris, Société Préhistorique Française, p.239-250.

Lemercier O. 2020. Is it possible to observe the demographic evolution from the Middle Neolithic to the Early Bronze Age in Mediterranean France (4500-1900 BCE) ? *In:* T. Lachenal, R. Roure, O. Lemercier (dir.), *Demography and Migration. Population Trajectories from the Neolithic to the Iron Age. Proceedings of the XVIIIth UISPP World Congress (4-9 June 2018, Paris, France) Sessions XXXII-2 and XXXIV-8.* Oxford, Archaeopress, p.5-20.

Lemercier O. (sous presse). Genèse du Bronze ancien dans le sud-est de la France : l'héritage campaniforme. *In:* S. Blanchet, T. Nicolas, B. Quilliec, B. Roberts (dir.), *Les sociétés du Bronze ancien atlantique du XXIVème au XVIIème s. av. J.-C.*, Actes du colloque de Rennes, 7-10 novembre 2018, Bordeaux, Ausonius.

Lemercier O., Strahm C. 2018. Nids de coucous et grandes maisons. L'habitat campaniforme, épicampaniforme et péricampaniforme en France dans son contexte européen. *In:* O. Lemercier, I. Sénépart, M. Besse& C. Mordant (dir.), *Habitations et habitat du Néolithique à l'âge du Bronze en France et régions voisines*, actes des 2e Rencontres Nord-Sud de Préhistoire Récente (Dijon, 19-21 novembre 2015), Toulouse, Archives d'Écologie Préhistorique, p.459-478.

Lemercier O., Blaise E., Cattin F., Convertini F., Desideri J., Furestier R., Gadbois-Langevin R., Labaune M. 2014a. 2500 avant notre ère : l'implantation campaniforme en France méditerranéenne. *In:* L. Mercuri, R.G. Villaescusa, F. Bertoncello (dir.), *Implantations humaines en milieu littoral méditerranéen : facteurs d'installation et processus d'appropriation de l'espace (Préhistoire, Antiquité, Moyen-Âge),* actes des XXXIVèmes Rencontres internationales d'Archéologie et d'Histoire d'Antibes (Antibes, 15-17 octobre 2013), Antibes, Éditions APDCA, p.191-203.

Lemercier O., Furestier R., Gadbois-Langevin R., Schulz Paulsson B. 2014b. Chronologie et périodisation des campaniformes en France méditerranéenne. *In:* I. Sénépart, F. Leandri, J. Cauliez, T. Perrin, E. Thirault (dir.), *Chronologie de la Préhistoire récente dans le sud de la France : Acquis 1992-2012. Actualité de la recherche.* Actes des 10e Rencontres Méridionales de Préhistoire Récente (Porticcio, 18-20 octobre 2012), Toulouse, Archives d'Écologie Préhistorique, p.175-195.

Lemercier O., Blaise E., Convertini F., Furestier R., Gilabert C. Labaune M. 2019. Beaker settlements in Mediterranean France in their cultural context. *In:* A. Gibson (dir.), *The Bell Beaker Settlement of Europe. The Bell Beaker phenomenon from a domestic perspective*, Oxford, Oxbow Books (Prehistoric Society Research Paper, 9), p.81-107.

Leonardi G., Cupito M., Baioni M., Longhi C., Martinelli N. 2015. Northern Italy around 2200 cal BC – From Copper Age to Early Bronze Age: continuity and/or discontinuity? *In:* H. Meller, H.W. Arz, R. Jung, R. Risch (dir.), *2200 BC – Ein Klimasturz als Ursache für den Zerfall der Alten Welt? - 2200 BC – A climatic breakdown as a cause for the collapse of the old world ?* 7. Mitteldeutscher Archäologen tag vom 23. bis 26. Oktober 2014 in Halle (Saale), Tagungen des Landesmuseums für Vorgeschichte Halle, 12/I, p.283-304.

Lespez L., Carozza L., Berger J.-F., Kuzucuoglu C., Ghilardi M., Carozza J.-M., Vanniere B. 2016. Rapid climatic change and social transformations: Uncertainties, adaptability and resilience. *In:* S. Thiébault, J.-P. Moatti (dir.), *The Mediterranean Region under Climate Change*, Paris, IRD Edition, p.35-45.

Lull V., Micó R., Rihuete Herrada C., Risch R. 2015. Transition and conflict at the end of the 3rd millennium BC in south Iberia. *In:* H. Meller, H.W. Arz, R. Jung, R. Risch (dir.), *2200 BC – Ein Klimasturz als Ursache für den Zerfall der Alten Welt? - 2200 BC – A climatic breakdown as a cause for the collapse of the old world ?* 7. Mitteldeutscher Archäologen tag vom 23. bis 26. Oktober 2014 in Halle (Saale), Tagungen des Landesmuseums für Vorgeschichte Halle, 12/I, p.365-408.

Magny M., Vannière B., Zanchetta G., Fouache E., Touchais G., Petrika L., Coussot C., Walter-Simonnet A.V., Arnaud F. 2009. Possible complexity of the climatic event around 4300–3800 cal. BP in the central and western Mediterranean, *The Holocene* 19, p.823-833.

Magny M., Combourieu Nebout N., De Beaulieu J.L., Bout-Roumazeilles V., Colombaroli D., Desprat S., Francke A., *et al.* 2013. North-south palaeohydrological contrasts in the central Mediterranean during the Holocene: tentative synthesis and working hypotheses, *Climate of the Past* 9, p.2043-2071.

Manning S.W. 2017. Comments on Climate, Intra-regional Variations, Chronology, the 2200 B.C. Horizon of Change in the East Mediterranean

Region, and Socio-political Change on Crete. *In:* F. Höflmayer (dir.), *The Late Third Millennium in the Ancient Near East: Chronology, C14 and Climate Change.* Papers from the Oriental Institute Seminar Held at the Oriental Institute of the University of Chicago, Oriental Institute seminars 11, Chicago, Oriental Institute, p.451-490.

Massa M., Sahoglu V. 2015. The 4.2 ka BP climatic event in west and central Anatolia: combining palaeoclimatic proxies and archaeological data. *In:* H. Meller, H.W. Arz, R. Jung, R. Risch (dir.), *2200 BC – Ein Klimasturz als Ursache für den Zerfall der Alten Welt? - 2200 BC – A climatic breakdown as a cause for the collapse of the old world ?* 7. Mitteldeutscher Archäologen tag vom 23. bis 26. Oktober 2014 in Halle (Saale), Tagungen des Landesmuseums für Vorgeschichte Halle, 12/I, p.61-78.

Mayewski, P.A., Rohling, E.E., Curt Stager, J., Karlén, W., Maasch, K.A., Meeker, L.D., Meyerson A., Gasse F., *et al.* 2004. Holocene Climate Variability, *Quaternary Research* 62, p.243–255.

Mejias Moreno M., Benitez De Lugo Enrich L., Pozo Tejado J. Del, Moraleda Sierra J. 2014. Los primeros aprovechamientos de aguas subterráneas en la Península Ibérica. Las motillas de Daimiel en la Edad del Bronce de La Mancha, *Boletín Geológico y Minero* 125, p.455-474.

Meller H., Arz H.W., Jung R., Rischr. (dir.) 2015. *2200 BC – Ein Klimasturz als Ursache für den Zerfall der Alten Welt? - 2200 BC – A climatic breakdown as a cause for the collapse of the old world?* 7. Mitteldeutscher Archäologen tag vom 23. bis 26. Oktober 2014 in Halle (Saale), Tagungen des Landesmuseums für Vorgeschichte Halle, 12(I/II).

Mercuri A.M., Florenzano A., Burjachs F., Giardini M., Kouli K., Masi A., Picornell-Gelabert L., Revelles J., Sadori L., Servera-Vives G., Torri P., Fyfe R. 2019. From influence to impact: The multifunctional land use in Mediterranean prehistory emerging from palynology of archaeological sites (8.0-2.8 ka BP), *The Holocene* 29, p.830-846.

Middleton G.D. 2012. Nothing Lasts Forever: Environmental Discourses on the Collapse of Past Societies, *Journal of Archaeological Research* 20, p.257-307.

Middleton G.D. 2018. Bang or whimper? The evidence for collapse of human civilizations at the start of the recently defined Meghalayan Age is equivocal, *Science* 361, 6408, p.1204-1205.

Mimokhod R.A. 2018. Paleoclimate and cultural genesis in Eastern Europe at the end of the 3rd millennium BC, *Rossiyskaya arkheologiya* 59, p.33-48.

Moreno García J. C. 2015. Climatic change or sociopolitical transformation? Reassessing late 3rd millennium BC in Egypt. *In:* H. Meller, H.W. Arz, R. Jung, R. Risch (dir.), *2200 BC – Ein Klimasturz als Ursache für den Zerfall der Alten Welt? - 2200 BC – A climatic breakdown as a cause for the collapse of the old world ?* 7. Mitteldeutscher Archäologen tag vom 23. bis 26. Oktober 2014 in Halle (Saale), Tagungen des Landesmuseums für Vorgeschichte Halle, 12/I, p.79-96.

Moulin B., Thirault E., Vital J., Bailly-Maître M.-C. 2012. Quatre années de prospection sur les extractions de cuivre de l'âge du Bronze ancien dans le massif des Rousses en Oisans (Isère et Savoie, France). *In:* T. Perrin, I. Sénépart, J. Cauliez, E. Thirault, S. Bonnardin (dir.), *Dynamismes et rythmes évolutifs des sociétés de la préhistoire récente. Actualité de la recherche.* Actes des 9èmes rencontres méridionales de Préhistoire Récente Saint-Georges-de-Didonne, 8-9 octobre 2010, Toulouse, Archives d'Écologie Préhistorique, p.341-369.

Müller J. 2015. Crisis – what crisis? Innovation: different approaches to climatic change around 2200 BC. *In:* H. Meller, H.W. Arz, R. Jung, R. Risch (dir.), *2200 BC – Ein Klimasturz als Ursache für den Zerfall der Alten Welt? - 200 BC – A climatic breakdown as a cause for the collapse of the old world ? 7.* Mitteldeutscher Archäologen tag vom 23. bis 26. Oktober 2014 in Halle (Saale), Tagungen des Landesmuseums für Vorgeschichte Halle, 12/II, p.651-669.

Nicolis F. 1998. Un nuovo aspetto ceramico tra età del Rame e età del Bronzo nell'Italia settentrionale, *Rivista di Scienze Preistoriche* XLIX, p.447-468.

Nocete F., Lizcano R., Peramo A., Gómez E. 2010. Emergence, collapse and continuity of the first political system in the Guadalquivir Basin from the fourth to the second millennium BC: The long-term sequence of Úbeda (Spain), *Journal of Anthropological Archaeology* 29, p.219-237

Pacciarelli M., Scarano T., Crispino A. 2015. The transition between the Copper and Bronze Ages in southern Italy and Sicily. *In:* H. Meller, H.W. Arz, R. Jung, R. Risch (dir.), *2200 BC – Ein Klimasturz als Ursache für den Zerfall der Alten Welt? - 200 BC – A climatic breakdown as a cause for the collapse of the old world ? 7.* Mitteldeutscher Archäologen tag vom 23. bis 26. Oktober 2014 in Halle (Saale), Tagungen des Landesmuseums für Vorgeschichte Halle, 12/I, p.253-282.

Poggiani Keller, R., Baioni, M., Leonini, V., Lo Vetro, D. 2006. Villanuova sul Clisi (BS) – Monte Covolo. Insediamento pluristratificato dal Neolitico tardo alla Media età del Bronzo. *In:* M. Baioni, R. Poggiani Keller (dir.), *Il bicchiere campaniforme : dal simbolo alla vita quotidiana. Aspetti insediativi nella Lombardia centro-orientale di un fenomeno culturale europeo del III millennio,* Annali del Museo, 20, Gavardo, Museo, p.79-115.

Pusztai Néfischl K., Kiss V., Kulcsár G., Szeverényi V. 2015. Old and new narratives for Hungary around 2200 BC. *In:* H. Meller, H.W. Arz, R. Jung, R. Risch (dir.), *2200 BC – Ein Klimasturz als Ursache für den Zerfall der Alten Welt? - 200 BC – A climatic breakdown as a cause for the collapse of the old world ? 7.* Mitteldeutscher Archäologen tag vom 23. bis 26. Oktober 2014 in Halle (Saale), Tagungen des Landesmuseums für Vorgeschichte Halle, 12/II, p.503-524.

Rahmstorf L. 2015. The Aegean before and after c. 2200 BC between Europe and Asia: trade as a prime mover of cultural change. *In:* H. Meller, H.W. Arz, R. Jung, R. Risch (dir.), *2200 BC – Ein Klimasturz als Ursache für den Zerfall der Alten Welt? - 2200 BC - A climatic breakdown as a cause for the collapse of the old world ?* 7. Mitteldeutscher Archäologen tag vom 23. bis 26. Oktober 2014 in Halle (Saale), Tagungen des Landesmuseums für Vorgeschichte Halle, 12/I, p.149-180.

Ran M., Chen L. 2019. The 4.2 ka BP climatic event and its cultural responses, *Quaternary International* 521, p.158-167.

Recchia G., Fiorentino G. 2015. Archipelagos adjacent to Sicily around 2200 BC: attractive environments or suitable geo-economic locations. *In:* H. Meller, H.W. Arz, R. Jung, R. Risch (dir.), *2200 BC – Ein Klimasturz als Ursache für den Zerfall der Alten Welt? - 2200 BC – A climatic breakdown as a cause for the collapse of the old world ?* 7. Mitteldeutscher Archäologen tag vom 23. bis 26. Oktober 2014 in Halle (Saale), Tagungen des Landesmuseums für Vorgeschichte Halle, 12/I, p.305-319.

Riquet R. 1976. L'anthropologie protohistorique française. *In:* J. Guilaine (dir.), *La Préhistoire française. II Les civilisations néolithiques et protohistoriques*, Paris, CNRS, p.135-152.

Roberts C.N., Woodbridge J., Palmisano A., Bevan A., Fyfe R., Shennan S. 2019 Mediterranean landscape change during the Holocene: Synthesis, comparison and regional trends in population, land cover and climate, *The Holocene* 29, p.923-937.

Roland T.P., Caseldine C.J., Charman D.J., Turney C.S.M., Amesbury M.J. 2014. Was there a '4.2 ka event' in Great Britain and Ireland? Evidence from the peatland record, *Quaternary Science Reviews* 83, p.11-27.

Schirrmacher J., Weinelt M., Blanz T., Andersen N., Salgueiro E., Schneider R.R. 2019. Multi-decadal atmospheric and marine climate variability in southern Iberia during the mid- to late-Holocene, *Climate of the Past,* 15, 2019, p.617-634.

Stockhammer P.W., Massy K., Knipper C., Friedrich R., Kromer B., Lindauer S., Radosavljevic J., Pernicka E., Krause J. 2015. Kontinuität und Wandel vom Endneolithikum zur frühen Bronzezeit in der Region Augsburg, *in* "H. Meller, H.W. Arz, R. Jung, R. Risch (dir.), *2200 BC – Ein Klimasturz als Ursache für den Zerfall der Alten Welt? - 2200 BC – A climatic breakdown as a cause for the collapse of the old world ?*" 7. Mitteldeutscher Archäologen tag vom 23. bis 26. Oktober 2014 in Halle (Saale), Tagungen des Landesmuseums für Vorgeschichte Halle, 12/II, p. 617-642.

Stoddart S., Woodbridge J., Palmisano A., Mercuri A.M., Mensing S.A., Colombaroli D., Sadori L., Magri D., Di Rita F., Giardini M., Mariotti M., Montanari C., Bellini C., Florenzano A., Torri P., Bevan A., Shennan S., Fyfe R., Roberts C.N. 2019. Tyrrhenian central Italy: Holocene population and landscape ecology, *The Holocene* 29, p.761-775

Toulemonde F., Auxiette G., Bouby L., Goude G., Peake R., collab. De Forest V. 2018. L'alimentation à l'âge du Bronze en France. *In:* J. Guilaine, D. Garcia (dir.), *La Protohistoire de la France*, Paris, Hermann (Histoire et Archéologie),   p.297-309.

Valera A.C. 2015. Social change in the late 3rd millennium BC in Portugal: the twilight of enclosures, *In:* H. Meller, H.W. Arz, R. Jung, R. Risch (dir.), *2200 BC – Ein Klimasturz als Ursache für den Zerfall der Alten Welt? - 200 BC – A climatic breakdown as a cause for the collapse of the old world ?* 7. Mitteldeutscher Archäologen tag vom 23. bis 26. Oktober 2014 in Halle (Saale), Tagungen des Landesmuseums für Vorgeschichte Halle, 12/II, p.409-428.

Vernet J.-L. 1999. Reconstructing Vegetation and Landscapes in the Mediterranean: the Contribution of Anthracology. *In:* P. Leveau, F. Trément, K. Walsh, G. Barker (dir.), *Environmental Reconstruction in Mediterranean Landscape Archaeology,* Oxford, Oxbow Books (The Archaeology of Mediterranean Landscapes 2), p.25-36.

Vital J. 2001. Actualités de l'âge du Bronze dans le sud-est de la France, *Documents d'Archéologie Méridionale* 24, p.243-252.

Vital J. 2004a. Du Néolithique final au Bronze Moyen dans le sud-est de la France : 2200-1450 av. J.-C., *Cypsela* 15, p.11-38.

Vital J. 2004b. Ruptures et continuités du Néolithique final au Bronze ancien dans la vallée du Rhône (France) : nouveaux éléments de compréhension, in « H.-J. Beier, R. Einicke (dir.), *Varia Neolithica III,* Langenweissbach, Beier and Beran. Archäologische Fachliteratur (Beitraäge zur Ur- und Frühgeschichte Mitteleuropas 37), p.251-277.

Vital J., Bouby L., Jallet F., Rey P.-J. 2007. Un autre regard sur le gisement du boulevard périphérique nord de Lyon (Rhône) au Néolithique et à l'âge du Bronze, *Gallia préhistoire* 49, p.1-126.

Vital J., Convertini F., Lemercier O. (dir.) 2012. *Composantes culturelles et Premières productions céramiques du Bronze ancien dans le sud-est de la France. Résultats du Projet Collectif de Recherche 1999-2009,* Oxford, Archaeopress (British Archaeological Reports, International Series 2446).

Walsh K., Berger J.-F., Roberts C.N., Vanniere B., Ghilardi M., Brown A. G., Woodbridge J., Lespez L., Estrany J., Glais A., Palmisano A., Finné M., Verstraeten G. 2019. Holocene demographic fluctuations, climate and erosion in the Mediterranean: A meta data-analysis, *The Holocene* 29, p.864-885.

Wang J., Sun L., Chen L., Xu L., Wang Y., Wang X. 2016. The abrupt climate change near 4,400 year BP on the cultural transition in Yuchisi, China and its global linkage, *Nature. Scientific Reports*, 6:27723 | DOI: 10.1038/srep27723.

Weinelt M., Schwab C., Kneisel J., Hinz M. 2015. Climate and societal change in the western Mediterranean area around 4.2 ka BP. *In:* H. Meller, H.W. Arz, R. Jung, R. Risch (dir.), *2200 BC – Ein Klimasturz als Ursache für den Zerfall der Alten Welt? - 2200 BC – A climatic breakdown as a cause for the collapse of the old world ?* 7. Mitteldeutscher Archäologen tag vom 23. bis 26. Oktober 2014 in Halle (Saale), Tagungen des Landesmuseums für Vorgeschichte Halle, 12/I, p.461-480.

Weiss H. 2015. Megadrought, collapse, and resilience in late 3rdmillenniumMesopotamia. *In:* H. Meller, H.W. Arz, R. Jung, R. Risch (dir.), *2200 BC – Ein Klimasturz als Ursache für den Zerfall der Alten Welt? - 2200 BC – A climatic breakdown as a cause for the collapse of the old world ?* 7. Mitteldeutscher Archäologen tag vom 23. bis 26. Oktober 2014 in Halle (Saale), Tagungen des Landesmuseums für Vorgeschichte Halle, 12/I, p.35-51.

Weiss H. 2016. Global megadrought, societal collapse and resilience at 4.2-3.9 ka BP across the Mediterranean and west Asia, *Pages Magazine* 24, p.62-63.

Weninger B. 2012. Réponse culturelle aux changements climatiques rapides de l'Holocène en Méditerranée orientale. *In:* J.-F. Berger (dir.), *Des climats et des hommes.* Paris, INRAP-La Découverte, p. 171-193.

Weninger B., Easton D. 2017. A Gap in the Early Bronze Age Pottery Sequence at Troy Dating to the Time of the 4.2 ka cal. BP. Event. *In:* F. Höflmayer (dir.), *The Late Third Millennium in the Ancient Near East: Chronology, C14 and Climate Change.* Papers from the Oriental Institute Seminar Held at the Oriental Institute of the University of Chicago, Oriental Institute seminars 11, Chicago, Oriental Institute, p.429-450.

Weninger B., Harper T. 2015. The geographic corridor for Rapid Climate Change in Southeast Europe and Ukraine. *In:* S. Hansen, P. Raczky, A. Anders, A. Reingruber (dir), *Neolithic and Copper Age between the Carpathians and the Aegean Sea. Chronologies and Technologies from the 6thto the 4thmillennium BCE.* International workshop Budapest, 30 March-1 April 2012. Bonn, Rudolf Habelt (Archäologie in Eurasien, 31), p.475-505.

Weninger B., Alram-Stern E., Bauer E., Clare L., Danzeglocke U., Jöris O., Kubatzki C., Rollefson G., Todorova H., Van Andel T. 2006. Climate forcing due to the 8200 cal.yr BP event observed at Early Neolithic sites in the eastern Mediterranean, *Quaternary Research* 66, p. 401-420.

Weninger B., Clare L., Rohling E.J., Bar-Yosef O., Böhner U., Budja M., Bundschuh M., Feurdean A., Gebel H.-G., Jöris O., Linstädter J., Mayewski P., Mühlenbruch T., Reingruber A., Rollefson G., Schyle D., Thissen L., Todorova H., Zielhofer C. 2009. The Impact of Rapid Climate Change on prehistoric societies during the Holocene in the Eastern Mediterranean, *Documenta Praehistorica* XXXVI, p.7-59.

Weninger B., Clare L., Gerritsen F., Horejs B., Krauss R., Linstädter J., Özbal R., Rohling E.J. 2014. Neolithisation of the Aegean and Southeast Europe during the 6600-6000 cal BC period of Rapid Climate Change, *Documenta Praehistorica* XLI, p.1-31.

Wiener M.H. 2013. "Minding the Gap": Gaps, Destructions, and Migrations in the Early Bronze Age Aegean. Causes and Consequences, *American Journal of Archaeology* 117, p. 581-592.

Wiener M.H. 2014 The interaction of climate change and agency in the collapse of civilizations ca. 2300–2000 BC, *Radiocarbon 56 / Tree-Ring Research* 70, p.S1-S16.

Zanchetta G., Regattieri E., Isola I., Drysdale R.N., Bini M., Baneschi I., Hellstrom J.C. 2016. The So-Called "4.2 Event" in the Central Mediterranean and its Climatic teleconnections, *Alpine and Mediterranean Quaternary* 29, 2016, p. 5-17.

Climat et sociétés à l'âge du Bronze en Europe occidentale

Cyril Marcigny[1]

Résumé

Le rythme et la nature des occupations post-néolithiques (entre le IIIe millénaire et le début de notre ère commune) dans un quart Nord-ouest de l'Europe, comprenant la France, le Benelux et le sud de la Grande-Bretagne, a fait l'objet ces vingt dernières années de nombreuses synthèses qui ouvrent sur des observations pertinentes à large échelle géographique. Au sein de cette zone, la Normandie est un échantillon représentatif grâce à de nombreuses fouilles préventives et programmées corrélées à des programmes de recherche sur le paléoenvironnement. Parmi les vestiges identifiés, sur le laps de temps couvrant l'âge du Bronze (compris dans son acceptation large, dès 2500 BCE) et le premier âge du Fer, les fossés, qu'ils appartiennent à des clôtures d'habitat (enceintes, espaces fortifiés ou enclos délimitant les établissements agricoles) ou des planimétries agraires (parcellaires), ont semblé être un bon indicateur de l'emprise humaine sur le milieu et la pérennité, toute relative, mais probablement plurigénérationnelle, des aménagements. Cet indicateur, que l'on pourrait qualifier de proxy pour reprendre un vocabulaire plus usité par nos collègues environnementalistes, est confronté ici à deux autres macro-indicateurs : la société et son degré de complexification observé via le prisme des données funéraires et le climat en prenant appui sur les graphes des variations de la teneur de l'atmosphère en ^{14}C.

Cette analyse prend appui sur les données robustes obtenues en Normandie (à partir de sites calés en chronologie grâce à un nombre important de datations isotopiques) mais ouvre sur l'ensemble de l'espace géographique Manche – Mer-du-Nord. Elle permet de proposer un scénario historique où différentes variables rentrent en jeu.

Abstract

The rhythm and nature of post-Neolithic occupations (between the 3rd millennium BC and the beginning of our common era) in North-Western Europe, including France, Benelux and southern Britain have been the focus of many studies over the past twenty years giving rise to pertinent observations on a large geographical scale. Normandy is representative of this research thanks to the many preventive and programmed excavations, which take into account analyses on the ancient

---

1 Inrap Normandie, Laboratoire d'Archéologie et Histoire Merlat (LAHM), UMR 6566-CReAAH (université de Rennes, Nantes, Le Mans, CNRS, MC), Le Chaos, 14400 Longues-sur-Mer, cyril.marcigny@inrap.fr

environment. Ditches, dating to the Bronze Age (from its beginning around 2500 BCE) and the Early Iron Age, that enclose settlements (enclosures, fortified sites or agricultural establishments) or agrarian plots, seem to be a good indicator of human influence on the environment with installations that probably lasted several generations. We aim to compare this type of indicator or environmental proxy with two other elements: society and its degree of complexity highlighted by funerary data and fluctuations in climate by studying the variations of ^{14}C in the atmosphere.

This analysis uses robust data from Normandy (from sites dated by a large number of isotopic dates) but also takes into account the entire Manche - Mer-du-Nord area. It makes it possible to propose a historical scenario with different variables coming into play.

Introduction

Un travail conduit, depuis une vingtaine d'années, sur le rythme et la nature des occupations protohistoriques post-néolithiques (entre le IIIe millénaire et le début de notre ère commune) de Normandie, a mis en évidence des cycles dans les formes de l'habitat, les structures agraires et plus spécifiquement les planimétries agraires (Marcigny, 2019). Cette lecture a été confrontée à un examen multiscalaire et multiproxy, multipliant les échelles d'études (dans l'Ouest de la France et le sud de l'Angleterre) et les indicateurs (paléoenvironnement, flux sédimentaires, climat), dans le cadre d'une modélisation résolument orientée sur l'observation de trois principales variables -le fossé, le climat et la société- et dans l'objectif de développer un (ou des) scénario(s) historique(s).

Dans le cadre de cet article nous livrons les premiers résultats de ce travail vu sous l'angle d'une restitution du comportement social des hommes, sur la période comprise entre 2500 et 600 BCE, soit des prémices de l'âge du Bronze au début du premier âge du Fer, face aux variations climatiques de la fin du Subboréal et les touts débuts du Subatlantique. Centrées sur la Normandie, où le travail a été conduit dans le détail, les données se veulent être de bons proxies heuristiques pour une partie de l'Europe du Nord-Ouest : Grand-Ouest français, Sud de l'Angleterre, Benelux, voire, mais dans une moindre mesure, le Nord-ouest de l'Allemagne (Figure 1).

Chronométrie

Avant de proposer notre lecture multi-proxy il n'est pas inutile de définir précisément les bornes temporelle et spatiale utilisées ici.

Pour la chronométrie, la plupart des sites normands a fait l'objet de datations radiocarbones pour préciser leurs ancrages chronologiques et confirmer les dates proposées grâce aux différents mobiliers (céramiques principalement mais aussi lithiques et plus rarement métalliques). Pour les sites les plus récemment étudiés (ceux après les années 2000), un protocole

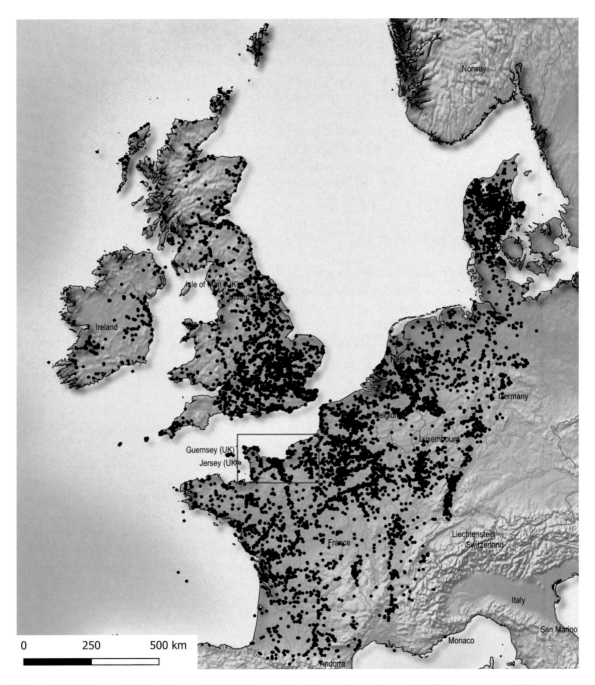

Figure 1. Cartographie des sites protohistoriques dans le quart nord-ouest de l'Europe, encadré en rouge la zone d'étude échantillon : la Normandie (SIG C. Marcigny/F. Audouit /datABronze/Inrap d'après les bases : https://archaeologydataservice.ac.uk/archives/view/prenorwesteurope_2014/ ; https://archaeologydataservice.ac.uk/archives/view/prebritire_2014/ ; https://www.inrap.fr/l-habitat-et-l-occupation-des-sols-l-age-du-bronze-et-le-debut-du-premier-age-du-11867).

similaire de prélèvement/datation a été élaboré dans le cadre de différents programmes de recherche. Ce protocole repose sur un choix strict des échantillons qui devaient correspondre aux processus anthropiques à renseigner. Ils remplissaient les critères d'une grille de fiabilité élaborée en interne et qui complètent l'évaluation de la qualité physico-chimique de la mesure selon les modalités préconisées par les laboratoires de datation. Cette grille prenait en considération la qualité des matériaux analysés (par exemple longévité, problème des effets réservoirs…), celle des contextes de prélèvement (par exemple position fonctionnelle vs détritique) et la qualité de l'association entre les matériaux datés et les assemblages archéologiques à dater. Les datations absolues se devaient aussi d'être nombreuses de manière à obtenir un nombre significatif de résultats permettant d'envisager, à terme, leur traitement statistique avec des méthodes adaptées à la distribution spécifique des densités de probabilité après calibration. Actuellement, c'est e 571 dates qui ont été obtenues sur l'âge du Bronze normand compris dans un sens chronologique large de la deuxième moitié du IIIe millénaire aux premiers siècles du premier âge du Fer.

Bien entendu, la très grande majorité de ces mesures d'âges ont été effectuées sur des échantillons uniques (graines, ossements, charbons, principalement) plutôt que sur des amalgames de matériaux provenant d'échantillons en réalité distincts. Ces mesures ont été réalisées presqu'exclusivement par AMS dans des laboratoires malheureusement distincts (Beta Analytic, Pozdan, Laval…) alors qu'initialement, nous ne voulions travailler qu'avec un seul laboratoire pour garder une homogénéité de traitement des prélèvements.

Ces résultats ont par la suite été utilisés de deux façons : d'abord bruts sans traitement statistique (au-delà de calculs de fréquences qui ont produit la mesure d'âge), puis par inférence bayésienne (à l'aide du logiciel ChronoModel). Pour ces dernières, il est important de garder à l'esprit que le résultat probabiliste qui est proposé a fait l'objet d'un calcul ayant pour base une ou des datations et des informations à priori (contexte, stratigraphie…).

Le modèle fait apparaître un découpage chronologique légèrement différent des chronologies en usage en France (figure 2), cinq grandes séquences peuvent ainsi être dégagées.
- séquence 1 : entre 2500/2300 et 2100 avant notre ère, elle vient coiffer la fin du Néolithique et une très grande partie du Bronze ancien I ;
- séquence 2 : entre 2100/2000 et 1800/1700 avant notre ère, elle couvre la fin du Bronze ancien I et la première moitié du Bronze ancien II ;
- séquence 3, entre 1800/1700 et 1500 avant notre ère, elle correspond aux derniers siècles du Bronze ancien II et au Bronze moyen I ;
- séquence 4, entre 1600/1500 et 1200/1150 avant notre ère, elle concerne le Bronze moyen II et l'étape ancienne du Bronze final (Bronze final I et IIa) ;

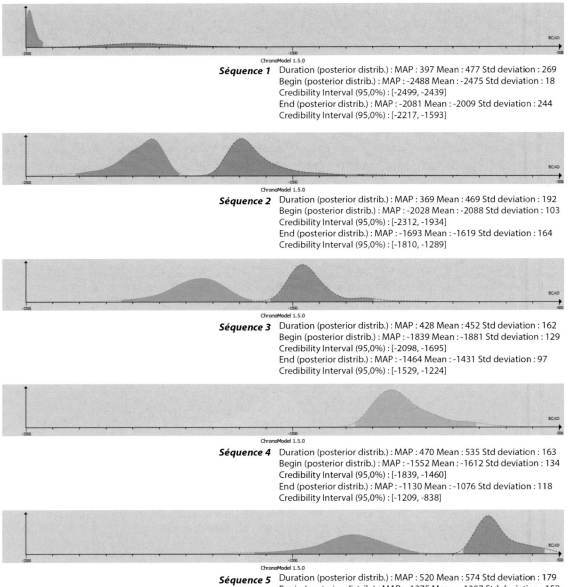

ChronoModel 1.5.0

Séquence 1   Duration (posterior distrib.) : MAP : 397 Mean : 477 Std deviation : 269
Begin (posterior distrib.) : MAP : -2488 Mean : -2475 Std deviation : 18
Credibility Interval (95,0%) : [-2499, -2439]
End (posterior distrib.) : MAP : -2081 Mean : -2009 Std deviation : 244
Credibility Interval (95,0%) : [-2217, -1593]

ChronoModel 1.5.0

Séquence 2   Duration (posterior distrib.) : MAP : 369 Mean : 469 Std deviation : 192
Begin (posterior distrib.) : MAP : -2028 Mean : -2088 Std deviation : 103
Credibility Interval (95,0%) : [-2312, -1934]
End (posterior distrib.) : MAP : -1693 Mean : -1619 Std deviation : 164
Credibility Interval (95,0%) : [-1810, -1289]

ChronoModel 1.5.0

Séquence 3   Duration (posterior distrib.) : MAP : 428 Mean : 452 Std deviation : 162
Begin (posterior distrib.) : MAP : -1839 Mean : -1881 Std deviation : 129
Credibility Interval (95,0%) : [-2098, -1695]
End (posterior distrib.) : MAP : -1464 Mean : -1431 Std deviation : 97
Credibility Interval (95,0%) : [-1529, -1224]

ChronoModel 1.5.0

Séquence 4   Duration (posterior distrib.) : MAP : 470 Mean : 535 Std deviation : 163
Begin (posterior distrib.) : MAP : -1552 Mean : -1612 Std deviation : 134
Credibility Interval (95,0%) : [-1839, -1460]
End (posterior distrib.) : MAP : -1130 Mean : -1076 Std deviation : 118
Credibility Interval (95,0%) : [-1209, -838]

ChronoModel 1.5.0

Séquence 5   Duration (posterior distrib.) : MAP : 520 Mean : 574 Std deviation : 179
Begin (posterior distrib.) : MAP : -1275 Mean : -1307 Std deviation : 158
Credibility Interval (95,0%) : [-1629, -1112]
End (posterior distrib.) : MAP : -767 Mean : -733 Std deviation : 78
Credibility Interval (95,0%) : [-840, -579]

• séquence 5 (la dernière), entre 1200/1150 (date légèrement faussée dans le modèle et étendue à 1300) et 750 avant notre ère, elle correspond à l'étape moyenne et finale du Bronze final et au début du premier âge du Fer (couvrant ainsi la transition subboréal/subatlantique).

Les cinq séquences se calent assez bien sur la chronologie néerlandaise et vient à partir du Bronze moyen se synchroniser avec les chronologies en usage (figure 3).

Figure 2. Les bornes chronologiques des 5 séquences temporelles proposées à l'issue de l'analyse bayésienne (logiciel Chronomodel v. 1.5., réal. C. Marcigny).

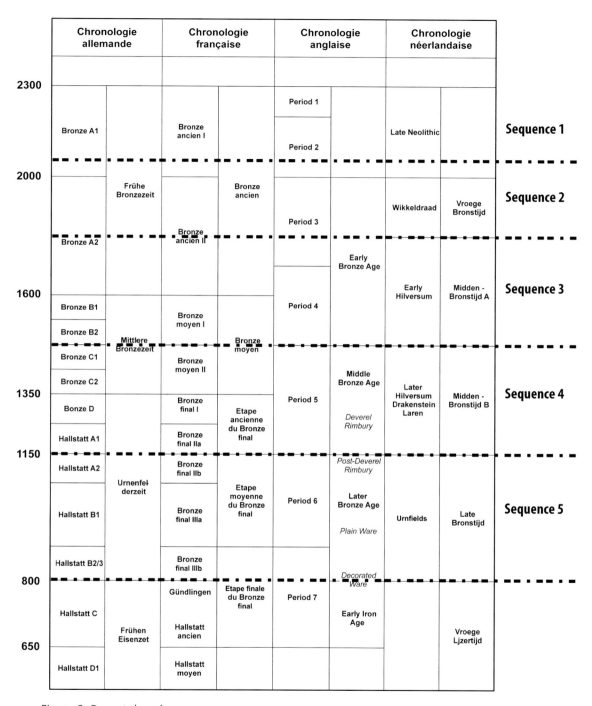

Figure 3. Report des séquences
déterminées par le modèle
bayésien sur le tableau synchronisé
des différentes chronologies
utilisées dans le Nord-Ouest de
l'Europe (DAO, C. Marcigny).

Les indicateurs

La clef d'entrée de notre modélisation est le fossé en tant que signature de l'appropriation foncière (via les fondations agraires : parcellaires ou clôtures) confronté à d'autres indicateurs qui pourraient paraître externes.

Le fossé : on creuse ou on ne creuse pas.

Le fossé nous permet de quantifier l'investissement (ou l'absence d'investissement) sur quatre types d'aménagements classiques en Protohistoire : les enceintes, les espaces fortifiés, les enclos délimitant les établissements agricoles et les planimétries agraires (ces deux derniers correspondant plus aux structures agraires). La profondeur des fossés, leurs longueurs, leurs présences ou non sont autant de critères mis à contribution pour interroger les données, en partant du présupposé que le fossé est le témoignage de la propriété. Seules les délimitations appartenant aux domaines funéraires -enclos de plan circulaire associés à des tombes ou enclos ovalaires et langgraben- n'ont pas été retenues ici, leurs significations sociales étant invariablement autres.

Rythme des installations

La chronologie de chacun de ces sites est renseignée sous forme numérique (borne_inf, borne_sup) dans une base de données interactive mise en place dans le cadre d'un projet d'enquête nationale sur l'âge du Bronze, élaborée et pilotée depuis 2005 par L. Carozza, C. Marcigny et M. Talon (Carozza *et al.* 2017). Elles sont ensuite projetées sur un diagramme de densités cumulées, via un calcul automatisé par une macro développée sous Excel (Marcigny et al., 2018 et 2020). Selon le principe du calcul de probabilité de densité, chaque occupation a la valeur 1, quelle que soit sa durée, son amplitude étant considérée comme constante. Cette valeur 1 est divisée par l'intervalle de temps qui constitue sa datation. La somme de toutes ces valeurs forme l'histogramme de densités cumulées (figure 4). Les datations précises sont bien visibles mais pondérées et les datations imprécises sont tout de même représentées, même si leur valeur est faible. Dans les règles de l'art, un lissage au moyen d'une moyenne mobile de rang 2 a été appliqué aux courbes pour éliminer les micro-variations sans valeur et pour améliorer la lecture des histogrammes.

La lecture sur le temps long de ces aménagements est similaire sur une vaste zone bordant la Manche et le sud de la Mer-du-Nord (figure 5, Bradley *et al.* 2016). Les courbes obtenues sur la Normandie peuvent donc être considérées comme un bon échantillon représentatif du rythme et de la nature des habitats contemporains du XXVIe au VIIe siècle avant notre ère sur le quart nord-ouest de l'Europe.

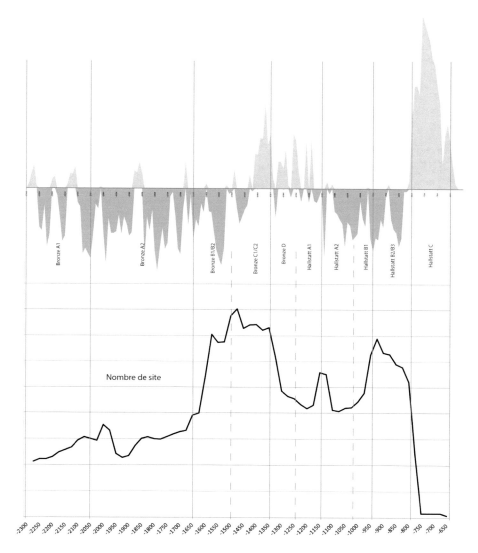

Figure 4. Représentation graphique des densités de sites (DHC) pour la Normandie corrélée avec l'indicateur climatique (variations du taux de 14C résiduel dans l'atmosphère) (DAO. F. Audouit/ datABronze/ Inrap).

Les planimétries agraires : de la ferme aux champs

Parmi les fossés, les planimétries agraires constituent un fait nouveau. En effet, à l'extrême fin du IIIe millénaire, entre 2200 et 2100 avant notre ère, une nouvelle forme d'organisation du paysage fait son apparition en Europe de l'Ouest : la planimétrie agraire.

Comme le soulignait J. Guilaine dans les années 1980, dans le domaine de l'histoire agraire « tout est à peu près à faire » (Guilaine, 1988). Il est vrai que les archéologues hexagonaux sont restés très longtemps en dehors du débat (Guilaine, 1991) et ce sont les découvertes de l'archéologie préventive à l'aube du XXIe siècle qui ont permis de faire débuter la thématique ; et ce encore bien timidement. En effet, l'âge du Bronze et le premier âge du Fer n'ont jamais été vraiment envisagés par les archéologues et les

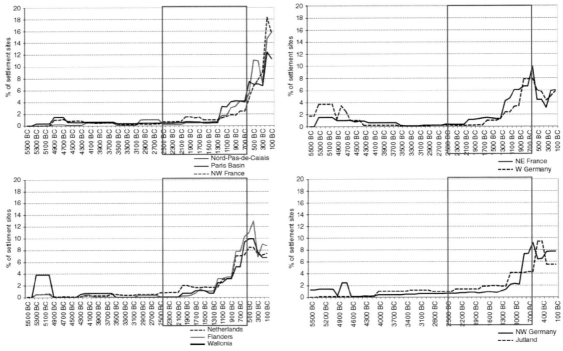

Figure 5. Courbes pour le quart Nord-Ouest de l'Europe (d'après Bradley *et al.* 2016, modifié).

archéogéographes travaillant en France comme un moment important en matière de construction des paysages (à l'exception notable de G. Chouquer ; Chouquer, 2007) et de restitution de l'histoire rurale. Les données sont toutefois conséquentes à l'échelle européenne (Bowen, 1961 ; Bradley, 1978 ; Drewett, 1978 ; Yates, 2007). Les collègues anglo-saxons, par exemple, intègrent, depuis les années soixante, l'âge du Bronze dans l'histoire des planimétries agraires (Johnston, 2000 ; Wickstead, 2007, 2008). Il en est de même des chercheurs des Pays-Bas (Louwe Kooijmans *et al.* 2005) ou de certaines régions d'Allemagne et d'Italie (Kristiansen, 1998 ; Kristiansen, Larson, 2005). En France, les données acquises dans l'Ouest, et plus particulièrement sur le littoral de la Manche (normand ou breton), sont aussi de plus en plus importantes et montrent un espace rural innovant, en pleine mutation, avec la fondation des premiers systèmes parcellaires, le développement des pratiques agraires, la constitution d'établissements agricoles, l'apparition de villages d'agriculteurs... Ces acquis illustrent bien cette modification des systèmes de production agricoles au cours du IIe millénaire, en rupture avec le Néolithique et amorçant d'une certaine manière le « boom » agraire de l'âge du Fer. Ils permettent, à l'instar des autres régions européennes, de dresser un portrait des paysages ruraux et des structures agraires de l'âge du Bronze de 2300 à 800 avant notre ère, à confronter aux incidences climatiques comme l'ont proposé récemment les archéologues œuvrant sur le second âge du Fer et l'Antiquité (Blancquaert *et al.* 2012).

Dans la zone qui nous intéresse ici, en Manche – Mer-du-Nord, en particulier dans le Sud de l'Angleterre et en Normandie, lors d'une plus forte densité d'occupation qui se confirmera au cours du Bronze moyen I (Marcigny, 2012a), de vastes espaces agraires vont être aussi mis en valeur et les premières planimétries agraires dressées (Fowler, 1971 ; Bradley 1978). Ces parcellaires sont peu étendus et forment des systèmes réticulés où se déplacent des établissements agricoles reliés par des chemins, qu'il est possible d'étudier comme un tout : de la ferme aux champs (Marcigny, Ghesquière, 2008 ; Marcigny, 2012b).

Il est ainsi aujourd'hui possible, grâce aux nombreuses fouilles de ces vingt dernières années en Normandie (et en élargissant la focale aux régions limitrophes : nord Bretagne et Sud de l'Angleterre), de dresser la trame chronologique de ces implantations agraires et d'en saisir l'organisation au fil du temps ; permettant même bien souvent, sur un même site, de mettre en phase les différentes modifications structurelles de l'organisation agraire, sur parfois presque un demi millénaire (comme à Tatihou, Marcigny, Ghesquière, 2003).

Dans les graphes proposés plus bas, sous le terme de planimétrie agraire, et à l'instar de ce que propose C. Reynaud (Reynaud, 2003), nous rassemblerons les différents aménagements liés à la mise en valeur et à l'exploitation du territoire (territoire devenu probablement terroir, si l'on considère la fixation de l'habitat comme un facteur décisif dans l'apparition de la notion de finage et de terroir, Lebeau 2000). Il s'agira principalement de systèmes fossoyés (les seules structures observables par les archéologues dans ces régions dont les sols sont érodés), de différentes natures et fonctions, certains étant liés à des voiries, d'autres à ce que nous regrouperons sous le terme parcellaire pour plus de commodité (sans pour autant préjuger dans un premier temps de la dimension foncière de ces structures).

La société

L'approche de la société porte exclusivement sur un critère qui est la complexification sociale. Elle est approchée grâce à une grille de lecture de la stratification sociale à partir des données funéraires via deux indicateurs : l'accès au monumentalisme (dimensions en plan des structures funéraires, volume des tertres, aménagements spécifiques –caveau, coffre, etc.-, …) et les viatiques (importance en nombre, types de matériaux, …).

Pour le Campaniforme et l'âge du Bronze ancien, la courbe prend principalement appui sur les résultats obtenus dans quelques secteurs géographiques ciblés : le Finistère breton, la Normandie occidentale et la région du Wessex en Grande-Bretagne (Needham, 2000). D'autres indicateurs montrent une plus forte hiérarchisation dans le Pays-de-Galles ou la Cornouaille (Needham, 2000) mais pour l'heure les régions plus orientales de notre zone d'étude, vers le Nord de la France ou le

Benelux, semblent moins dissertes, à l'exception notable des grands enclos funéraires (Desfossés *et al.* 2000 ; connus aussi dans le Kent, Grinsell, 1992) qui pourraient être un des indices de cette forme de complexification. Dès le Bronze moyen, au détour du XVIIe siècle avant notre ère, les données sont moins disparates et unissent à nouveau la zone Manche - Mer-du-Nord. On retrouve ainsi sur l'ensemble de ce secteur des sites funéraires où l'accès aux monuments devient la norme et où les distinctions sociales semblent effacées. Ils réapparaitront et sous une forme moins forte qu'au Bronze ancien 2 à la fin de l'âge du Bronze, à partir des XIe-Xe siècles.

Climat

En l'absence de données régionales exploitables sur le laps de temps retenu ici, le choix s'est porté sur l'utilisation des variations de la teneur de l'atmosphère en [14]C résiduel (Stuiver *et al.* 1998), considérée comme un enregistrement des fluctuations de l'activité solaire (Hoyt et Schatten, 1997, Bond *et al.*, 2001) et donc, comme un bon indicateur empirique des variations du climat (Magny, 1993). Cette première estimation météorologique a été confrontée lorsque c'était possible aux données issues des recherches paléoenvironnementales (récemment modélisées pour le Bassin parisien et le Massif armoricain, David, 2014), de manière à valider les variations climatiques proposées.

L'objet de notre propos n'est bien entendu pas de s'engager dans une démarche déterministe, mais les oscillations climatiques, au même titre que d'autres marqueurs, peuvent être un des éléments moteurs des mutations sociales de ces sociétés essentiellement agro-pastorales dont les pratiques agraires et donc, les moyens de subsistance, peuvent être fortement impactés par les conditions météorologiques. Cette approche fait écho aux travaux bien connus de E. LeRoy-Ladurie ou V Schnirelman (LeRoy-Ladurie, 1967 ; Schnirelman, 1992) pour des périodes plus récentes qui ont bien mis en exergue les dynamiques de peuplements mais aussi les conséquences politiques des périodes de disette ou de famine.

Si le fait politique échappe à l'analyse pour les périodes qui nous préoccupent, nous partons du postulat, comme déjà souligné dans une précédente étude (Marcigny, 2012b), que les crises, et en particulier les crises environnementales, ont eu des répercussions sur les systèmes techniques et sociaux et qu'à ce niveau, le protohistorien a une certaine visibilité qui lui permet, pour peu que l'on examine ces systèmes sur la longue durée, de sentir les évolutions conjointes des sociétés et de leurs contextes   environnementaux.

La confrontation des cinq séquences chronologiques déduites du modèle bayésien avec la courbe des variations de la teneur de l'atmosphère en [14]C résiduel (figure 6) montre un certain synchronisme avec de probables

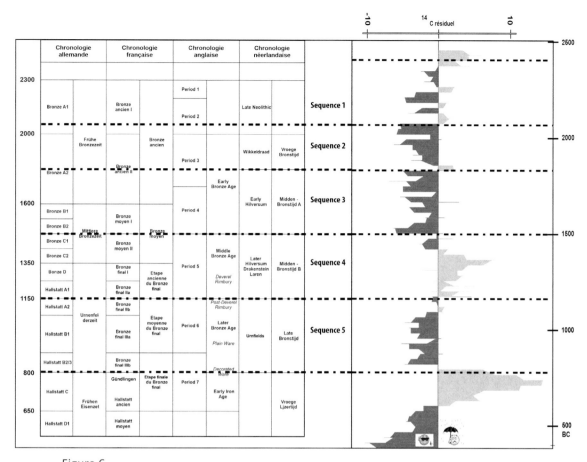

Figure 6. Corrélations entre le climat et les différentes séquences chronologiques (d'après Stuiver et al. 1998 ; DAO M. Magny modifié C. Marcigny).

fluctuations climatiques. La première variation, la fin du Néolithique (vers 2500/2300 avant notre ère, début de la séquence 1), est plutôt favorable. La seconde est centrée autour d'un épisode défavorable (le fameux *4200 BP event*, début de la séquence 2), avant une longue période où l'activité solaire est plutôt bonne (entre 2100 et 1500 avant notre ère, séquences 2 et 3), juste ponctuée d'un court épisode de péjoration autour de 1800 avant notre ère (au début de la séquence 3). La séquence 4 correspond peu ou prou à la péjoration climatique du Bronze moyen II et du début du Bronze final (Magny *et al.* 2007). La dernière séquence correspond à une période favorable avant la péjoration de la transition Subatlantique/Subboréal (Van Geel, Magny, 2002) qui vient coiffer les premiers siècles du premier âge du Fer.

Confrontation des indicateurs

La confrontation de ces trois indicateurs/proxies permet de faire des constats à partir des structures construites grâce à des fossés (figure 7).

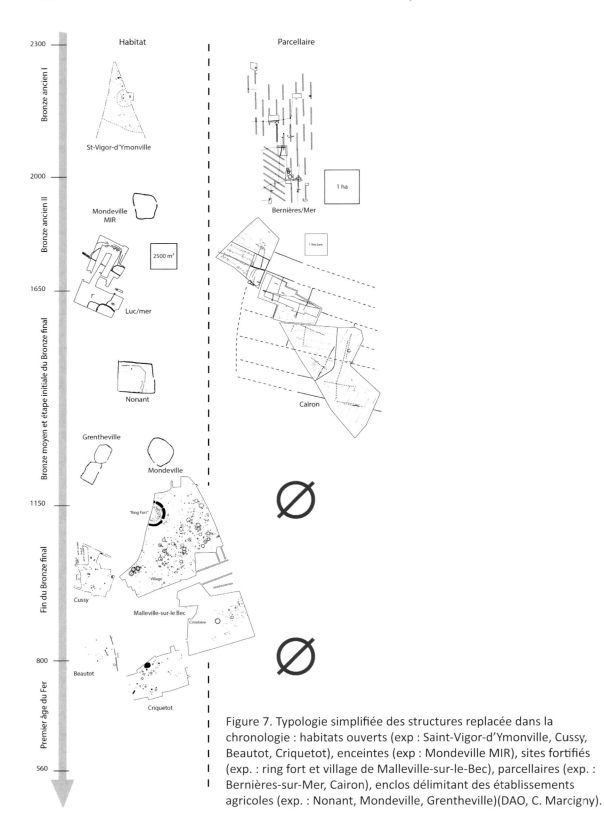

Figure 7. Typologie simplifiée des structures replacée dans la chronologie : habitats ouverts (exp : Saint-Vigor-d'Ymonville, Cussy, Beautot, Criquetot), enceintes (exp : Mondeville MIR), sites fortifiés (exp. : ring fort et village de Malleville-sur-le-Bec), parcellaires (exp. : Bernières-sur-Mer, Cairon), enclos délimitant des établissements agricoles (exp. : Nonant, Mondeville, Grentheville)(DAO, C. Marcigny).

Les enceintes et les sites fortifiés

Les enceintes et sites fortifiés forment une première classe de site, souvent associée à une plus forte stratification sociale. Pour l'âge du Bronze et le premier âge du Fer, ce postulat paraît valide au regard des vestiges souvent collectés dans les comblements des fossés (mobilier de « prestige », artisanat spécialisé, ...) et de l'investissement collectif nécessaire pour concevoir ces habitats délimités par des fossés souvent larges (voire très larges pour les sites fortifiés) et profonds à très profonds.

Confrontés à notre lecture de la hiérarchisation de la société, pas de grandes surprises, ces sites sont identifiés dans il phases contemporaines de la présence des grands tumuli du Bronze ancien 2 ou de tombes privilégiées des étapes moyenne et surtout finale du Bronze final (figure 8).

Confronté à l'estimation du climat, le graphe des enceintes montre des courbes qui font écho aux phases favorables ou défavorables. Les enceintes les plus anciennes apparaissent ainsi à la sortie de l'évènement 4200 BP (autour des XXIIe-XXIe siècles), à la fin du « phénomène » campaniforme. Il s'agit d'une première vague de construction qui semble marquer le pas au début du XXe siècle, pour prendre véritablement son essor entre la deuxième moitié du XXe siècle et le XVIIIe siècle. Ce deuxième moment est compris entre deux petites péjorations climatiques dont il est difficile de savoir si elles ont eu un impact véritable sur la société ; bien qu'il est séduisant de penser celle du XVIIIe/XVIIe siècle comme une phase qui aurait accompagnée (voire provoquer) la fin des enceintes. Ce n'est toutefois pas l'hypothèse que nous retiendrons ici (cf. en conclusion).

Les fortifications (de hauteur ou en méandre, auxquelles on rajoutera les ring forts), qui forment le deuxième groupe de cette classe de site, apparaissent bien plus tard à partir de la deuxième moitié du XIIe siècle et deviendront très présents dans le paysage du Xe siècle au tout début du premier âge du Fer (première moitié du VIIIe siècle). Ici la variable climatique semble avoir eu une incidence sur la mise en place de ce type de sites. Ils se développent après la forte péjoration comprise entre la fin du XVe et le XIIe siècle (très forte en début de séquence, au Bronze moyen II) et disparaitront lors du passage entre le Subboréal et Subatlantique. D'autres évènements sont connus sur ces deux laps de temps et montrent les tensions politiques et économiques qui teintent les rapports sociaux du Bronze final : la présence d'armement en très grands nombres (défensifs et offensifs), de conflits interpersonnels lors des XIIIe et XIIe siècles et les changements dans les pratiques agro-pastorales des IXe et VIIIe siècles.

Les établissements agricoles

Les établissements agricoles et plus particulièrement la présence ou non de systèmes de clôture fondés dans le sol (sous la forme de fossés) pour délimiter l'habitat est aussi une structuration qui peut être interrogée.

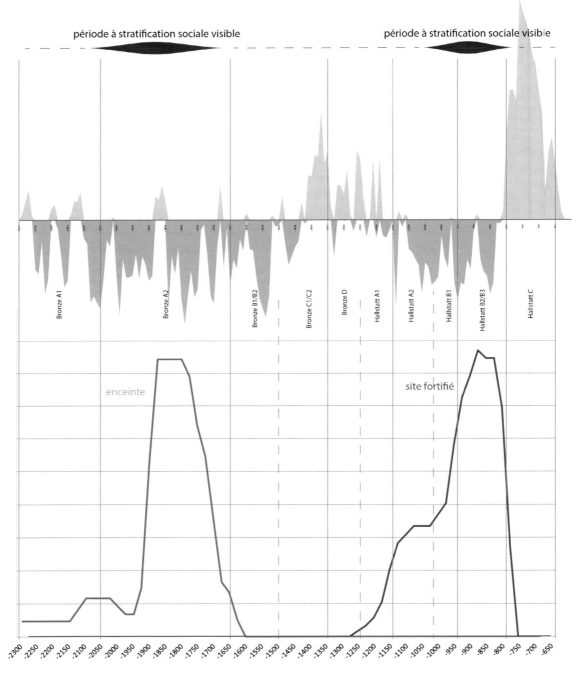

Figure 8. Confrontation entre les courbes des densités de sites –enceinte et site fortifié- et le climat, en haut évaluation des séquences à stratification sociale (DAO, C. Marcigny).

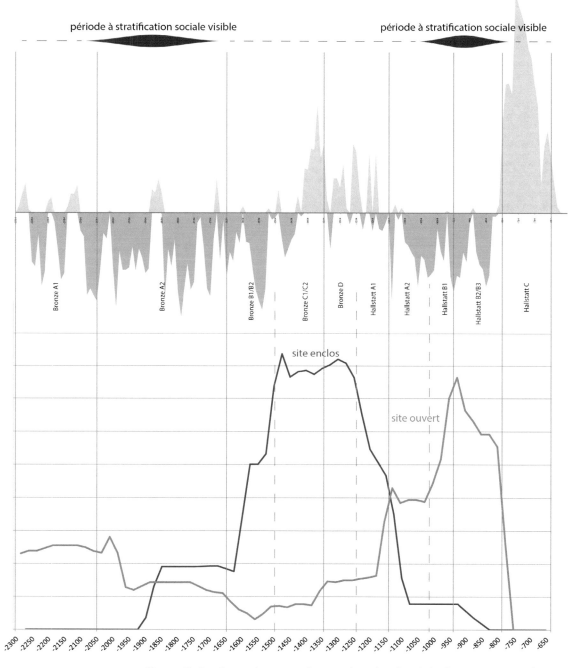

Figure 9. Confrontation entre les courbes des densités de sites -site enclos, site ouvert- et le climat, en haut évaluation des séquences à stratification sociale (DAO, C. Marcigny).

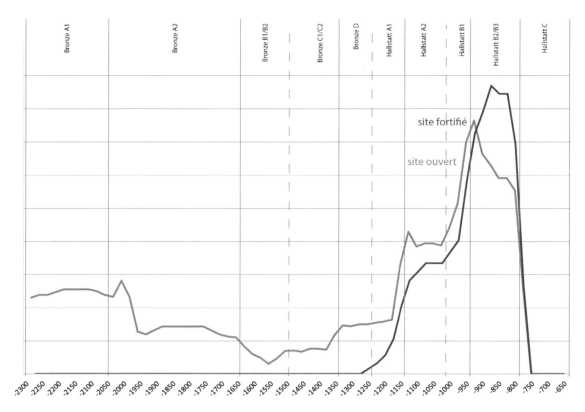

Figure 10. Confrontation entre les courbes des densités de sites : site fortifié, site ouvert (DAO, C. Marcigny).

Les habitats ouverts, sans clôture fondée, sont présents sur deux plages de temps principalement : en début de séquence (dans le prolongement du Néolithique) jusqu'au XXe siècle, puis durant l'âge du Bronze final et le début de l'âge du Fer, à partir des XIIIe-XIIe siècles (figure 9). On retrouve ici des phases déjà évoquées précédemment ; les établissements agricoles non clos répondant assez bien aux périodes à stratification sociale, le rythme de la courbe obtenue étant d'ailleurs très proches de celle des sites fortifiés (figure 10).

Les fermes pourvues de clôture présentent une autre signature ; leur rythme d'implantation correspond à une séquence qui s'ouvre après la disparition des enceintes (figure 11). Elles sont très nombreuses à la fin du Bronze ancien II et au Bronze moyen I, mais elles amorcent un déclin en phase avec la péjoration climatique du Bronze moyen II (au XVe et XIVe siècle), pour reprendre de la vigueur ensuite avant de disparaître au XIIe siècle lorsque les habitats ouverts prennent le relais.

La clôture ou l'absence de clôture est ici un élément social important et marque probablement une volonté de marquer sa propriété à des périodes où il est possible pour les fermiers de le faire (et d'y consacrer un investissement humain).

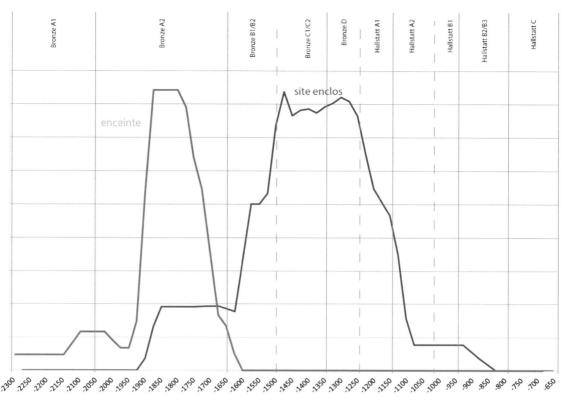

Figure 11.
Confrontation
entre les courbes
des densités de
sites : enceinte,
site enclos (DAO,
C. Marcigny).

Le rôle de la clôture et du soin apporté à cette matérialisation de l'espace sont en effet deux faits suffisamment importants pour être soulignés. Claude Mordant s'est déjà posé la question en 2008 (Mordant, 2008, fig. 74) et a bien montré, à travers les exemples de Nonant (Marcigny, 2005) et de Florémont dans les Vosges (Buzzi *et al.* 1994), les similitudes structurelles de ces établissements ruraux, au-delà de ces systèmes de délimitation si caractéristiques de l'Ouest de la France et du Sud de l'Angleterre. Il propose d'y voir ici une sorte de prémices de l'enbocagement, caractéristique du domaine armoricain. Sans réfuter cette proposition, nous proposons de voir dans cette volonté de clore son espace domestique, un phénomène plus en lien avec la pleine propriété et une certaine forme d'émancipation des fermiers du Bronze moyen et de l'étape ancienne du Bronze final par rapport à leurs prédécesseurs. Bien entendu, cette assertion historique est plus que difficile à mettre en évidence sans d'autres sources que l'archéologie, mais il est par contre certain que cette forme d'établissement va disparaître aux étapes moyenne et finale du Bronze final pour laisser la place à un établissement rural plus en phase avec ce que l'on connaît plus à l'est de la France. La presque stricte exclusion de ces deux formes d'habitat est donc un élément fort qui méritera d'être examiné sous l'angle social, ce que nous tenterons en fin d'article.

Les planimétries agraires

Dernière classe de site, les planimétries agraires fondées dans le sol à l'aide de fossés forment un volet particulièrement intéressant de l'étude de l'âge du Bronze car elles constituent une véritable innovation dans les pratiques agraires européennes, comme souligné plus haut, et permettent d'envisager un rapport des sociétés au « foncier » totalement différent des pratiques du Néolithique (Brun, Marcigny, 2012). En Normandie, une politique très volontariste de datations isotopiques sur ce type d'aménagement permet d'avoir une bonne vision du rythme de création et de restructuration de ces parcellaires (figure 12). La courbe obtenue montre deux séquences. La première est centrée sur le Bronze ancien II, elle s'affaisse à partir du XVIIIe siècle, pour reprendre ensuite dans la deuxième moitié du XVIIe siècle (séquence centrée sur le Bronze moyen I). La création de nouveaux parcellaires est progressivement abandonnée au Bronze moyen II, pour totalement disparaître durant le XIIe siècle. Durant cette phase des processus de condamnation ont même été identifiés sur certains sites (Omonville-la-Petite, Bernières-sur-Mer, Marcigny, 2017), accréditant l'hypothèse d'une refonte des pratiques liées au « foncier » et la propriété (changement social et économique probablement très important).

Confrontée aux indicateurs sociaux et climatiques, la courbe montre aussi de bonnes correspondances. La première séquence à parcellaire se cale peu ou prou sur la période à stratification sociale visible. La seconde s'achève alors que débute la péjoration du Bronze moyen II. Cet abandon généralisé de la construction parcellaire creusée trouve aussi des parallèles outre-Manche avec la désertion des terres du Dartmoor centrée sur les dates 1395 et 1155 avant notre ère, bien illustrée sur ce site grâce aux horizons organiques (tourbes) qui recouvrent les anciennes limites parcellaires (Fyfe *et al.* 2008).

Les planimétries agraires étant un très fort investissement pour les communautés agricoles (Marcigny, 2017), nous avons tenté d'évaluer cet effort en distinguant deux volumes de creusement à partir de la surface topographique des sites : des profonds (entre 1,50 et 2,50 m de profondeur pour autant d'ouverture) et des plus légers, correspondant à la fondation de nouvelles planimétries ou l'entretien des limites parcellaires déjà existantes (entre 1,00 m et 1,20 m de profondeur). Ces données réintroduites dans la base de données montrent très clairement deux moments différents. Des creusements profonds en début de séquence avec un pic de fréquence qui colle à celui des enceintes (figure 13), puis des creusements plus légers dans une phase où la stratification sociale semble moins complexe (développement en phase avec l'apparition des fermes encloses).

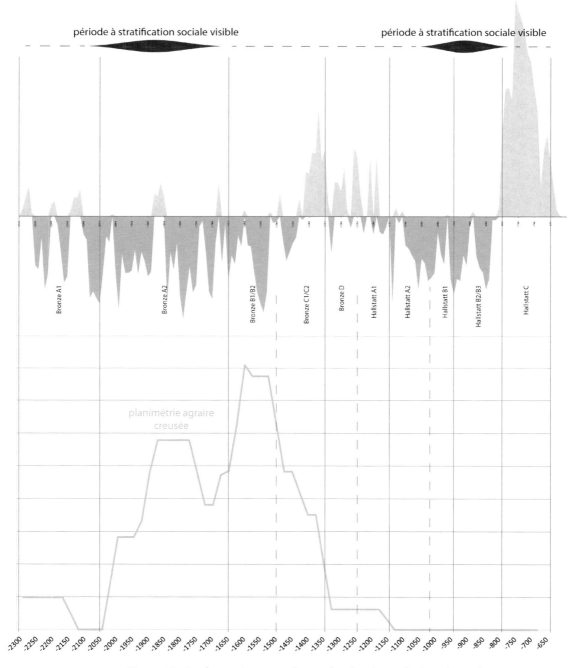

Figure 12. Confrontation entre la courbe des densités des planimétries agraires et le climat, en haut évaluation des séquences à stratification sociale (DAO, C. Marcigny).

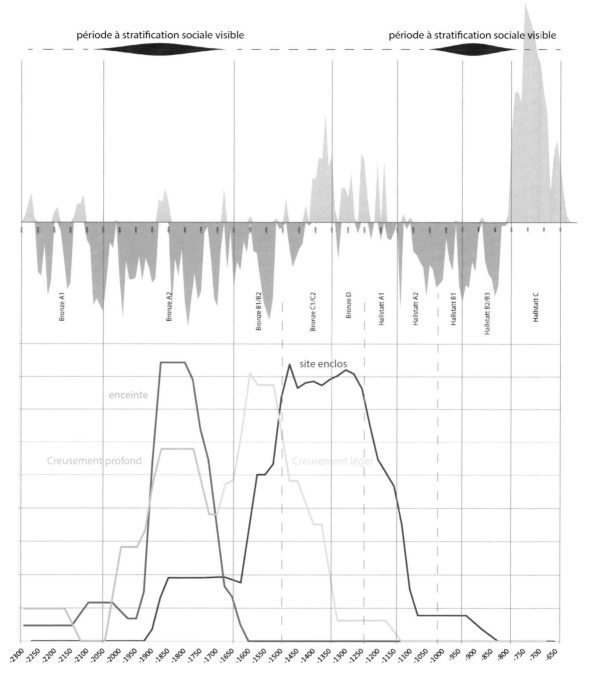

Figure 13. Confrontation entre les courbes des densités de sites -enceinte, site enclos- et celles des planimétries agraires creusées profondément ou non, en haut le climat et l'évaluation des séquences à stratification sociale (DAO, C. Marcigny).

Conclusion : un récit historique

À l'issue de ce travail, liminaire et encore imparfait, les trois indicateurs/ proxies retenus et leur lecture conjointe permetent de faire des premiers constats et de proposer quelques pistes de réflexions visant à la restitution d'un récit historique. Ce dernier prend invariablement appui sur une problématisation simple (schéma : faits vs résultats) et qui intègre des explications historiques justifiées par les preuves dégagées par l'analyse archéologique. Bien entendu, l'exercice reste périlleux, l'étude des faits sur des périodes sans sources écrites étant bien souvent sujette à caution mais, à notre sens, les premiers résultats exposés ici peuvent être diserts et être mis à contribution pour peu qu'ils soient examinés à l'aune de l'ensemble des facteurs disponibles (paramètres sociaux, économiques, voire politiques) dont le climat (sans cacher pour autant la complexité de la dynamique du climat et de l'évolution des sociétés).

Ce bilan s'appuie sur les propositions faites autour de trois échelons territoriaux emboîtés (Marcigny, 2008), formant autant de niveaux d'appropriation de l'espace : le territoire immédiat (de la ferme au village), le territoire biologique (du terroir à la chefferie) et le territoire culturel (sphère des relations économiques et politiques). Cette notion d'appropriation ou de propriété est en effet un des éléments fondateurs de l'âge du Bronze à la transition Bronze ancien I et II. Il apparaît en effet que la fondation de larges planimétries agraires (profondément creusées dans le sol, rappelons-le), à l'échelle du territoire biologique, signe la mainmise sur le territoire d'une nouvelle classe sociale qui se met en place au sortir des péjorations climatiques de la fin du Néolithique. Ces élites pourraient fonder, pour partie, leur pouvoir sur le contrôle des terres et sur des conditions climatiques très favorables à l'agriculture. Dans ce contexte, le lien avec les planimétries agraires devient évident, puisque ces grands travaux visent très certainement à s'accaparer et à verrouiller certains terroirs reconnus encore aujourd'hui pour leur forte potentialité agricole (Dubreuil, 1992). Les enceintes, implantées sur des points topographiques bien visibles dans le paysage et systématiquement en marge des trames parcellaires, pourraient alors être le siège de ces élites ; les zones dédiées à l'agriculture étant plutôt occupées par de petites unités domestiques sans clôture, sans propriété (mise en fermage des terres ?).

À partir de cette date, certains territoires vont être comme figés et perdurer sur un laps de temps conséquent (près de cinq cents ans pour Tatihou, par exemple, Marcigny, Ghesquière, 2003). Seule l'occupation des espaces change de nature. On passe ainsi d'un habitat principalement ouvert au Bronze ancien à un habitat fermé, délimité par de profonds fossés au cours du Bronze moyen I, où le sentiment d'appropriation doit être plus fort. Là encore, on a affaire à un élément historique important qui va probablement voir une classe de fermiers (sans doute propriétaires) s'installer au sein de parcellaires antérieurs. Cette plus forte « démocratisation » (ou ce tassement de la hiérarchisation sociale) au cours du Bronze moyen I se ressent aussi à la lecture des données funéraires avec une plus forte accession aux tumulus, jusqu'alors réservés à des personnages importants, formant ainsi de vastes ensembles tumulaires. Ce mouvement voit les élites disparaître et les enceintes abandonnées au profit de réseaux de fermes formant de petits hameaux. La trame générale des planimétries agraires n'est toutefois pas remise en question. Il semble que les valeurs identificatrices et structurantes de cette forme de gestion du terroir agricole restent de mise, assurant alors sa pérennité au-delà de la disparition des élites, conservant ainsi ses fonctions stabilisatrices, de permanence et d'identification. Ces espaces sont alors utilisés sur plusieurs générations sans modifications importantes de leurs plans d'ensemble qui devient, au-delà de l'outil de gestion et de production, un socle instituant et constituant pour la communauté. Durant la même phase, de nouvelles terres sont occupées et mises en valeur selon des procédés différents (fossés parcellaires peu profonds, discontinus) mais qui visent aussi à cette stabilité sur le long terme.

Ces territoires vont perdre de leur unité à la fin du Bronze moyen et durant l'étape initiale du Bronze final, lors de la péjoration climatique centrée sur le XIVe siècle. À cette époque, on va assister à l'abandon de secteurs densément occupés et à la fondation de nouveaux territoires, sans le lien durable que constitue la trame parcellaire.La pérennité des zones occupées n'est ainsi plus assurée. L'ensemble de cette phase de transition, entre Bronze moyen et final, est marqué par une instabilité climatique, sociale et un probable stress économique. Ce stress trouverait son incarnation, à partir du XIIIe siècle, à travers le développement de l'armement défensif et offensif (Uckelmann, 2008) et les traces de violence qui impactent la société du XIIIe siècle avant notre ère. Ces dernières, moins évidentes à saisir en archéologie, ont toutefois été identifiées en Grande-Bretagne à Tormarton, (Osgood, 2006), à Tollense, en Allemagne (Jantzen *et al.* 2011) ou en Norvège avec les charniers de Sund « Nord-Trøndelag » (Fyllingen, 2003).

Cette plus grande instabilité se concrétise par de petits habitats ouverts dont la durée d'occupation est courte, et qui se déplacent dans le territoire, peut-être au gré de l'épuisement des sols. À côté de ces établissements qui forment un premier niveau d'intégration, on retrouve des habitats groupés (villages, Rathbone, 2013), des sites de hauteur fortifiés, des ring forts et des territoires fortifiés (dike), dont la durée de fréquentation est bien plus importante, (sur plusieurs générations) qui polarisent une partie des activités artisanales. L'ensemble de ces composantes constituent alors un territoire mouvant, bien moins réifié qu'au cours du Bronze ancien/moyen, où les notions de protection (au sein de l'habitat groupé, par exemple) et de défense (derrière le rempart des sites de hauteur ou les dikes) deviennent importantes. Ce mouvement, qui s'amorce au cours du Bronze final IIa, va bien souvent s'effondrer durant les premiers temps du premier âge du Fer, lors de la péjoration climatique accompagnant le passage Subboréal/ Subatlantique.

On assiste alors à une recomposition de la société au cours du Hallstatt ancien, avec de nouveaux habitats (ouverts) une densité moindre des sites et de nouvelles pratiques agricoles qui collent bien aux conditions environnementales des VIIIe et VIIe siècles. C'est ainsi que, dans l'Ouest de la France, l'orge nue est peu à peu abandonnée au profit de l'orge vêtue (Matterne, 2001). Outre-Manche, dans le Sud de l'Angleterre, le même phénomène se produit, à la même époque, où le déclin de l'orge nue coïncide avec l'enregistrement du seigle dans les spectres carpologiques (Harding, 1989). Ces mutations des paysages ruraux se concrétisent à la fin du IXe siècle mais s'amorcent déjà au cours de l'étape moyenne du Bronze final. Elles semblent prendre une nouvelle ampleur au début de l'âge du Fer, peut-être lors de troubles économiques importants liés pour partie à des conditions de production agricole moins efficientes. Il faudra attendre la deuxième moitié du VIIe siècle et le VIe siècle pour que les campagnes normandes reprennent de la vitalité autour de nouveaux établissements agricoles enclos (Ghesquière, Marcigny, 2019), c'est durant cette phase, où le climat devient plus clément, que l'on va voir réapparaitre des réseaux viaires et de nouvelles planimétries agraires.

Bibliographie

Bond G., Kromer B., Beer J., Muscheler R., Evans M.-N., Showers W., Hoffmann S., Lotti-Bond R., Hajdas I., Bonani G. 2001. Persistent solar influence on North Atlantic climate during the Holocene, *Science*, 294, p. 2130–2136.

Bowen H.C., 1961. Ancient Fields, A tentative analysis of vanishing earthworks and landscape, *British Association for the Advancement of Science, Research Committee on Ancient Fields,* London, 1961, 80 p.

Blancquaert G., Leroyer C., Lorho T., Malrain F., Zech-Matterne V 2012. Rythmes de créations et d'abandons des établissements ruraux du second âge du Fer et interactions environnementales, in « F. Bertoncello, F. Braemer, *Variabilités environnementales, mutations sociales. Nature, intensités, échelles et temporalités des changements* », Actes des 32e rencontres internationales d'archéologie et d'histoire d'Antibes (Antibes, 20-22 octobre 2011), Antibes, Éditions de l'Association pour la promotion et la diffusion des connaissances archéologiques, p. 233–245.

Bradley R., Haselgrove C., Vander Linden M., Webley L. (eds) 2016. *The Later Prehistory of North-West Europe: The Evidence of Development-Led Fieldwork*, Oxford University Press, Oxford, 480 p.

Brun P., Marcigny C., 2012. Une connaissance de l'âge du bronze transfigurée par l'archéologie préventive ,in « Nouveaux champs de la recherche archéologique », *Archéopages*, hors-série 10 ans, p. 132-139.

Buzzi P., Dreidemy C., Guillaume C., Koenig M. P., Mervelet P., 1994. La déviation de la RN 57 en Lorraine, Bilan des recherches archéologiques, *Revue Archéologique de l'Est,* 45, 1, p. 15-90.

Carozza, L., Marcigny C., Talon, M., 2017. *L'habitat et l'occupation des sols à l'âge du Bronze et au début du premier âge du Fer,* Recherches Archéologiques, 12, Paris : Inrap/CNRS Éditions., Paris, 2017, 374 p.

Chouquer G., 2007. Transmissions et transformations dans les formes parcellaires en France. Esquisse d'un schéma général d'interprétation, in « *La mémoire des forêts* », Office National des Forêts / institut National de la Recherche Agronomique / Direction Régionale des Affaires Culturelles de Lorraine, Paris, 2007, p. 21-33.

David R. 2014. Modélisation de la végétation holocène du Nord-Ouest de la France : Reconstruction de la chronologie et de l'évolution du couvert végétal du Bassin parisien et du Massif armoricain, thèse de doctorat, université de Rennes 1, Rennes, 284 p.

Desfosses Y. (dir.), 2000. *Archéologie préventive en vallée de Canche, les sites protohistoriques fouillés dans le cadre de ma réalisation de l'autoroute A16,* Nord-Ouest Archéologie 11, Berck-sur-Mer, 427 p.

Drewett P., 1978. Field systems and land allotment in Sussex, 3rd millennium BC to 4th century AD, *in.* "Bowen H.C., Fowler P.J. (dir.), *Early land allotment in the British Isles",* British Archaeological Reports, 48, p. 67–80

Dubreuil V., 1992. Typologie des paysages ruraux de l'Ouest de la France à partir de classifications d'images du satellite NOAA, *Norois,* n°155, p.283-296.

Fowler, P.J., 1971. Early prehistoric agriculture in western Europe : some archaeological evidence, *in.* "Simpson D.D.A. (ed.) *Economy and Settlement in Neolithic and Early Bronze Age Britain and Europe",* Leicester University Press, Leicester, p. 153-182.

Fyfe, R.M., Brück J., Johnston, R., Lewis, H., Roland, T.P., Wickstead, H., 2008. Historical context and chronology of Bronze Age land enclosure on Dartmoor, UK. *Journal of Archaeological Science,* 35(8), p. 2250-2261.

Fyllingen H., 2003. Society and Violence in the Early Bronze Age: An Analysis of Human Skeletons from Nord-Trøndelag, Norway, *Norwegian Archaeological Review,* 36:1, p. 27-43.

Ghesquiere E.; Marcigny C., 2019. Rythme et nature des occupations protohistoriques en Normandie, de l'âge du Bronze final eu début du second âge du Fer, *in.* « E. Leroy-Langelin et Y. Lorin (dir.), *L'habitat des Hauts-de-France et ses marges à la Protohistoire ancienne* », HABATA 1, Revue du Nord, Hors série, Collection Art et Archéologie, n° 27, 2019, p. 81-99.

Grinsell, L. 1992. The Bronze Age Round Barrows of Kent, *Proceedings of the Prehistoric Society,* 58(1), p. 355-384

Guilaine J., 1988. Le Néolithique et l'Âge du Bronze en France : regards sur une recherche et un patrimoine, *Espacio, Tiempo y Forma,* Serie I, Prehistoria, t. I, 1988, p. 213-225.

Guilaine J. (dir.), 1991. Pour une archéologie agraire. Paris, Armand Colin, 572 p.

Harding A.F., 1989. Interpreting the evidence for agricultural change in the Late Bronze Age, *in "Northern Europe, Bronze Age Study group",* Colloquium in Stockholm (10-11 may 1985), éd. The Museum of National Antiquities, Stockholm Studies, 6, 1989, p. 173-181.

Hoyt D.-V., Schatten M. 1997. *The role of the Sun in climate change*, University Press, Oxford, 279 p.

Jantzen D., Brinker U., Orschiedt J., Heinemeier J., Piek J., Hauenstein K., Krüger J., Lidke G., Lübke H., Lampe R., Lorenz S., Schult M., Terberger T., 2011. A Bronze Age battlefield ? Weapons and trauma in the TollenseValley, north-eastern Germany, *Antiquity*, 85, 2011, p. 417-433.

Johnston R., 2000. Field Systems and the Atlantic Bronze Age : thoughts on a regional perspectives, *in.* "Henderson J.C. dir., *The Prehistory and Early History of Atlantic Europe",* Papers from a session held at the European Association of Archaeologists Fourth Annual Meeting in Göteborg (1998), British Archaeological Reports, International Series. Oxford, Archaeopress, p. 47-55.

Kristiansen K., 1998. *Europe before History.* Cambridge UniversityPress, Cambridge, 505 p.

Kristiansen, K., Larson, T., 2005. *The Rise of Bronze Age society*, Travels, transmissions and transformations, Cambridge, 449 p.

Lebeau R., 2000. *Les grands types de structure agraire dans le monde*, Paris, Armand Colin

Le Roy-Ladurie E. 1967. *Histoire du climat depuis l'An Mil*, Paris, Flammarion, 377 p.

Louwe Kooijmans L.P., Van Den Broeke P.W., Fokkens H., Van Gijn A.L. (éd.), 2005. *The Prehistory of the Netherlands*, Amsterdam University Press, 2 volumes, 844 p.

Magny M. 1993. Solar influences on Holocene climatic changes illustrated by correlations between past lake-level fluctuations and the atmospheric 14C record, *QuaternaryResearch*, 40, p. 1–9.

Magny M., Bossuet G., Gauthier E., Richard H., Vanniere B., Billaud Y., Marguet A., Mouthon J., 2007. Variations du climat pendant l'âge du Bronze au centre ouest de l'Europe : vers l'établissement d'une chronologie à haute résolution, in « Mordant C., Richard H., Magny M. (dir.). *Environnements et cultures à l'Age du Bronze en Europe occidentale »,* Actes du 129e Congrès national des sociétés historiques et scientifiques (CTHS). Besançon, 19-21 avril 2004. Paris : Editions du CTHS (Documents préhistoriques, 21), p. 13-28.

Marcigny C., 2005. Une ferme de l'âge du Bronze à Nonant (Calvados)., *in* « Marcigny C., Colonna C., Ghesquière E. et Verron G. (dir.), *La Normandie à l'aube de l'histoire, les découvertes archéologiques de l'âge du Bronze 2300-800 av. J.-C »,* Somogy Editions d'art, 2005, p. 48–49.

Marcigny C., 2008. Les territoires de l'âge du Bronze : du territoire immédiat au territoire culturel. Quelques exemples de l'Ouest de la France, *Archéopages*, 21, p. 22-29.

Marcigny C., 2012a. Les paysages ruraux de l'âge du Bronze, structures agraires et organisations sociales dans l'Ouest de la France, *in.*

« Carpentier V. et Marcigny C. (dir.), *Des Hommes aux Champs, Pour une archéologie des espaces ruraux du Néolithique au Moyen Age* », Actes de la table ronde de Caen (octobre 2008), Presses Universitaires de Rennes, p. 71-80.

Marcigny C., 2012b. Au bord de la mer. Rythmes et natures des occupations protohistoriques en Normandie (IIIe millénaire - fin de l'âge du Fer), in. « Honegger M. et Mordant C. (éd.), *Au bord de l'eau, Archéologie des zones littorales du Néolithique à la Protohistoire* », 135e Congrès CTHS (Neuchâtel, Suisse, 2010), Cahier d'Archéologie Romande, p. 365-384.

Marcigny C., 2017. Les choses changent. Les modifications de la structure agraire au IIe millénaire sur les rives de la Manche, in. « Lachenal T., Mordant C., Nicolas T., Véber C., *Le Bronze moyen et l'origine du Bronze final en Europe occidentale (XVIIe-XIIIe siècle av J.C.)* », Colloque APRAB (2014, Strasbourg), Monographie d'Archéologie du Grand-Est, MAGE 1, Strasbourg, p. 645-658.

Marcigny C., 2019. *L'âge du Bronze en Normandie (2300-800 avant notre ère). Paysans et métallurgistes,*coll. Archéologies normandes, Orep éditions, 124 p.

Marcigny C., Ghesquière E., 2003. *L'île Tatihou à l'âge du Bronze (Manche), Habitats et occupation du sol*, Documents d'Archéologie Française (DAF), n° 96, 192 p.

Marcigny C., Ghesquiere E., 2008. Espace rural et systèmes agraires dans l'ouest de la France à l'âge du Bronze : quelques exemples normands, *in* « Guilaine J. (dir.), *Villes, villages, campagnes de l'Âge du Bronze* », séminaires du Collège de France, Editions Errance, p. 256-278.

Marcigny, C., Nere, E. Peake, R., Riquier,V., Le Den Mat, G., 2018. Rythme et nature des occupations du IIIer millénaire à l'aube de l'âge du Fer en France septentrionale, *in.* « O. Lemercier, I. Senepart, M. Besse, C. Mordant (dir.), *Habitations et habitat du Néolithique à l'âge du Bronze en France et ses marges* », Actes des rencontres Nord/Sud de Préhistoire récente, Toulouse : Archives d'Ecologie Préhistorique, Toulouse, p. 513-524.

Marcigny C., Riquier V., Audouit F., Frénée E., Néré E., Peake R., Talon M., 2020. Dynamique de peuplement de la fin du Néolithique à la fin de l'âge du Bronze en France, *in* "T. Lachenal, R. Roure et O. Lemercier (dir.), *Demography and Migration, Population trajectories from the Neolithic to the Iron Age"*, Proceedings of the XVIII UISPP World Congress (4-9 june 2018, Paris), sessions XXXII-2 and XXXIV-8, Archaeopress Archaeology, p. 59-70.

Matterne V. 2001. *Agriculture et alimentation végétale durant l'âge du Fer et l'époque gallo-romaine en France septentrionale, Archéologie des Plantes et des Animaux*, vol. 1, ed. M. Mergoil, 310 p. 105 pl.

Mordant C., 2008. L'habitat à l'âge du Bronze en France orientale, *in.* « Guilaine J. (dir.), *Villes, villages, campagnes de l'Âge du Bronze* », séminaires du Collège de France, Editions Errance, p. 204-219.

Needham S. P. 2000. Power Pulse Across a Cultural Divide: Cosmologically Driven Acquisition Between Armorica and Wessex, *Proceedings of the Prehistoric Society*, 66, p. 51-207

Osgood H., 2006. The dead of Tormarton°: Bronze Age combat victims ? *In.* "Otto T., Thrane H., Vandkilde H. (ed.), *Warfare and society°: archaeological and social anthropological perspectives"*, Aarhus UniversityPress, Aarhus, p.331–40.

Rathbone S., 2013. A consideration of villages in Neolithic and Bronze Age Britain and Ireland, *Proceedings of the Prehistoric Society*, 79, p. 39-60.

Reynaud C., 2003. Les systèmes agraires antiques : Quelle approche archéologique ?, *in* « Lepetz S., Matterne V. (dir.), *Cultivateurs, éleveurs et artisans dans les campagnes de Gaule romaine* », Actes du VIème Colloque AGER (Compiègne, 5 au 7 juin 2002), Revue Archéologique de Picardie, n°1-2, p. 281-298.

Shnirelman V. A. 1992. Crises and Economic Dynamics in Traditional Societies, *Journal of Anthropological Archaeology*, 11, p.25–46.

Stuiver M., Reimer P. J., Bard E., Beck J.-W., Burr G.-S., Hughen K.-A., Kromer B., Mccormac G., Van Der Plicht J., Spurk M. 1998. Intcal98 radiocarbon age calibration, 24 000–0 cal BP, *Radiocarbon,* 40, p.1041–1083.

Uckelmann M., 2008. Irlando der Iberien – Überlegungen zum Ursprungeiner Ornament form der Bronzezeit, in. „Verse F., Knoche B., Graefe J., Hohlbein M., Schierhold K., Siemann S., Uckelmann M., Woltermann G. ed., *Durch die Zeiten ... Festschrift für Albrecht Jockenhövel"*, zum 65, Geburtstag, Rahden, Marie Leidorf, p. 259–268.

Van Geel B., Magny M., 2002. Mise en évidence d'un forçage solaire du climat à partir de données paléoécologiques et archéologiques : la transition Subboréal-Subatlantique, *in* « Richard H., Vignot A. (dir.), *Equilibres et ruptures dans les écosystèmes depuis 20000 ans en Europe de l'Ouest* », Actes du colloque international de Besançon, 18-22 septembre 2000, Annales littéraires : 730, série « Environnement, sociétés et archéologie » n°3, Besançon, PUFC 831, p. 107-122.

Wickstead H., 2007. *Land Division and identity in Later Prehistoric Dartmoor, South-West Britain : Translocating Tenure,* PhD thesis. University College, London, 347 p.

Wickstead H., 2008. *Theorising Tenure : A Landscape Analysis of Later Prehistoric Dartmoor,* British Archaeological Reports, British Series. Oxford, Archaeopress, 231 p.

Yates D.T., 2007. *Land, Power and Prestige. Bronze Age Field Systems in Southern England,* Oxford, 204 p.

Climat et société à l'âge du Fer

Olivier Buchsenschutz [1]

Résumé

Si l'analyse des temps préhistoriques a toujours mis en relation étroite climatologues et archéologues, les spécialistes de l'âge du Fer ont longtemps privilégié le mobilier indigène ou importé pour analyser et dater les vestiges de sépultures, puis d'habitats, sans apporter une attention sérieuse au climat et à l'environnement. L'étude de l'alimentation et des végétaux les a conduits à analyser l'environnement naturel, sa modification par les hommes, et l'influence éventuelle du climat sur les évènements historiques et l'évolution des sociétés. La collaboration de spécialistes variés suppose une démarche rigoureuse, une confrontation objective et critique, une grande prudence vis-à-vis des tentations déterministes. Les différences d'échelle temporelles, climatiques et spatiales doivent être cohérentes et adaptées à la problématique historique visée.

Summary

If the analysis of prehistoric times has always linked climatologists and archaeologists, specialists in the Iron Age have long favored indigenous or imported furniture to analyze and date the remains of burials, then habitats, without providing any serious attention to the climate and the environment. The study of food and plants has led them to analyze the natural environment, its modification by humans, and the possible influence of climate change on historical events and the evolution of societies. The collaboration of various specialists presupposes a rigorous approach, an objective and critical confrontation, a great prudence vis-à-vis deterministic temptations. Temporal, climatic and spatial differences must be consistent and adapted to the historical problem.

Introduction

Les sciences environnementales, la philologie et l'archéologie contribuent chacune à expliquer les évènements ou les évolutions historiques de la protohistoire récente. Elles ont toutes des lacunes et des points de vue différents, leur échelle du temps et de l'espace n'est pas la même. Il s'agit ici de voir comment on peut corréler l'évolution de l'environnement avec celle de la société sans verser dans un déterminisme naïf. Les géographes ne

1 Archéologie et Philologie d'Orient et d'Occident (AOROC). CNRS : UMR8546 - Ecole Normale Supérieure de Paris - ENS Paris - 45 Rue d'Ulm 75230 PARIS CEDEX 05 - France. olivier.buchsenschutz@ens.psl.eu

croient plus au dicton vendéen qui disait « le calcaire produit l'instituteur, le granit produit le curé ». Le déterminisme a été critiqué et abandonné dès le début du XXe siècle, mais il réapparait de temps en temps aujourd'hui, peut-être parce que la dégradation de l'environnement menace nos conditions de survie dans l' « anthropocène ».

Le développement du dialogue entre environnementalistes, géographes, historiens et archéologues en France

Les travaux d'Emmanuel Le Roy Ladurie, et particulièrement *L'histoire du climat depuis l'an mil*, parue en 1967, méritent d'être lus ou relus à la lumière des recherches récentes sur l'environnement et des craintes actuelles sur l'évolution du climat. A partir de recherches archivistiques sur les forêts, les vendanges et les glaciers, il découvre les travaux des climatologues et détaille l'histoire chaotique des dialogues interdisciplinaires. Tant que les spécialistes des sciences naturelles environnementales d'une part, les historiens d'autre part, construisent chacun avec leurs sources une histoire du climat, les connaissances progressent normalement. Mesurer ensuite comment la perception de l'évolution climatique dans les textes se raccorde ou non avec des mesures physiques de plus en plus précises exige un dialogue interdisciplinaire approfondi. Souvent on se contente, - aujourd'hui encore-, de l'évoquer sommairement en conclusion d'un article pour ouvrir des pistes de recherches qui ne sont pas validées. Ma contribution voudrait seulement attirer l'attention sur les difficultés de ces recherches interdisciplinaires et sur la tentation, à laquelle j'ai parfois moi aussi cédé, de s'appuyer sur une autre discipline pour valider une hypothèse que nos sources habituelles ne permettent pas de prouver.

J'ai déjà été étonné de constater, lors d'une réunion informelle au Comité national du CNRS au début des années 1980 sur le développement des recherches environnementales, que les historiens envisageaient d'analyser l'influence du climat sur la société toujours à partir de textes et de récits des catastrophes naturelles. Pour les archéologues des périodes protohistoriques ou historiques, l'exemple des préhistoriens avait déjà suscité des contacts étroits avec les sciences environnementales.

Un numéro spécial de la revue « *Les Nouvelles de l'Archéologie* » en 1992 rend compte d'une table-ronde réunie à Bibracte (altitude 800 m) par Hervé Richard et Michel Magny (1993). L'étincelle qui a déclenché cette réunion, expliquent les auteurs dans la conclusion, avait été provoquée par le questionnement des archéologues « qui, subissant brouillard et pluie durant deux à trois semaines en juillet, se demandaient « comment était-il possible de supporter un tel climat à l'époque ?». Les contributions soulignent les difficultés que rencontrent les spécialistes quand on leur impose une fourchette de datation précise, ici 500 BC à 500 AD, étroite pour les environnementalistes, large pour les historiens. L'identification de phases de climat « favorables » ou « défavorables » définies par les hydrologues et les palynologues est déjà biaisée par les altérations anthropiques ; la dendrochronologie est encore préoccupée par des problèmes de méthode ; le bilan dressé par Ch. Goudineau de l'apport des sources littéraires est limité à la perception du temps court plutôt que du climat. L'introduction et la conclusion de ce volume définissent déjà clairement l'intérêt du dialogue approfondi entre les différentes disciplines et met en lumière le problème de leur avancement respectif : toutes ne sont pas prêtes au même moment à contribuer efficacement aux questions posées par l'histoire du climat et l'histoire des hommes. Le volume consacré aux recherches sur l'environnement du mont Beuvray quant à lui montre l'intérêt de réunir une large équipe de spécialistes autour d'un site majeur soumis à de fortes contraintes climatiques (Buchsenschutz, Richard 1996).

En 1991 est créé une formation de troisième cycle « Archéologie et environnement », réunissant plusieurs universités et départements « scientifiques ou littéraires ». Elle visait à enseigner aux étudiants le dialogue entre environnementalistes, archéologues et historiens. Son champ de recherche était ouvert au monde entier et privilégiait l'évolution humaine depuis le Néolithique - les préhistoriens avaient déjà depuis longtemps des enseignements et des équipes de recherche pluridisciplinaires - jusqu'à l'époque moderne. Cette démarche n'existait pas en France pour la protohistoire, l'Antiquité et les périodes plus récentes. Les étudiants sélectionnés avaient aussi bien une formation « historique ou archéologique » que « de sciences naturelles ». Les cours, les stages, les mémoires, étaient conçus, accompagnés et évalués par des géologues, des botanistes, des ostéologues, des archéologues. Le but n'était pas de former des chercheurs polyvalents, mais plutôt de favoriser le dialogue entre disciplines, d'apprendre à expliquer, à comprendre, à pratiquer une méthode d'analyse en connaissant ses possibilités et ses limites. Les excursions sur le terrain et les soutenances, qui suscitaient des commentaires et une concurrence parfois passionnée entre les différentes disciplines sont un de mes meilleurs souvenirs. Cette formation continue aujourd'hui à apprendre aux chercheurs à dialoguer avec des spécialistes d'autres disciplines pour confirmer ou écarter une hypothèse avec une approche critique. La protohistoire européenne bénéficie non seulement des méthodes des préhistoriens pour analyser l'environnement, mais aussi des plus anciens témoignages écrits des contemporains sur le climat de cette période.

Les témoignages des auteurs grecs et latins sur l'Europe moyenne à l'âge du Fer

Les variations climatiques ont-elles pu influencer l'évolution des sociétés anciennes? Comment les sociétés percevaient-elles les variations et l'action du climat ? Nous connaissons plusieurs témoignages des auteurs grecs et latins sur le climat de l'Europe moyenne, au Nord et à l'Ouest des Alpes : ils ont une vision géographique très sommaire de ces régions, ils connaissent des itinéraires, des fleuves, certaines montagnes, mais ils n'ont pas une vision précise de l'organisation générale des régions qu'ils ont traversées. Ils n'ont aucune idée de l'évolution du climat sur le long terme. Cependant ils se posent déjà des questions assez proches des nôtres, que nous évoquons ici brièvement, les textes ayant été déjà présentés ailleurs (Buchsenschutz, 2009).

Le différences entre les climats des rives de la Méditerranée et de l'Europe moyenne sont connues et sommairement décrites, par exemple par Strabon ou Tacite. L'absence de la culture de la vigne et de l'olivier caractérise les pays « barbares ». Mais plusieurs auteurs font allusion à la facilité de recueillir du fourrage, qui contraste avec la difficulté d'obtenir des moissons précoces, surtout dans les zones les plus septentrionales. Pline l'Ancien, en louant les pâturages de la Germanie, relève un effet bénéfique de ce climat : il affirme dans ce passage que le climat autant que le sol est responsable de la pousse de l'herbe. (Pline l'Ancien, *Histoire naturelle*, XVII, 4, 26). La différence entre les pays occidentaux océaniques et le climat plus contrasté de l'Europe centrale est bien identifiée par Tacite, et plus tard par Ausone.

Strabon associe volontiers les mœurs des populations à la rudesse relative de leur environnement géographique en Espagne : les zones côtières favorisent les formations de sociétés urbaines, les plaines portent des campagnes parsemées de villages, les zones montagneuses abritent des tribus de pillards. Ce déterminisme grossier issu d'une vision idéologique simpliste contraste avec des observations beaucoup plus crédibles que le géographe a puisées dans des récits de voyageurs.

Les auteurs anciens parlent rarement du temps (climat) long. Ils n'ont aucun outil pour enregistrer les évolutions à long terme. Mais toutefois, dans la tradition des observations réalisées par Hipparque sur les mouvements périodiques des astres, ils conçoivent que le climat peut changer à long terme, et que « *les contrées qui jadis ne pouvaient, à cause de la longue rigueur de l'hiver, conserver un seul des pieds de vigne ou d'olivier qu'on leur avait confiés, abondaient maintenant, grâce à l'adoucissement et à l'attiédissement des anciens froids, en larges récoltes d'olives, en productives vendanges* » (Columelle, De re rustica, I. 1.5).

Pour les périodes historiques où les textes sont plus abondants, les corrélations entre le climat et des évènements historiques sont exceptionnelles, mais elles font déjà l'objet de discussions entre les auteurs. Strabon (Géographie, VII, 2) conteste ainsi le rôle d'une grande marée dans la migration des Cimbres, partis de la côte Sud de la mer du Nord à la fin du IIe siècle BC, et la relation de modification de l'habitat des Celtes avec la montée des eaux avancée par d'autres auteurs. Les modifications les plus spectaculaires du climat sont considérées soit comme exagérées, soit comme un paramètre marginal d'évènements politiques importants.

Quelques cas particuliers révélés par l'archéologie

Rappelons ici pour mémoire quelques références à des études combinant archéologie et environnement dans l'Europe de l'âge du Fer, sans citer les travaux sur les lacs alpins qui sont à la fois nombreux et bien connus. C'est souvent la contrainte de l'eau par rapport aux habitats qui a motivé le développement d'une problématique environnementale.

La côte méridionale de la mer du Nord

Les travaux réunissant des équipes capables de gérer des études environnementales et archéologiques croisées ont été développés avant la deuxième guerre mondiale à Gröningen aux Pays-Bas (van Giffen 1936 et successeurs), puis en Allemagne à Wilhelmshaven (Haarnagel 1979 et successeurs). Ils montrent comment les hommes résistent à la montée de la mer par la construction de buttes artificielles, les *Wurten*, répondant en quelque sorte aux questions de Strabon. L'histoire des variations fines du niveau marin du début de l'âge du Fer jusqu'à l'époque actuelle constitue un chronomètre très précis et à chaque mouvement, on peut constater comment réagissent les groupes humains : ils profitent des vestiges de moraines fossiles, des canaux d'évacuation des marées (*Priel*), ou construisent des buttes artificielles (*Wurten, Terpen*) pour chaque maison, et bien tôt pour tout un village. Au Moyen-âge les digues protègent peu à peu toute la contrée submersible. Cette stratégie correspond à des structures domestiques qui subsistent jusqu'au XIXe siècle, avec des « maisons-étables » à trois nefs en bois. La fouille de la Feddersen Wierde (Ie siècle. BC – Ve siècle AD, Haarnagel 1979) a révélé une évolution sociale qui manifeste une spécialisation des activités et une hiérarchisation progressive, avec des maisons où l'atelier artisanal remplace les vaches, d'autres où la présence d'importations révèle une famille privilégiée. L'évolution du niveau de la mer a été maîtrisée jusqu'au Ve siècle AD, l'organisation sociale ou les éventuelles migrations ne semblent pas, dans la longue durée, avoir été provoquées par ce paramètre.

La Wetterau et l'évolution des campagnes

Plus ponctuellement, l'analyse du paysage de la plaine de la Wetterau au Nord de Francfort-sur-le-Main a été confrontée au développement des sépultures et des résidences princières de la fin du 1er âge du Fer dans la région (Stobbe, Airie 2002). On observe un déboisement entre les 7e et 5e siècle, avec une intensité maximum au 6e siècle. Au moment où est construit le complexe princier du Glauberg, avec ses riches sépultures et son sanctuaire, on est plutôt dans une phase de recul de cette exploitation de la plaine ; elle ne reprend qu'au 3e siècle BC. Dans l'état actuel des connaissances, la corrélation entre l'évolution de la société, celle de la production agricole et la courbe climatique ne sont pas avérées.

L'évolution des parcellaires pendant la Protohistoire est beaucoup trop complexe pour être décrite ici. La construction du paysage, quand on a la chance d'avoir des éléments de datation, reflète très tôt plusieurs paramètres contradictoires : l'initiative individuelle à l'échelle d'une propriété, la procédure de partage des terres à l'occasion de mariages, une organisation collective à plus petite échelle de la distribution des terres dans un terroir donné, la répartition entre les jardins, l'*infield* et l'*outfield*, la rotation des cultures, le choix de nouvelles productions etc. Les contraintes environnementales, liées au sol, à l'exposition, à l'eau, définissent un cadre général à l'intérieur duquel les choix des groupes humains reposent sur des stratégies complexes, parfois à la limite des possibilités du milieu naturel et du climat.

Je suis bien conscient que ces exemples isolés ne prouvent rien, sinon qu'il faut trouver un autre moyen pour comparer les courbes diachroniques longues des environnementalistes avec les données archéologiques des milieux tempérés. Notre expérience des recherches sur le Berry avec H. Richard, A. Maussion et B. Vannière ont buté au départ sur l'éparpillement des échantillons environnementaux : pratiquement aucun gisement archéologique ne livrait ni des pollens, ni ces charbons sur lesquels B. Vannière a développé son analyse (Batardy *et al.* 2000). L'archéologie de cette région pour les périodes protohistoriques et romaine ne livrait presque aucune stratigraphie. C'est seulement après quelques mois de recherches communes sur le terrain que des parallélismes ont pu enfin être établis.

Cette rapide revue à travers quelques exemples montre que les régions où nous disposons de bonnes courbes d'évolution environnementale sont précisément celles où l'installation humaine était la plus difficile et la plus sensible aux légères variations climatiques. Dans des terroirs contrastés mais généralement très favorables à l'agriculture comme en Berry, on lit aujourd'hui plus facilement des variations de stratégies d'exploitation agricole que des traces de crises du climat.

Comment confronter les données archéologiques et historiques aux données environnementales ?

P. Brun et P. Ruby par exemple (Brun, Ruby, 2008, p. 55), mettent en parallèle l'évolution de la teneur en carbone 14 résiduel avec les changements culturels et historiques.

De nombreux auteurs ont rapproché la disparition des villes du 5e s. BC et la baisse de l'occupation du territoire de la Gaule, des migrations celtiques en Italie. Or Tite-Live rappelle qu'avant la migration du 4e s. BC, il y avait eu d'autres invasions au début du 6e s. Elles auraient été provoquées par une croissance excessive de la population celtique : « sous son [Ambigat roi des Bituriges] règne, les productions de la Gaule et son peuplement étaient tels, que la prolifération de la multitude se révélait à peu près ingouvernable. » Il faut faire la part du mythe dans ce texte, et se poser des questions sur sa chronologie. Il n'en reste pas moins que Tite-Live ne parle pas de migration provoquée par une crise alimentaire ou climatique, mais rappelle qu' « autrefois » c'est un excès de richesse qui aurait poussé les Celtes à migrer en Italie (Peyre, Buchsenschutz, 2008).

Le travail synthétisé par C. Marcigny en Normandie (Marcigny, 2012) mériterait à lui seul une longue discussion. Nous le suivons volontiers quand il met en parallèle les variations climatiques avec la densité de la population. En revanche, je ne vois pas très bien en quoi il a pu jouer un rôle dans l'évolution des formes de l'habitat. L'exemple de l'Allemagne du nord évoqué plus haut montre le contraire.

Quantifier les données archéologiques

Il ne s'agit donc pas ici de critiquer les courbes fournies aujourd'hui par les analyses environnementales, mais plutôt de se demander comment les archéologues pourraient mettre en regard des données pertinentes et une chronologie aussi fine.

Une tentative dans ce sens a été proposée dans notre laboratoire à travers la « Basefer ». Il s'agit de réunir dans un même fichier les données générales (contrôlées par une publication dans une revue scientifique ou une thèse universitaire) sur les gisements de l'âge du Fer en France. Conçue avant tout pour établir des cartes de répartition, cette base se limite au signalement par présence/absence de structures (habitat, fosses, rempart massif, silo..), de mobilier (parure, arme, épée, os animaux..) et à une évaluation de la période d'occupation du site (pour les sites stratifiés une nouvelle fiche est établie quand il y a une solution de continuité ou un changement de fonction du gisement). Cet atlas n'entre pas en concurrence avec les bases spécialisées qui existent déjà pour l'âge du Fer, et qui prennent en charge l'analyse fine des données environnementales.

Le tableau synthétique ci-dessus montre ainsi une raréfaction de tous les types de sites aux 4e/3e s. BC. Elle correspond bien à la période défavorable aux cultures, isolée par Chr Petit et N. Bernigaud, et aux migrations vers l'Italie, mais il serait imprudent de proposer dès aujourd'hui une corrélation entre les deux phénomènes.

Il faut souligner en effet qu'une datation des occupations reste très difficile pour les archéologues. Sur environ 15000 gisements présents dans la base, nous n'en avons retenu ici que 1700 qui ont été classés en fonction de leur date d'abandon ; nous avons éliminé ceux qui avaient une occupation longue et insuffisamment documentée, parce qu'il s'agit alors d'une incertitude sur la datation plutôt que d'une occupation longue. Cette incertitude sur les dates d'occupation réelle, particulièrement pour cette documentation publiée mais relativement ancienne, est un handicap

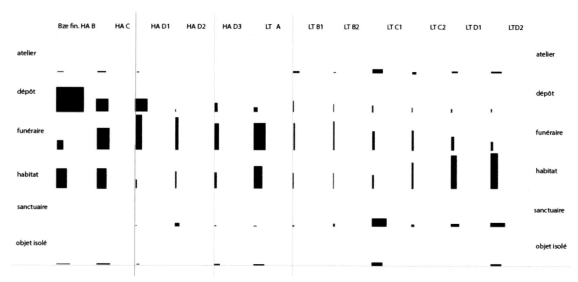

Pourcentages croisés des gisements datés, Basefer

1706 ont été retenus, chacun n'étant occupé que pendant la période typologique affichée.
largeur du rectangle noir dans chaque case : % en ligne
hauteur du rectangle noir dans chaque case : % en colonne
valeur de côté du carré blanc : 70 %   24 février 2014, 15567 sites

Figure 1. Effectifs des sites de l'Age du Fer en France (données extraites de la base de données "BaseFer"), corrélés avec la périodisation des cultures de Hallstatt et de La Tène

sérieux ; il ne sera corrigé que lorsque les fouilles récentes seront à leur tour publiées et inscrites dans la base. Il appartient donc aux archéologues de publier et de cataloguer les découvertes récentes, notamment des fouilles de sauvetage, dont les résultats abondants et souvent spectaculaires sont trop rarement aboutis et validés par une publication examinée par un comité de lecture.

En ce qui concerne le début de La Tène moyenne, on constate bien que le nombre de gisements diminue en France. C'est en même temps une période où s'affirme la culture celtique avec le développement d'un art et d'un artisanat indépendant des modèles méditerranéens, et la capacité à coloniser une partie de l'Italie septentrionale. La mise au point très lente de la faux, l'adoption rapide des meules rotatives de Catalogne, la diversification de l'outillage agricole en général, montrent un souci d'améliorer la productivité, qui semblerait un peu paradoxal s'il était contemporain d'une détérioration du climat.

Vers une méthode d'analyse interdisciplinaire

Comment passer de la confrontation brutale d'une courbe de mesures physiques à une appréciation de l'anthropisation et de sa capacité à surmonter les variations climatiques ?

Il faut d'abord réunir des données chronologiques et spatiales environnementales adaptées à l'échelle de la période historique concernée.

Pour le millénaire que recouvre l'âge du Fer européen, les variations de l'énergie solaire ou la précession des équinoxes ne signifient pas grand-chose. Même l'analyse des extraordinaires calottes polaires répond aux questions chronologiques mais ne rend pas compte des variations régionales. On recherche plutôt, à l'échelle des régions, l'enchaînement des années sèches ou pluvieuses, des étés torrides ou tardifs, en l'absence des témoignages archivistiques des périodes modernes. Enfin il ne faut pas oublier que les grands domaines agricoles, de Varron jusqu'au début du 20e s. en passant par O. de Serres, ont su répartir leurs cultures dans leur vaste domaine en fonction des capacités naturelles de chaque parcelle, avant l'arrivée des méthodes modernes de transformation des sols qui unifient les paysages.

G. Bertrand et son équipe ont mis en évidence des analyses qui cherchent à concilier les approches environnementale, géographique et historique (Bertrand 2000, 2002). Chacune de ces disciplines apporte une contribution incomplète à l'étude du temps qui passe et du temps qu'il fait, privilégient le rôle du climat, de l'espace ou de la chronologie comme fondement des stratégies humaines. Pour éviter cet écueil, il propose une analyse à différentes échelles : l'**hétérostasie** est définie comme la prise en compte des différentes échelles de temps et d'espace ; on considère chaque cas particulier comme un cube dans une mosaïque à trois dimensions avec une infinité de paramètres. Le **géosystème** représente alors le temps de la Source (des composants et des mécanismes biophysiques plus ou moins anthropisés), le **territoire** est le temps de la Ressource (exploitation des différentes ressources par les sociétés), le **paysage** est le temps du Ressourcement (multiples temporalités des vécus et des représentations, des symboles, des mythes, et des rêves).

Conclusion

La corrélation directe entre un évènement historique (une modification brutale dans le mode de vie des humains) et une crise climatique majeure est exceptionnelle. Les mauvaises récoltes des années 1780 contribuent sans doute au déclenchement de la Révolution française, mais le bouleversement social et idéologique que celle-ci provoque n'a rien à voir avec elles. Les réflexions du géographe G. Bertrand méritent d'être relues à la lumière des nouvelles données sur l'environnement des derniers millénaires. Il insiste notamment sur les différentes échelles de travail, qui ne répondent ni aux mêmes méthodes d'analyse, ni aux mêmes questions. Les stratégies économiques, sociales ou politiques, qui intéressent historiens et protohistoriens, sont développées à partir d'un cadre de connaissances limité dans l'espace et le temps, le reste étant occupé par des mythes. L'évolution du climat contribue certainement au succès ou à l'échec de ces stratégies. Elle en est rarement la cause dans les derniers millénaires de l'histoire humaine.

Bibliographie

Batardy C., Buchsenschutz O, Dumasy F. (dir.) 2000. *Le Berry antique : atlas 2000*. Tours : RACF, 2001, 190 p. (*Supplément à la Revue archéologique du centre de la France* ; 21).

Bertrand G. 2002. La discordance des temps, *in* « Richard H., Vignot A., *Equilibres et ruptures dans les écosystèmes depuis 20 000 ans en Europe de l'Ouest* », Actes du colloque international de Besançon 18-22 septembre 2000, Coll. Annales Littéraires Série Environnement, Sociétés et Archéologie n°3, p. 15-23.

Bertrand C. et G. 2000. Le géosystème : un espace-temps anthropisé. Esquisse d'une temporalité environnementale, *in* « *Les temps de l'environnement*, Barrué-Pastor M., Bertrand G. (éds.) », Journées du Pirevs 1977, Toulouse, Presses universitaires du Mirail, p. 65-78.

Brun P., Ruby P. 2008. *L'âge du Fer en France : Premières villes, premiers Etats celtiques*, La Découverte, Paris, 179 p.

Buchsenschutz O. 2000. Histoire et environnement, le temps retrouvé, *in* « *Les temps de l'environnement*, Barrué-Pastor M., Bertrand G. (éds.) », Journées du Pirevs, 1977, Toulouse, Presses universitaires du Mirail, p.117-121.

Buchsenschutz O., Richard H. (dir.) 1996. L'environnement du Mont Beuvray. Glux-en-Glenne, Bibracte, 1996, 208 p., 114 ill. (Bibracte; 1)

Buchsenschutz O. 2009. Note sur la perception du climat de l'Europe nord-alpine par les Celtes et par les auteurs grecs et latins, *in* « Bagley J.-M., Eggl Ch., Neumann D., Schefzig M. (dir.), *Alpen, Kult und Eisenzeit* », Festschrift für Amei Lang zum 65. Geburtstag, Leidorf 2009, p. 317-322.

Buchsenschutz O. et al. 2015. *L'Europe celtique à l'âge du Fer (VIIIe - Ier siècles)*, *Nouvelle Clio*, 2015, PUF, Paris. p. 86-91.

Buchsenschutz O. 2018. Les habitats de l'âge du Fer en France du IVe au Ie s. avant notre ère, in « Guilaine J., Garcia D., *La protohistoire de la France* », Hermann, Paris, p. 403-418.

Giffen Van A. E. 1936. Der Warf in Ezinge, Provinz Groningen, Holland, und seine westgermanischen Häuser, *Germania*, 20/1, p. 40-47.

Haarnagel W. 1979. *Die Grabung Feddersen Wierde : Methode, Hausbau, Siedlungs- und Wirtschaftsformen sowie Sozialstruktur*, 2 vol., Wiesbaden, Fr. Steiner.

Leroy-Ladurie E. 1967. *Histoire du climat depuis l'an mil*, Paris, Flammarion, Nouvelle bibliothèque scientifique.

Marcigny C., 2012. Au bord de la mer. Rythmes et natures des occupations protohistoriques en Normandie (IIIe millénaire - fin de l'âge du Fer), in « M. Honegger, C. Mordant éd., Congrès CTHS », *Cahier d'Archéologie Romande*, p. 365-384.

Peyre Ch., Buchsenschutz O. 2008. Tite-Live, Bourges, et les premiers processus d'urbanisation à l'âge du Fer en Europe septentrionale, *Germania*, 86, 2008, 1. Halbband, p.29-46.

Redde M., Barral Ph., Favory F. (dir.) *et al.* 2011. *Aspects de la romanisation dans l'Est de la Gaule*, (Collection Bibracte; 21), Glux-en-Glenne, Bibracte, Centre archéologique européen, Clamecy, 2 vol., 966 p.

Richard H., Magny M. 1993. Le climat à la fin de l'âge du Fer et dans l'Antiquité (500 BC - 500AD), Méthodes d'approche et résultats, *Les Nouvelles de l'Archéologie*, 50, hiver 92, p. 5-60.

Discussion et conclusions sur les sociétés humaines face aux changements climatiques des premiers 9000 ans de l'Holocène

François Djindjian

La fin de la dernière glaciation et le début de l'interglaciaire Holocène ont été une formidable avancée dans l'évolution de l'Humanité, qui va passer d'une population de l'ordre du million d'individus (d'un même ordre de grandeur que la population de nombreuses espèces animales) à sept milliards en environ 12 000 ans.

Ce constat amène à se poser plusieurs questions :

- Pourquoi cette révolution technologique, économique et démographique ne s'est-elle pas faite plus tôt au MIS 5, c'est-à-dire pendant l'avant-dernier interglaciaire ?
- En particulier, pourquoi le processus mésolithique (microlithisation des armatures et invention de l'arc) ne s'est-il pas fait en Europe et au Proche-Orient pensant cette période interglaciaire qui a connu le même environnement forestier ?

C'est que nos connaissances sur les peuplements humains au MIS 5 sont encore notoirement insuffisantes, et cela pour des raisons d'abord taphonomiques, car les niveaux d'occupation en grottes et en abris ont subi des phénomènes d'érosion à la fin du MIS 5. En outre, les sites de plein air, situés plus en profondeur, sont plus rarement découverts par les travaux modernes d'aménagement. C'est pourquoi, une meilleure connaissance du MIS 5 nous aiderait à mieux comprendre les mécanismes d'innovations des débuts du MIS 1.

L'adaptation au changement climatique de la fin de la dernière glaciation ne s'est pas faite de façon synchrone dans toutes les parties du monde. Plusieurs facteurs géographiques entrent en jeu, dont principalement la latitude, mais aussi la proximité océanique et la variation rapide d'altitude qui multiplie et diversifie les biotopes. Le Proche-Orient a anticipé cette adaptation dès la fin du maximum glaciaire avec un processus d'épipaléolithisation à partir de 17 000 BP (et sans doute plus tôt). Ce processus progresse à partir de 14 000 BP en Europe méditerranéenne avec l'Epigravettien récent puis final, en Afrique du Nord avec l'Ibéromaurusien et au Caucase avec l'Imérétien, et enfin dans toute l'Europe moyenne à l'Alleröd (Azilianisation). En Extrême-Orient, un phénomène comparable s'observe avec la « microblade industry ». La microlithisation des industries est un processus encore mal connu. Certains y voient le résultat de la diminution du territoire des chasseurs cueilleurs épipaléolithiques et mésolithiques, et donc la restriction à l'accès aux sources d'approvisionnement de la matière première (Kozlowski ce volume), tous y voient le changement des pratiques cynégétiques avec le développement de la forêt à l'origine de l'invention de l'arc (dont les microlithes sont les pointes de flèches). Le débitage change également, avec le retour du percuteur dur, le développement de la technique du microburin pour fragmenter les supports et en Extrême-Orient le débitage par pression, tous convergent à faciliter l'obtention de supports lamellaires et la taille de pièces microlithiques.

Nous connaissons bien en Europe les processus qui modifient la culture matérielle du tardiglaciaire européen, depuis les épisodes à lamelles à dos du Magdalénien, du Mézinien et de l'Epigravettien méditerranéen, les épisodes à pièces à dos anguleux du Magdalénien final, les épisodes à pointes de Cresswell et à pointes hambourgiennes des plaines du Nord (Bölling, Dryas II) des grandes plaines du Nord, les épisodes à pointes à dos courbe (Azilien, Federmesser) de l'Alleröd, les épisodes à pointes pédonculées (Dryas III, Préboréal), et enfin les armatures microlithes du Mésolithique holocène. Elles sont liées à des changements de pratiques cynégétiques induits par le changement de végétation et le changement d'espèces animales chassées. On retrouve des processus analogues en Afrique du Nord et en Amérique du Nord.

Les groupes de chasseurs-cueilleurs épipaléolithiques puis mésolithiques vivent sur des territoires plus restreints, mais restent mobiles pour pouvoir exploiter des ressources alimentaires diversifiées qui changent suivant les saisons.

Ce n'est pas le cas au Proche-Orient, où la pratique de la cueillette, connue dès le dernier maximum glaciaire, complète les ressources alimentaires de la chasse, et permet de franchir une étape supplémentaire mais essentielle, celle de la sédentarisation (Natoufien). Le passage de la cueillette à la culture et de la chasse à la domestication s'est sans doute fait progressivement et parallèlement, de façon à permettre des retours en arrière en cas de mauvaises récoltes ou d'épidémies animales. L'archéologie nous apprend cependant que ce passage fut réussi au Proche-Orient il y a onze mille ans environ, grâce à l'exceptionnel climat de l'Holocène humide et dans une région à la latitude idéale pour cette révolution économique et sociale.

Ailleurs, l'économie des chasseurs-cueilleurs holocènes va persister en s'adaptant aux environnements les plus variés, dans les zones périglaciaires du Nord de l'Amérique et de l'Eurasie, dans les taïgas, dans les steppes eurasiatiques, dans la grande plaine nord-américaine et sur ses côtes occidentales et orientales, dans les forêts tropicales humides, dans les environnements semi-désertiques du centre de l'Australie ou de l'Afrique australe et sous les climats méditerranéens et de savanes de l'hémisphère Sud (Afrique australe, Argentine, Australie). Les derniers de ces peuplements verront l'arrivée des Européens, à partir du XVI° siècle, qui les découvriront non sans étonnement ni sans interrogations (« sont-ils des humains ? sont-ils des créatures de Dieu ? ») puis des ethnologues, d'abord missionnaires puis explorateurs et enfin scientifiques jusqu'au moment où les exterminations, les extinctions et les acculturations firent disparaitre ou dénaturer l'objet de leurs études, comme l'ont si bien décrit, entre autres, les ethnologues américains (comme A. Kroeber (Kroeber, 1961) avec Ishi, le dernier des Yahi de Californie), français (comme Cl. Lévi-Strauss ou J. Emperaire), anglais, sans oublier les ethnologues russes de la Sibérie au XIXᵉ et XXᵉ siècle.

Un des phénomènes les plus spectaculaires des activités de ces chasseurs, pécheurs et cueilleurs installés non loin des rivages marins, est l'existence d'amas coquilliers, témoins d'une consommation importante de coquillages marins. Ces amas coquilliers (ou *shell mound*), sont connus sous des noms différents suivant les endroits où ils sont trouvés : Scandinavie (kokkenmodding), Cantabrie (conchero), Brésil (sambaqui), etc. On les retrouve également sur la côte d'Afrique du Nord, au Sénégal, en Asie du Sud-est et au Japon. Ce sont ces groupes qui ont pu, grâce aux ressources de la pêche et des coquillages, se sédentariser comme les Indiens de la côte Ouest au Nord de l'Amérique et les Jomons au Japon, et entrer dans une économie de stockage qui conditionne la naissance des processus d'échanges, comme l'a bien décrit A. Testart (Testart, 2014).

Nous connaissons aussi des cas où l'économie de chasseurs-cueilleurs holocènes a échoué : les colonisations des iles où la faune endémique est insuffisante pour alimenter une occupation humaine pérenne. Il en résulte généralement l'extinction des plus grandes espèces endémiques et le départ des groupes humains, avant que de nouveaux groupes néolithiques, apportant graines et animaux domestiques, ne viennent s'y installer (iles de la Méditerranée, iles du Pacifique, iles de l'océan indien).

Comme l'avait bien compris le grand généticien et agronome N. Vavilov (Vavilov, 1926), les différents centres d'invention de l'agriculture et de l'élevage, de l'ancien monde comme du nouveau monde, diffusèrent leurs innovations par des migrations de populations dans des régions encore peuplées par des chasseurs-cueilleurs holocènes. Nous connaissons encore insuffisamment les processus de cette cohabitation ni leur durée réelle mais nous savons qu'à terme l'économie néolithique l'emporta dans la plupart des cas, tandis que l'anthropologie physique et la génétique nous révélèrent le métissage des deux populations.

Nous ne connaissons pas cependant les cas où l'économie néolithique échoua, car l'archéologue ne peut découvrir, sauf exception, que les systèmes qui ont réussi. Nous connaissons aussi des cas d'acculturation où les groupes de chasseurs-cueilleurs empruntèrent des solutions techniques aux agriculteurs, comme la céramique dans le Mésolithique d'Europe orientale, non pas pour stocker les graines mais comme récipient de cuisine pour confectionner des soupes et des bouillies. Nous connaissons, par contre, les cas où l'économie néolithique disparut après avoir réussi, suite aux changements climatiques de l'Holocène, et tout particulièrement à la fin de l'Holocène humide avec les débuts de l'aridification.

Avec les progrès des recherches archéologiques, nous en savons plus sur les lieux d'origine et les dates d'apparition de ces foyers d'agriculture. Au Proche-Orient, en Chine, en Amérique centrale, en Amérique du Sud, ces foyers apparaissent pendant l'Holocène humide il y a onze mille ans au Proche-Orient, dix mille ans en Amérique du Sud, huit mille cinq cent ans en Chine, huit mille ans en Amérique centrale. C'est donc l'Holocène humide qui a donné l'environnement favorable à des sociétés qui avaient inventé l'agriculture et l'élevage, de mettre en œuvre de façon pérenne ces innovations.

Au Proche-Orient, les débuts de l'agriculture sont marqués par un néolithique acéramique : PPNA (10 040-8 940 BC), PPNB (8 940-6 400 BC) dont les séquences ont été établies sur le site de Jéricho et confirmées sur de nombreux sites du croissant fertile et en Anatolie. La communauté archéologique est unanime pour considérer que la fin des méga-sites PPNB est lié à l'événement climatique 8200 BP (Guilaine, Nowak, ce volume) dont les conséquences furent le passage au néolithique céramique (Yarmoukien du Levant) et à des habitats plus réduits. J. Guilaine observe des lacunes de peuplement dans les iles de la Méditerranée orientale et M. Nowak observe une continuité voire des conditions favorables de peuplements dans les Balkans créant des colonisations opportunistes en Europe, preuve du rôle essentiel de la latitude dans l'évolution de l'agriculture. Il faut également faire remarquer que les différences observées entre le signal climatique des carottages glaciaires ou marins et les changements de paléoenvironnement sur les sites archéologiques sont rendues plus complexes par la transformation anthropique de l'environnement due à la pratique agropastorale, deux processus différents qu'il faut pouvoir séparer.

Pastoralisme et épisode 8200 BP

Le pastoralisme nomade (qu'il faut distinguer de la transhumance à la recherche de pâturages saisonniers souvent en altitude), est une économie spécialisée dans laquelle l'agriculture ne joue aucun rôle ou seulement un rôle résiduel (qui est souvent compensé par des échanges avec des populations d'agriculteurs). L'étude archéologique du pastoralisme nomade est difficile car les traces laissées au sol par les populations pastorales sont peu nombreuses : il s'agit d'enclos à bétail et d'accumulations de déjections animales des troupeaux (Biagetti *et al.* 2003). Elles sont donc surtout visibles dans des endroits à sédimentation inexistante comme les régions désertiques et steppiques. Les structures funéraires sont souvent les seules architectures conservées (comme les célèbres kourganes scythes). Aux périodes historiques, le pastoralisme s'est développé dans des environnements désertiques et semi-désertiques, dans des steppes et sur des haut-plateaux froids. En Afrique du Nord, ce sont les Peuls aujourd'hui au Sahel, probablement issus d'une longue tradition qui remonte au 6ème millénaire BC ; en Afrique australe, ce sont les Hottentots ; En Afrique orientale, les Masaï dans la région des grands lacs, et les Somalis dans la corne de l'Afrique. Au Proche-Orient, ce sont les bédouins, dans la péninsule arabique et à l'intérieur du croissant fertile ; En Asie centrale et en Mongolie, ce sont les indo-européens depuis le 4° millénaire (Cimmériens, Scythes, Sarmates, Alains, Parthes), et aux époques historiques, depuis le IV° siècle jusqu'au XIII° siècle, Turcs et Mongols (Huns, Avars, Kazars, Seldjoukides, Azeris, Mongols, etc.).

La domestication des espèces animales est connue hors contexte néolithique avec l'élevage du renne (Samis de Carélie, Nenetses et Chukchis de Sibérie). C'est également le cas dans les steppes de Mongolie et du Nord Caucase (cheval, ovicaprinés). Elle a donc pu être acquise directement des économies de chasseurs-cueilleurs dans des régions impropres à l'agriculture. Mais généralement, la domestication fait partie du « package » néolithique, des agriculteurs-éleveurs, dès les débuts de chaque foyer de néolithisation.

La question se pose alors, dans chaque cas, des origines des pastoralismes.

C'est au Proche-Orient et dans le Nord de l'Afrique que les débuts du pastoralisme nous sont les mieux connus. Pour le Proche-Orient, une publication collective (Bar-Yosef, Khazanov, 1992) sous le titre « *Pastoralism in the Levant* » nous fournit un ensemble de contributions sur le sujet, avec notamment celle de J. Zarins (Zarins, 1992) qui émet le premier la thèse du lien entre l'émergence d'un nomadisme pastoral et de l'événement 8200 BP. Il est en effet probable que la fin de l'Holocène humide au Proche-Orient a obligé les agriculteurs-éleveurs à s'adapter au coup de froid-sec de l'épisode 8200 BP : abandon des terres les plus arides, accélération des migrations vers une Europe moins sèche (Nowak, ce volume) et passage au pastoralisme. Ce serait alors le temps du développement d'un pastoralisme bédouin dans la péninsule arabique et au centre du croissant fertile.

En Afrique, il n'existe pas encore de datations sur le début du développement du pastoralisme (ovicaprinés, chameaux) des Somalis dans la corne de l'Afrique, ni du pastoralisme Masai (bovins) dans la région des grands lacs (Kenya, Tanzanie) ni des Hottentots dont nous connaissons surtout le temps historique de leur arrivée en Afrique australe.

Mais il est également probable, qu'indépendamment des variations climatiques, le pastoralisme se soit naturellement développé dans des régions impropres à l'agriculture mais favorable à un élevage, qui s'est naturellement spécialisé en pastoralisme nomade. Les deux processus ne s'opposent pas, ils se complètent. B. Barich (Barich, ce volume) nous retrace ainsi les pastoralismes successifs du Sahara en relation avec l'aridité croissante de la région : bovins et ovicaprinés (à partir de 8500/8200 BP), cheval (vers 1 500 BC), âne (au moins avant 1 200 BC), dromadaire (réintroduit au IV° siècle BC après son extinction au Pléistocène) tout en en cherchant les origines de ces troupeaux et de ces nomades à l'Est (Levant, vallée du Nil), et en tenant compte d'hybridations avec des espèces sauvages subsahariennes. Le pastoralisme du Sahara ne se serait donc pas issu d'une agriculture locale mais d'une agriculture distante qui se serait adaptée par un pastoralisme nomade aux nouveaux territoires du Sahara, à partir de l'épisode 8200 BP. C'est l'épisode 4200 BP qui aurait mis un terme au pastoralisme bovin au Sahara, à l'origine de la longue pérégrination des pasteurs peuls à travers le Nord de l'Afrique, cherchant des pâturages au Sahel et laissant aux Touaregs un pastoralisme d'ovicaprinés au Sahara.

Un épisode 5 900 BP a été proposé par M. Claussen (Claussen *et al.* 1999), basé sur des travaux de simulation, rapporté au 4° évènement de Bond, et montrant une accélération de l'assèchement du Sahara et du Proche-Orient. Cet événement a été mis en relation avec la fin de la période d'Obeid en Mésopotamie et avec la fin de la culture chalcolithique de Cucuteni-Tripolié (Roumanie, Ukraine) liée à l'arrivée de nomades d'Asie centrale (culture des tombes en fosse), première introduction du cheval domestiqué en Europe, souvent invoqué comme l'arrivée des premiers locuteurs indo-européens.

Ainsi l'adaptation des agriculteurs-éleveurs à l'aridification progressive de l'Holocène à partir de 8200 BP, s'est traduit par la mise en œuvre plusieurs types de solutions :

- Le changement d'équilibre entre ressources de l'agriculture et ressources de l'élevage,
- Les débuts d'une irrigation artisanale,
- Le passage au pastoralisme,
- Le changement de pastoralisme, comme au Sahara, où le pastoralisme d'ovicaprinés succède au pastoralisme de bovidés,
- La migration vers des territoires moins arides comme la néolithisation de l'Europe à partir du Proche-Orient,
- L'abandon de terres incultivables ou demandant trop d'efforts pour les cultiver.

Les débuts de l'Hydraulique

De quand datent les innovations techniques de l'hydraulique pour l'irrigation des espaces cultivés ? Une remarquable synthèse a été publiée par un ingénieur P.L. Viollet (Viollet, 2004) dont le premier chapitre concerne les travaux hydrauliques des sociétés anciennes de l'Eurasie, notamment en Egypte, en Mésopotamie, en Asie centrale, en Arabie, dans le monde égéen et en Grèce. L'ancienneté des travaux hydrauliques y est démontrée remontant au IV° millénaire BC, et concerne des travaux d'irrigation, mais aussi d'aménagements divers (barrages, digues, tunnels, canaux, etc.) facilitant le transport fluvial ou maritime, la défense d'une ville, l'assainissement des zones marécageuses, le contrôle de la dangerosité des crues et l'approvisionnement en eau (comme son complément, l'évacuation des eaux usées).

Concernant l'irrigation des champs cultivés, plusieurs processus peuvent être invoqués pour le recours à cette irrigation. Le premier et sans doute le plus important est l'amélioration des rendements, le second l'extension des terres cultivables et le troisième la réponse à une aridité croissante du climat. Il n'est donc pas irrationnel de considérer que l'irrigation ait pu commencer avant l'épisode 4200 BP. Cette première irrigation était sans doute modeste, opportuniste, locale, recourant à des techniques rudimentaires et ne demandait que les ressources de la cellule familiale ou villageoise. Nous en avons des preuves en Mésopotamie remontant au VII° millénaire BC dans la culture de Samarra par des canaux d'irrigation et les premiers puits qui apparaissent au VI° millénaire.

Cette première agriculture est une agriculture des vallées alluviales. Aussi les dangers sont-ils d'abord ceux des crues destructrices et des changements de lits des rivières et des fleuves (comme l'exemple célèbre du fleuve Jaune en Chine) avant d'être ceux de l'aridité croissante du climat, qui sera marqué par les chocs des événements 8200 BP et 4200 BP.

En outre, un facteur lié à l'origine de l'irrigation, trop négligé, est celui de l'activité des fleuves. En phase de sédimentation, ils présentent une morphologie « en tresse » favorable à une première agriculture. Mais en phase de creusement dans ses alluvions, ce qui est le cas notamment pour l'Euphrate à partir du VI° millénaire, des terrasses se forment, au-dessus du lit du fleuve, hors d'atteinte des crues : il faut donc les irriguer ! Cela se fait en s'alimentant en amont et en remontant l'eau par un canal de dérivation à très faible pente (qui est d'une grande difficulté technique de réalisation sans théodolites) pour atteindre la terrasse, au bout de quelques kilomètres. Cette technique a été particulièrement pratiquée en Asie centrale.

En Egypte, les premiers travaux hydrauliques dateraient de l'Ancien Empire vers 3200 BC (Manning, 2002). W. Schenkel a cependant émis en 1978 la théorie de la révolution de l'irrigation, situant celle-ci au cours de la première période intermédiaire (2160-2055 BC), période qui a suivi l'effondrement de l'Ancien Empire, souvent liée aux conséquences économiques et sociales de l'épisode aride 4200 BP (Shenkel, 1994). Ces deux propositions ne sont pas contradictoires car la vraie question est à quel moment se généralise la construction de bassins de retenue des eaux de décrue du Nil ? L'aménagement hydraulique (digue d'al Lahoun, canal, barrages, réservoirs) de la dépression du Fayoum (lac Moeris) qui date de la XII° dynastie vers 1890 BC (sous Sesostris II) révèle la capacité technique impressionnante déjà atteinte par les Egyptiens à cette époque. Il faut également mentionner les travaux d'amélioration de la navigation sur le Nil à la première cataracte (Merenrê I vers 2 400 BP, puis Sesostris III et Thoutmosis III) et à la deuxième cataracte (passe de Semna et site de Mirgissa : fortifications, plan d'eau et glissière pour bateaux) par Sesostris III vers 1870 BP. La technologie était donc déjà présente et la crise aride de 4200 BP a du accélérer les travaux d'irrigation du Nil. Cependant, le passage à deux récoltes annuelles nécessite une irrigation plus régulière dans le cycle annuel que lors de la seule période de crue. C'est alors qu'apparait, au Nouvel Empire vers 1500 BC, le seau à balancier ou chadouf (seau, perche et contrepoids), puis, à l'époque ptolémaïque, le manège (saquiya) qui est constitué d'une roue horizontale entrainé par des animaux et une roue verticale qui puise l'eau par des godets et l'élève.

En Asie centrale, les recherches archéologiques de la période soviétique sous la direction de S.P. Tolstov (expédition du Khorezm, avec la contribution essentielle sur l'irrigation de B.V. Andrianov dans les années 1950) puis de V. Masson ont permis de proposer ces débuts à l'œuvre des sociétés chalcolithiques de la première moitié du IIIème millénaire BC, puis à des travaux importants au IIᵉ millénaire œuvre des sociétés de l'âge du Bronze (Lisitsina, 1981). Les travaux de la DAFA en Bactriane sous la direction de J. Cl. Gardin sont arrivés aux mêmes conclusions. Cependant, dans la littérature soviétique, les facteurs climatiques comme la croissance de l'aridité interviennent peu comme cause première du schéma marxiste de l'évolution sociale par stade. Un volume des Annales a été consacré en 2002 (57ème année, volume 3) à la question des travaux d'irrigation, en lien avec le développement d'un état centralisé, ayant seul l'autorité de mobiliser les ressources manuelles collectives nécessaires à ces grands travaux d'aménagement. Ce sujet, qui intéresse plus ici l'historien des idéologies que l'ingénieur hydrologue, est la critique des théories du « despotisme asiatique » chère à Marx, Engels et aux générations marxistes du monde académique, enthousiastes ou disciplinés, qui leur ont succédés depuis, de la « bureaucratie hydraulique » de M. Weber, à « l'hypothèse hydraulique » de K. Wittfogel (Wittfogel, 1957) et de sa remise en cause pour l'Egypte par K. Butzer (Butzer, 1976, 1996). Les estimations « marxistes » de S.P. Tosltov sur les travaux d'irrigation de l'oasis de Geoksyur en Turkménistan - les 500 000 jours-hommes soit 25 000 hommes pendant 20 jours soit plus de 20 fois la population estimée (donc beaucoup d'esclaves !) - ont été réduit par G. Lisitsina à 2500 jours-hommes soit cent personnes pendant 25 jours compatibles avec une population de 1000 à 1200 personnes sur 8 hectares d'habitats (Francfort et Lecomte, 2002, note 24)

Dans les régions comprises entre les montagnes et le désert très aride, les seules ressources en eau sont les oueds, à sec en temps normal, mais qui sont sujets à des crues violentes lorsque, deux ou trois fois par an, entre mars et août, tombent de fortes pluies sur les hautes montagnes dont ils sont issus. La technique utilise des murets déflecteurs, des déversoirs et des petites digues, plus rarement de véritables barrages, construits dans le lit des oueds. Ils permettent de guider une partie de l'eau de la crue, chargée de limons, vers un système ramifié de canaux en terre, munis de dispositifs de répartition en pierres, avec des vannes. Le courant est calmé par la pente des canaux, moindre que celle de l'oued, de telle sorte que seuls les sédiments fins sont entraînés vers les cultures. Deux ressources contribuent à la fertilité : l'eau elle-même, qui permet de semer dès l'inondation terminée, et les sédiments qui se déposent et permettent de reconstituer des couches de terre arable. Un des exemples les mieux étudiés est le système hydraulique construit dès le III° millénaire BC au royaume de Qataban et Saba sur le piémont oriental des montagnes du Yémen, qui fonctionnera jusqu'à sa destruction au VIème siècle évoquée dans le Coran.

D'autres techniques, plus élaborées encore, émergent à partir du 1er millénaire av. J.C., en relation probable avec le développement de la métallurgie du fer et des techniques de forage de mines. Il s'agit de la construction de galeries de drainage à partir de puits verticaux, appelées qanats en Iran et karez en Asie centrale. Cette technique a été retrouvée en Iran achéménide comme au Proche-Orient ; il fait également partie de la tradition urartéenne dans le Sud-Caucase (Briant, 2001). C'est également au I° millénaire qu'apparaissent les techniques de remontée d'eau comme la vis d'Archimède actionnée par manivelle, les roues à godet et les pompes.

4200 BP et après : collapses et accélération des travaux hydrauliques

L'événement climatique 4200 BP a fait l'objet d'une médiatisation importante, corrélant cet épisode avec l'effondrement des premiers grands états agricoles : Empire d'Akkad en Mésopotamie, Ancien Empire en Egypte, Civilisations de l'Indus, Empire minoen. Dans ce volume, J. Guilaine en étudie les effets dans les iles de la Méditerranée tandis qu'O. Lemercier analyse les processus de la fin des sociétés chalcolithiques en Europe occidentale en liaison avec le phénomène campaniforme et les débuts de l'âge du Bronze.

Cet événement climatique est sans doute le premier dans l'histoire de l'Humanité à nous amener à distinguer entre corrélation directe (à court terme ou à long terme) et corrélation systémique (l'événement climatique créant ou contribuant à un affaiblissement d'une société avec des conséquences en chaine : crise économique, famines, perte d'autorité, querelles dynastiques, affaiblissement de l'armée face à une menace intérieure ou extérieure attendant un moment propice pour se libérer ou pour envahir, etc.). Reprenant Tocqueville, sur les causes de la révolution française de 1789, faut-il les chercher dans (1) l'explosion d'un volcan islandais à l'origine d'étés pluvieux entraînant de mauvaises récoltes, des disettes, des spéculations sur la farine et une augmentation insoutenable des prix du pain, (2) la faillite financière de la France conséquence de l'échec des réformes de Turgot et de l'endettement de Necker obligeant à convoquer les Etats Généraux, (3) du manque d'autorité du roi Louis XVI, (4) d'un changement d'état d'esprit de la société du siècle des Lumières face à une aristocratie figée dans la défense de ses privilèges et à une intelligentsia demandant l'égalité pour tous, (5) d'une gestion de crise particulièrement mal effectuée ? Ou plutôt dans tous ces processus à la fois qui ont eu le mauvais goût de survenir au même moment (pour répondre à O. Buchsenschutz, dans ce volume) ?

Laissons à l'historien ces difficultés à résoudre, car l'archéologue ne peut généralement observer que la fin du processus, c'est-à-dire les résultats d'un effondrement qui se traduisent par l'éclatement d'un Empire, l'abandon des villes (et leur remplacement pas des petits habitats), une crise économique de longue durée qui se révèle dans la baisse de la production agricole et du commerce international, l'arrêt de la construction de monuments de prestige (palais, tombes), des invasions de peuples extérieurs et une dépopulation.

Un cas moins connu mais très instructif d'effondrement après 4200 BP, est la culture Kouro-Araxe, le Bronze ancien du Sud Caucase dont l'apogée se situe entre 3500 et 2400 BC. Cette civilisation développe une métallurgie précoce à partir d'un substrat chalcolithique déjà présent au V° millénaire en relation étroite avec la Mésopotamie (influences Halaf et Obeid). La phase récente de cette culture (2 600- 2400 BC) révèle une expansion géographique importante vers le Nord Caucase, vers l'Iran, vers l'Anatolie orientale et la Palestine (Bronze ancien III de Khirbet Kerak). Vers 2100, l'abandon des grands villages (dont la superficie est souvent supérieure à l'hectare) met en évidence un effondrement et son remplacement dans le Sud Caucase par un Bronze moyen (2100-1550 BC) qui a fourni aux archéologues de grands kourganes marquant l'arrivée de populations nomades. En Anatolie centrale, les riches tombes d'Alaca-Höyük et Horoztepê, d'inspiration nettement steppique, et datées de la fin du III° millénaire, ont été également interprétés comme l'arrivée d'une vague de peuplement (souvent interprétée aussi comme indo-européenne à l'instar de la culture des tombes en fosse d'Ukraine et Russie méridionale marquant la fin vers 3000 BC du chalcolithique Cucuteni-Tripolié (qui apparait au début du IV° millénaire BC)). Il faut faire remarquer au passage les décalages chronologiques importants entre Proche-Orient et Europe occidentale sur le néolithique, le chalcolithique et l'âge du Bronze (bien que les terminologies utilisées soient les mêmes) et l'acquisition rapide de la métallurgie du Caucase par les peuples nomades des steppes dans la seconde moitié du IV° millénaire.

L'Europe occidentale

C'est à la fin du III° millénaire BC que disparaissent les grandes sociétés du Néolithique final et du Chalcolithique d'Europe occidentale et qu'un Bronze ancien leur succède, dont le nombre et la taille des sites est significativement en retrait par rapport à la période précédente, traduisant un recul démographique qui a été souligné par de nombreux protohistoriens depuis près de 50 ans (R.P. Riquet, J. Guilaine notamment). Il est vrai que lier un effondrement socio-économique avec une innovation technologique majeure (métallurgie du bronze) peut déconcerter un archéologue plus au fait des évidences typologiques à la Montélius que des complexités systémiques des changements sociétaux. Ici aussi, le passage d'une économie agricole à une économie plus pastorale a été évoqué. Les points de vue différents d'O. Lemercier (ce volume) et de J.F. Berger (Berger *et al.* 2000 ; 2019) sont instructifs des difficultés à surmonter pour pouvoir développer et valider un discours systémique sur les transitions culturelles et leurs causalités. Il n'en reste pas moins que l'épisode 4200 BP a dû jouer un rôle significatif sinon majeur dans cette transition, sans doute accentuée par le développement important des sociétés du néolithique final que leur taille rendait plus fragile à l'arrivée d'une crise climatique.

Un exemple particulièrement intéressant d'adaptation à la crise 4200 BP est celui des Motillas dans la péninsule ibérique (région de La Mancha), où le creusement de puits profonds (près de 20 m) jusqu'à la nappe phréatique de la rivière, a permis de résister à l'aridité. Ici aussi, ce sont les temps des débuts de l'âge du Bronze (Blanco-Gonzalès *et al.* 2018).

Mais comme le fait remarquer O. Lemercier (ce volume), il faut tenir compte d'un gradient de cette aridité en fonction de la latitude. Elle se fait sentir plus particulièrement dans les latitudes méditerranéennes, au Moyen-Orient, plus encore au Sahara (assèchement des grands paléolacs), mais aussi dans les latitudes tropicales humides avec le recul de la forêt (comme en Amazonie). Par contre, elle se fait moins sentir en Europe moyenne où les transitions semblent moins brutales. L'Europe, là encore, amortit l'amplitude des variations climatiques favorisant les migrations de populations et accélérant les innovations technologiques.

Une bonne information sur la variabilité du climat à l'Holocène en Europe moyenne nous est donnée par les fluctuations du niveau des lacs alpins (Magny, 2004, 2005), qui révèlent les périodes d'humidité excessive, avec pour conséquence la remontée des eaux qui engloutit les villages installés sur les rives des lacs (dont les pieux des fondations encore conservés sous l'eau ont longtemps fait croire au mythe des palafittes), notamment le retour d'humidité après l'épisode 4200 BP, l'épisode 1500-1100 BC et l'épisode 800-400 BC (quand le niveau du lac Léman monte de 10 mètres). Ces fluctuations ont également concerné l'avancée des glaciers dans les Alpes, l'expansion des lacs, des tourbières, et les crues de la Vistule en Pologne (Starkel, 1966).

Car par ailleurs, nous manquons d'informations multiples sur les variations climatiques de ces périodes. En effet, nous ne disposons le plus souvent que des variations de la teneur de l'atmosphère en [14]C résiduel, interprétée comme un enregistrement des fluctuations de l'activité solaire, et qui est généralement considérée, faute de mieux, comme un indicateur empirique des variations du climat. Trois épisodes de péjoration climatique sont identifiables: l'épisode 2200 BC (4200 BP), la période 1400-1150 BC et la période 800-600 BC. C. Marcigny (ce volume) met bien en évidence la corrélation des variations climatiques avec l'emprise humaine sur le milieu (fortifications, habitats, parcellisations) en Europe occidentale sur la période 2600-600 BC., par rapport à l'amélioration 1600-1400 BC et les conséquences négatives de ces trois péjorations climatiques.

Mais qu'en est-il pour l'âge du Bronze sur le reste de l'Europe ?

Nous savons que les développements précoces de la métallurgie du Bronze apparaissent dans les zones riches en minerais de cuivre et en minerais d'étain (qui a la propriété de faire baisser la température de fusion du cuivre et de produire un bon alliage). Le Caucase est le plus ancien centre de métallurgie vers 3500 BC (culture Kouro-Araxe du Sud Caucase, culture de Maikop du Nord Caucase), puis la Bulgarie (chalcolithique de la culture de Varna dès le V° millénaire BC), la période minoenne en Crête (à partir de l'Asie mineure et du cuivre de Chypre) vers 2700 BC, puis l'Europe centrale vers 2300 BC (culture d'Unetice) avec les minerais des Monts métallifères, le Sud-ouest de la péninsule ibérique (dès le début du III° millénaire BC, nous remémorant le mythe de Tartessos, l'approvisionnement des phéniciens et la mine réelle de cuivre de Rio Tinto). C'est après l'épisode 4200 BP, que débute l'âge du Bronze (Bronze ancien), en Espagne vers 1800 BC (culture d'El Argar), en Italie vers 2200 BC (culture de la Polada de la vallée du Pô), en France vers 2000 BC, en Grande-Bretagne vers 2100 BC (culture du Wessex) après les phénomènes chalcolithiques campaniformes et épicampaniformes. Les débuts de ce Bronze ancien sont donc synchronisés avec l'épisode 4200 BP.

L'amélioration climatique 1600-1400 BC voit le grand développement du Bronze moyen, bien noté par C. Marcigny (ce volume). Mais cette période voit également l'épanouissement de grands état du monde méditerranéen comme l'empire mycénien (1650-1100), l'empire hittite (1625-1220), le nouvel empire égyptien (1590-1085), la période moyenne de l'empire assyrien (1521-911) et sans oublier la culture des Terramares en Italie (1700-1150), la culture Nuragique en Sardaigne (à partir de 1800 BC). Ces empires vont s'effondrer presque simultanément entre 1250 et 1150 BC. Cet effondrement a-t-il des causes climatiques, comme l'ont proposé B.L. Drake (Drake, 2012) et E. Cline (Cline, 2015), ce dernier sous le titre volontairement provocateur « *1177 BC, the year civilisation collapsed* » ?

Ces « siècles obscurs » de la Grèce (XIIème-IXème siècles) méritent cependant une analyse plus approfondie, comme celle qu'a tentée A. Schnapp-Gourbeillon (Schnapp-Gourbeillon, 2002). L'effondrement des pouvoirs centraux, la destruction des palais, l'essor de la piraterie, l'insécurité, la dégradation du commerce international, la baisse démographique, notamment, peuvent être mieux mis en relation avec les conséquences systémiques d'une péjoration climatique que par l'invasion des « Doriens » (le retour des Héraclides et la fondation de Sparte) des historiens de l'Antiquité et des philologues du XIX° siècle. Les Phéniciens sortiront cependant plus vite et mieux des destructions attribuées aux « peuples de la mer », résultant d'un effet domino qui est une conséquence de l'effondrement systémique plutôt que la cause. Les Phéniciens prendront alors des positions qui leur donneront une prééminence sur le commerce méditerranéen (succédant aux Mycéniens), position que leur disputeront plus tard les Grecs, après leur sortie plus tardive des siècles obscurs.

La pénétration de la métallurgie du Fer (par réduction du minerai et invention du bas-fourneau) à partir d'un centre anatolien probable, se fait en Europe au début du premier millénaire BC et arrive en Europe occidentale vers 800 BC. La typologie des trois âges technologiques (pierre, Bronze, Fer) a longtemps et aujourd'hui encore obscurci la connaissance des changements des processus internes socio-économiques des sociétés protohistoriques au cours des transitions Néolithique/Chalcolithique/Bronze et Bronze/Fer.

C'est en effet un paradoxe de constater que les deux transitions Néolithique/Age du Bronze et Age du bronze/Age du fer qui apportent chacune une technologie révolutionnaire s'effectuent dans un contexte de déstabilisation voire d'effondrement des sociétés. Ces deux révolutions technologiques ont souvent été considérées comme la cause de ces déstabilisations ? Ne faut-il pas plutôt les considérer comme la conséquence et la concrétisation de l'élaboration d'un nouveau système ?

Dans le premier cas, la métallurgie se diffuse véhiculée par des migrations de sociétés pastorales qui les empruntent au contact des foyers de métallurgie et s'infiltrent ou prennent le contrôle des sociétés chalcolithiques et néolithique final. C'est à l'Est de l'Europe, l'arrivée des pasteurs Yamnaya, faisant l'acquisition de la métallurgie du Nord Caucase, à l'origine de ce qui sera la culture Cordés, puis la culture Campaniforme. Ce modèle semble être confirmé par les études génétiques récentes (Olalde *et al.* 2018, 2019). Dans le second cas, à la fin de l'âge du Bronze, sont bien enregistrés des mouvements de population d'origines diverses, animées par la déstabilisation des Etats circumméditerranéens, comme ceux connus historiquement et regroupés sous le nom de « peuples de la mer » ou l'expansion vers l'Ouest et le Sud de cultures centre-européennes (culture des Champs d'Urnes (1350-1150) et Rhin-Suisse-France orientale ou RSFO (1150-950). Ils diffusent la métallurgie du Fer en moins de deux siècles sur l'Europe et le Proche-Orient.

Les variations climatiques des derniers 1500 av. J.C. ont été estimées par les variations de la teneur en ^{14}C résiduel dans l'atmosphère (figure 1), qui met en évidence une alternance d'épisodes plus chauds et des épisodes plus froids (Brun, Ruby, 2008). Trois épisodes de péjoration climatique sont enregistrés : 1400-1150 av. J.C. (déjà cité précédemment), 800-600 av. J.C et 400-300 av. J.C. Le VIII° siècle av. J.C. correspond aux débuts du premier âge du Fer (Hallstatt : 730-460 av. J.C.) dans la chronologie française et le V° siècle av. J.C., celui des débuts du second âge du Fer (La Tène : 460-25 av. J.C.). L'apogée de la civilisation de Hallstatt correspond à l'amélioration 580-430 av. J.C. Les invasions celtes en Italie, en Grèce, et jusqu'en Asie Mineure (Galates) s'échelonnent de 390 av. J.C. à 250 av. J.C. Elles correspondent à la période de péjoration climatique 400-300 av. J.C. Y-a-t-il une relation de causalité entre les deux processus ? O. Buchsenschutz (ce volume) en doute. Gilbert Kaenel (voir introduction) y pense. P. Brun et P. Ruby (Brun,

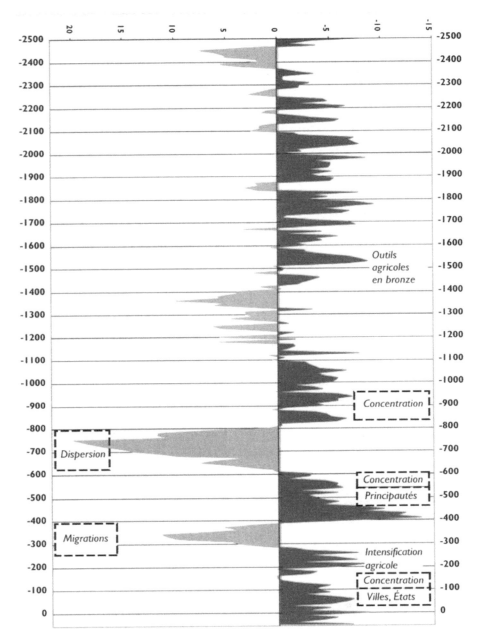

Figure 1.
Variations de la
teneur en 14C
résiduel dans
l'atmosphère
entre 2 500 et 0
av. J.C. (d'après
Brun, Ruby,
2008, p.55)

Ruby, 2008) voient un net développement économique et sociétal à partir de 250 av. J.C., « *un changement d'échelle* » pour reprendre leur expression, marqué par la vulgarisation de l'outillage agricole en fer et l'invention des haut-fourneaux vers 150 av. J.C., l'anthropisation du paysage, la reprise démographique, l'émergence des bourgs, le développement du commerce, la monétarisation, que l'amélioration climatique accélère.

Avec l'invention de l'agriculture et de l'élevage, puis de l'urbanisation et enfin des premiers Etats, les sociétés humaines ont développé des systèmes qui les rendent plus fragiles aux évènements climatiques, et cela à différentes échelles : temps long (aridité de l'Holocène récent), temps moyen (changements séculaires liés aux variations de l'activité solaire), épisodes brefs comme le 8200 BP et le 4200 BP largement évoqués et les conséquences climatiques multi-annuelles des éruptions volcaniques. Les progrès de l'archéologie et des Sciences auxquelles elle fait appel, permettent de proposer avec une accélération rapide depuis une cinquantaine d'années, des corrélations et des débuts d'explications sur l'influence des changements climatiques sur les sociétés humaines. Le déterminisme climatique prôné par plusieurs études spectaculaires mais aussi parfois quelque peu caricaturales a pu agacer les archéologues qui, aux prises avec les difficultés de synthèse qu'apportent aujourd'hui l'hyperspécialisation et la prolifération des données, se méfient des solutions simples. Et de fait, avec la complexification des sociétés humaines, les conséquences d'un changement climatique sont systémiques et non mono-causales : baisse des rendements agricoles, diminution des ressources alimentaires, épidémies, déstabilisation sociale, insécurité, baisse des échanges commerciaux, spéculations, crise économique et financière, guerres et invasions, baisse démographique, etc. L'archéologie peut en mesurer les effets sur un temps assez long pour qu'il en reste des traces matérielles. Les premières sources écrites nous sont d'un grand secours. L'archéologie est cependant démunie face à des phénomènes courts de plusieurs années d'origine météorologique dont les conséquences sur la production agricole créent des disettes, voire des famines, dont les conséquences dramatiques sont pourtant bien connues aux périodes historiques jusqu'à des dates très récentes comme E. Le Roy Ladurie (Le Roy Ladurie, 2004, 2006, 2009) l'a bien montré.

Bibliographie

Bar-Yosef O., Khazanov A. 1992. *Pastoralism in the Levant. Archaeological material. Anthropologic perspectives*, Monographs in Word archaeology, n°10, Madison prehistory Press

Berger J.-F., Magnin F., Thiébault S., Vital J. 2000. Emprise et déprise culturelle à l'Age du Bronze : l'exemple du Bassin Valdainais (Drôme) et de la moyenne vallée du Rhône, *Bulletin de la Société préhistorique française*, 97, 2000, p. 95-119.

Berger J.-F., Shennan S., Woodbridge J., Palmisano A., Mazier F., Nuninger L., Guillon S., Doyen E., Begeot C., Andrieu-Ponel V., Azuara J., Bevan A., Fyfe R., Roberts C. N. 2019. Holocene land cover and population dynamics in Southern France, *The Holocene*, 29, 2019, p.776-798

Biagetti S., Di Lernia S. 2003. Vers un modèle ethnographique-écologique d'une société pastorale préhistorique saharienne. *Sahara*, 14, p.6-30

Blanco-González A., Lillios K. T., López-Sáez J. A., 2018. Cultural, demographic and environmental dynamics of the Copper and early Bronze Age in

Iberia (3300–1500 BC): Towards and interregional multiproxy comparison at the time of the 4.2 ky BP event, *Journal of World Prehistory*, 31, 2018, p. 1-79.

Briant P. (dir.) 2001. *Irrigation et drainage dans l' Antiquité, qanats et canalisation souterraines en Iran, en Egypte et en Grèce.* Paris, Collège de France/Thotm Édition,

Brun P., Ruby P. 2008. *L'âge du Fer en France.* Paris, La Découverte

Butzer , K. W. 1976. *Early Hydraulic Civilization in Egypt. A Study in Cultural Ecology.* Chicago, University of Chicago Press.

Butzer K. W. 1984. Long-Term Nile Flood Variations and Political Discontinuities in Pharaonic Egypt, in "J. Desmond Clark et Steven A. Brandt (dir.), *From Hunters to Farmers. The Causes and Consequences of Food Production in Africa*", Berkeley, University of California Press, p.102-112

Butzer K.W 1996. Irrigation, Raised Field and State Management: Wittfogel Redux ? *Antiquity*, 70, p.200-204

Clausen M., Kubatski Cl., Brovkin V., Ganopolski A. *et al.* 1999. Simulation of an abrupt change in Saharan vegetation in the mid-Holocene. *Geophysical research Letters*, 26, 14, p.2037-2040

Cline E., 2015. *1177 BC, the Year Civilization collapsed* Princeton University Press

Drake B.L. 2012. The influence of climate change to the Late Bronze age collapse and the Greek Dark Age, *Archaeological Science*, 39, 2012, p.1862-1870

Ghesquière E., Marchand G. 2010. *Le Mésolithique en France.* Paris, La Découverte

Kroeber, Th. 1961. *Ishi*, (version française 1968) Paris, Plon, Terre Humaine

Le Roy-Ladurie E. 2004, 2006, 2009. *Histoire humaine et comparée du climat.* Paris, Fayard, 3 vol.

Lisitsina, G.N., 1981. The History of Irrigation Agriculture in Southern Turkmenia, *in* P. L. Kohl (dir.), *The Bronze Age Civilization of Central Asia. Recent Soviet Discoveries*, New York, Sharpe, 1981, p. 350-358.

Manning, J.G., 2002. Irrigation et État en Égypte antique. *Annales. Histoire, Sciences Sociales.* 57° année, 3, 2002, p.611-623

Magny, M. 2004. Holocene climatic variability as reflected by mid-European lake-level fluctuations, and its probable impact on prehistoric human settlements. *Quaternary International*, 113, p.65–80.

Magny M., Honegger M., Chalumeau L. 2005. Nouvelles données pour l'histoire des fluctuations holocènes duniveau du lac de Neuchâtel (Suisse): la séquence sédimentaire deMarin-Les Piécettes, *Eclogae Geologicae Helvetiae*, 2005, p.1-11

Olalde I, Brace S, Allentoft M.E. *et al.* 2018. The Beaker phenomenon and the genomic transformation of northwest Europe. *Nature,* 2018, 555 (7695), p.190-196

Olalde I., Mallick S., Patterson N., Rohmand N. et al. 2019. The genomic history of the Iberian Peninsula over the past 8000 years. *Science,* 2019, 363(6432), p.1230-1234

Schenkel W., 1978. *Die Bewdsserungs revolution im Alien Àgypten*, Mayence, Verlag Philipp von Zabern

Schenkel W., 1994. Les systèmes d'irrigation dans l'Egypte ancienne et leur genèse, *Archéo-Nil* 4, S. p.27-35

Schnapp-Gourbeillon A. 2002. *Aux origines de la Grèce (XIIIème-VIIIème siècles avant notre ère), la genèse du politique.* Paris, Les Belles-Lettres

Starkel L. 1966. Post-glacial climate and the moulding of European relief. In *"World climate from 8 000 to 0 BC"*, Royal Meteo. Soc., London, p.15-33

Testart A. 2014. *L'amazone et la cuisinière. Anthropologie de la division sexuelle du travail.* Paris, Gallimard

Vavilov N. 1926. Études sur l'origine des plantes cultivées. Revue de botanique appliquée et d'agriculture coloniale, 6° année, bulletin n°60, août 1926. p. 476-484

Viollet P.L. 2004. Histoire de l'Hydraulique 5 000 ans d'Histoire, Paris Presses de l'ENPC

Wittfogel K. 1957. *Oriental despotism.* New Haven, Yale University Press

Zarins J. 1992. Nomadisme pastoral en Arabie: ethnoarchéologie et archives archéologiques. In « Bar-Yosef O., Khazanov A. edts, *Pastoralism in the Levant"*, Madison Wisconsin, Prehistory Press, monographs in world archaeology, 10